Praise

The Rough Guide to Climate Change

The first edition of
*The Rough Guide to Climate
Change* was shortlisted
for the 2007 Royal Society
Prize for Science Books

Shortlisted

Royal Society
PRIZES FOR
SCIENCE
BOOKS 07

Credits

The Rough Guide to Climate Change
Editing & design: Tom Cabot and Duncan Clark
Picture research & illustration: Robert Henson,
Duncan Clark, Hannah Stuart & Tom Cabot
Proofreading: Jason Freeman
Production: Rebecca Short

Rough Guides Reference
Director: Andrew Lockett
Editors: Kate Berens,
Tom Cabot, Matthew Milton,
Tracy Hopkins, Joe Staines

Cover Images

Front cover: melting ice in the Arctic (Paul Nicklen/National Geographic). **Back cover**: Solar Power Plant, Nellis Air Force Base, Nevada (Proehl Studios/Corbis). **Flaps**: Greenland ice sheet (NASA); forest fire (John McColgan, Bureau of Land Management, Alaska Fire Service); European heat wave, July 2003 (NASA); golden toad (Charles H. Smith/US Fish and Wildlife Service); New Orleans after Katrina (Jocelyn Augustino/FEMA); arid earth (Patrik Giardino/Corbis); Himalayan melt lakes (NASA).

Publishing Information

This third edition published 2011 by
Rough Guides Ltd, 80 Strand, London WC2R 0RL
11 Community Centre, Panchsheel Park, New Delhi 110017, India
Email: mail@roughguides.com

Distributed by the Penguin Group:
Penguin Books Ltd, 80 Strand, London WC2R 0RL
Penguin Group (USA), 375 Hudson Street, New York 10014, USA
Penguin Group (Australia), 250 Camberwell Road, Camberwell, Victoria 3124, Australia
Penguin Group (NZ), 67 Apollo Drive, Mairangi Bay,
Auckland 1310, New Zealand

Rough Guides is represented in Canada by Tourmaline Editions Inc.,
662 King Street West, Suite 304, Toronto, Ontario, M5V 1M7

Printed in Singapore by Toppan Security Printing Pte. Ltd.

Typeset in FF Nexus and Myriad

416 pages; includes index

A catalogue record for this book is available from the British Library

ISBN 13: 978-1-84836-579-7

THE ROUGH GUIDE to

Climate Change

Robert Henson

Contributing editor
Duncan Clark

www.roughguides.com

ASKHAM BRYAN
COLLEGE
LEARNING RESOURCES

Acknowledgements

My deep gratitude goes to the experts who graciously provided interviews and text reviews for this book. Listed here with their affiliations at the time of their input, they are: Sim Aberson (NOAA Atlantic Oceanographic and Meteorological Laboratory); Lisa Alexander (Monash University); David Anderson and Connie Woodhouse (NOAA Paleoclimatology Program); Myles Allen, Ian Curtis, Tina Jardine, Lenny Smith and David Stainforth (Oxford University); Caspar Ammann, Aiguo Dai, James Done, Mary Hayden, Greg Holland, Jim Hurrell, Jeff Kiehl, Joanie Kleypas, Linda Mearns, Jerry Meehl, Susi Moser, Bette Otto-Bliesner, David Schneider, Claudia Tebaldi, Tom Wigley and Jeff Yin (NCAR); Vicki Arroyo, Elliot Diringer, Judi Greenwald, Katie Mandes, Namrata Rastogi, Truman Semans and John Woody (Pew Center on Global Climate Change); Mustafa Babiker (Arab Planning Institute); Richard Baker (UK Central Science Laboratory); Dan Becker (Sierra Club); Lance Bosart (University at Albany, State University of New York); Harold Brooks (NOAA Storm Prediction Center); Shui Bin (University of Maryland); Andy Challinor (University of Reading); John Christy (University of Alabama in Huntsville); Hugh Cobb (NOAA National Hurricane Center); Judith Curry (Georgia Institute of Technology); Lisa Dilling, Roger Pielke Jr. and Konrad Steffen (University of Colorado); Nikolai Dotzek (Deutsches Zentrum für Luft- und Raumfahrt); David Easterling and Tom Karl (NOAA National Climatic Data Center); Kerry Emanuel (Massachusetts Institute of Technology); Ben Felzer (Marine Biological Laboratory); Jeff Fiedler (UCAR); Piers Forster (University of Leeds); Hayley Fowler (Newcastle University); Greg Franta and Joel Swisher (Rocky Mountain Institute); Jonathan Gregory (University of Reading/UK Met Office Hadley Centre); James Hansen and Cynthia Rosenzweig (NASA Goddard Institute for Space Studies); Eric Haxthausen (Environmental Defense); Brian Huber (Smithsonian National Museum of Natural History); Tim Johns, Richard Jones and Jason Lowe (UK Met Office Hadley Centre); Phil Jones and Nathan Gillett (University of East Anglia); David Karoly (University of Oklahoma); David Keith (University of Calgary); Georg Kaser and Thomas Mölg (University of Innsbruck); Paul Kench (University of Auckland); Eric Klinenberg (New York University); Chuck Kutscher (US National Renewable Energy Laboratory); Beate Liepert (Lamont-Doherty Earth Observatory); Diana Liverman (University of Arizona); Glenn Milne (Durham University); Andy Parsons (UK Natural Environmental Research Council); Greg Pasternack (University of California, Davis); Rick Piltz (ClimateScienceWatch.org); John Church and James Risbey (CSIRO); Alan Robock (Rutgers University); Vladimir Romanovsky (University of Alaska); William Ruddiman (University of Virginia); Stephen Schneider (Stanford University); Peter Schultz (US Climate Change Science Program); Glenn Sheehan (Barrow Arctic Science Consortium); Scott Sheridan (Kent State University); Thomas Smith (NOAA); Susan Solomon (NOAA/IPCC); Eric Steig (University of Washington); Peter Stott (UK Met Office Hadley Centre); Jeremy Symons (National Wildlife Federation); David Thompson (Colorado State University); Lonnie Thompson (Ohio State University); Isabella Velicogna (University of California, Irvine); David Travis (University of Wisconsin–Whitewater); Chris Walker (Swiss Re); Chris West (UK Climate Impacts Programme); S. Joseph Wright (Smithsonian Tropical Research Institute); Gary Yohe (Wesleyan University).

Special thanks go to the friends and colleagues who read large portions of the book and provided much-needed moral support along the way: Joe Barsugli, Brad Bradford, Andrew Freedman, Zhenya Gallon, Roger Heape, Theo Horesh, Matt Kelsch, Richard Ordway, John Perry, Catherine Shea, Wes Simmons, Stephan Sylvan and Ann Thurlow. And every writer should have an agent and supportive friend like Robert Shepard in her or his corner.

The enthusiasm of Rough Guides for this project smoothed the way on many occasions. I'm grateful to Mark Ellingham and Andrew Lockett for championing the book and to Richard Trillo, Katy Ball and colleagues for getting the word out about it. The multitalented Duncan Clark provided superb editing, designing and illustration for the first and second editions and valuable consultation on the third. Tom Cabot skilfully edited text and updated the design for the third edition.

No one could ask for a better springboard for writing about climate change than working at the University Corporation for Atmospheric Research, which operates the National Center for Atmospheric Research. It's impossible to adequately convey my appreciation for the opportunity I've had as a writer and editor at UCAR and NCAR to learn about climate change from many dozens of scientists over more than two decades. This book couldn't have been written without two intervals of half-time leave made possible by Lucy Warner and Jack Fellows, plus the support of my colleagues in UCAR Communications. The opinions expressed herein are my own (as are any errors), and the content does not necessarily reflect the views of my UCAR and NCAR colleagues or the institution as a whole.

The choices we make in the next few years will shape the planet we bequeath to the youth of this century. This book is dedicated to them, including those in my own circle of life – David, Chloe and Ava Henson, Renée BeauSejour and Natasha Kelsch Mather.

Contents

Part 4: Debates & solutions

From spats and spin to saving the planet

Part 5: What can you do?

Reducing your footprint & lobbying for action

Part 6: Resources

Climate change books and websites

Foreword

by James Lovelock

author of *The Revenge of Gaia*

When the Intergovernmental Panel on Climate Change published its third assessment report in 2001 it received only brief attention from the world's media. Yet it was the most thorough statement ever made by climate scientists about Earth and to me the scariest document ever written. It talked laconically but authoritatively about the probability of the world being three degrees Celsius hotter by the end of the century and of a distinct possibility of it being almost six degrees hotter.

Three degrees does not sound like much but it represents a rise in temperature comparable with the global heating that occurred between the last ice age, some 15,000 years ago, and the warmth of the eighteenth century. When Earth was cold, giant glaciers sometimes extended from the polar regions as far south as St Louis in the US and the Alps in Europe. Later this century when it is three degrees hotter glaciers everywhere will be melting in a climate of often unbearable heat and drought, punctuated with storms and floods. The consequences for humanity could be truly horrific; if we fail to act swiftly, the full impact of global heating could cull us along with vast populations of the plants and animals with whom we share Earth. In a worst case scenario, there might – in the twenty-second century – be only a remnant of humanity eking out a diminished existence in the polar regions and the few remaining oases left on a hot and arid Earth.

The Rough Guide to Climate Change makes the arcane science of climate change comprehensible and sets the scene for the apocalyptic events that may lie ahead. You may well ask why science did not warn us sooner; I think that there are several reasons. We were from the 1970s until the end of the century distracted by the important global problem of stratospheric ozone depletion, which we knew was manageable. We threw all our efforts into it and succeeded but had little time to spend on climate change. Climate science was also neglected because twentieth-century science failed to recognize the true nature of Earth as a responsive self-regulating entity. Biologists were so carried away by Darwin's great vision that they failed to see that living things were tightly coupled to their material environment and that evolution concerns the whole Earth system with living organisms an integral part of it. Earth is not the Goldilocks planet of the solar system sitting at the right place for life. It was in this favourable state some two billion years ago but now our planet has to work hard, against ever increasing heat from the Sun, to keep itself habitable. We have chosen the worst of times to add to its difficulties.

We are fortunate to have this readable yet accurate book by Robert Henson at the prestigious National Center for Atmospheric Research in Colorado. For me it is a treasured source and map that will guide my thoughts as the climate changes. I recommend it to everyone as a truthful reference point in an ocean of misinformation and special pleading.

James Lovelock, July 2006

Introduction

When strangers meet at a bus stop or in a coffee shop, weather is the universal icebreaker. Yesterday's sweltering heat, the storm predicted for this weekend: it's all fair game. Even longer-term climate shifts find their way into chitchat. "It used to snow harder when I was a kid" is a classic example – and one explicable in part by the fact that any amount of snow looks more impressive from a child's height.

Today, however, such clichés have an edge to them, because we know that humans play a role in determining the course of climate. When we hear about Arctic tundra melting or a devastating hurricane, we're now forced to consider the fingerprints of humanity – and that's going well beyond small talk. Indeed, climate change is as much a divider as weather has traditionally been a unifier. Weather has always seemed to transcend politics, but human-induced climate change is wedded to politics: it's an outgrowth of countless decisions made by local, regional and national governments, as well as individuals and corporations. Sadly, it's also become – particularly in the US – a polarized subject, linked to other issues so frequently that it often serves as shorthand for one's entire world view.

It might come as a surprise, then, how much of the basic science behind global climate change is rock-solid and accepted by virtually all parties. Aside from a handful of sceptics and a few headline-grabbing controversies, most of the debate among experts these days revolves around interpretation. Just how warm will Earth get? Which computer projections for the year 2050 are likely to be the most accurate? How should we go about trying to reduce the blanket of greenhouse gases that's getting thicker each year? How can we best adapt to unavoidable changes? These are difficult questions – but they're about the *nature* of global climate change, not its mere existence.

Ever since Louis Agassiz introduced the notion of ice ages in the 1830s, we've known that prehistoric climates differed markedly from the present. And it's been clear for many decades that carbon dioxide and other greenhouse gases keep our planet warmer than it would otherwise be. But it's taken a long time to truly connect the dots, to fully understand that we can push our planet's climate into dangerous, uncharted territory just by the way we go about living our lives.

Although more and more people recognize the risks of climate change, not everyone is convinced of the danger, in part thanks to a battlefield of rhetoric. Industries heavily invested in the status quo, such as oil and coal, have contributed to the spin. So has the blogosphere, where anyone can cast himself or herself as an expert on climate. The true experts – those who have spent decades studying climate – faced the unauthorized release in 2009 of hundreds of emails exchanged among scientists. Though none of the messages threatened the foundations of global warming science, their content was furiously spun by sceptics. By 2010, the near-collapse of global climate negotiations made any solution to the **climate crisis** seem further away than ever.

Still, if global warming is one of the most daunting challenges humanity has faced, it's also a unique opportunity. Fossil fuels do more than power our cars and homes. They fuel crisis – from instability in the oil-drenched Middle East to smog-choked skies across the great cities of the developing world. But the era of cheap, plentiful oil is drawing to an end, and the difficult steps needed to deal with global warming could hasten the world's transition to cleaner, sustainable forms of energy. And, as many who have written on this subject point out, we may yet emerge from that transition with new ways of achieving progress on other tough issues. In that light, a warming world could affect our interconnected global society as much as it shapes the climate itself.

How this book works

Whether you're alarmed, sceptical or simply curious about climate change, this book will help you sort through the many facets of this sprawling issue. **The basics** lays out some key questions and answers, explains how global warming actually works, and examines the sources of the greenhouse gases heating up our planet. **The symptoms** provides an in-depth look at how climate change is already affecting life on Earth and how these changes may play out in the future.

The science describes how the global warm-up has been measured and puts the current climatic changes in the context of Earth's distant history and future. It also takes a look at the computer models that tell us what we can expect over the next century. **Debates & solutions** surveys the global-warming dialogue and explores the ways in which we might be able to eliminate or reduce the threat of climate change. These include political agreements, such as the Kyoto Protocol, as well as cleaner energy sources and sci-fi-esque geoengineering schemes.

For solutions on an individual or family scale, turn to **What can you do?**, which provides tips on reducing your carbon footprint at home and on the road. Finally, **Resources** provides details of other books on the subject and a list of recommended websites.

PART 1

THE BASICS

Global warming in
a nutshell

Smoke stacks at the Bao Steel mill, Baotou, Inner Mongolia.

Climate change: a primer

Key questions and answers

Before exploring the various aspects of climate change in depth, let's quickly answer some of the most frequently asked questions about the issue. The following twenty pages will bring you up to speed with the current situation and the future outlook. For more information on each topic, follow the reference to the relevant chapter later in the book.

The big picture

Is the planet really warming up?

In a word, yes. Independent teams of scientists have laboriously combed through more than a century's worth of temperature records (in the case of England, closer to 300 years' worth). These analyses all point to **a rise of more than 0.7°C (1.3°F)** in the average surface air temperature of Earth over the last century (see the graph on the inside front cover). The chapter Keeping track (p.189) explains how this average is calculated. The map overleaf shows how warming since the 1970s has played out regionally.

In recent years global temperatures have spiked dramatically, reaching a new high in 1998 and similar levels in 2010. Strong **El Niño** events (see p.130) helped boost global temperatures in both years, but it's not as if readings were chilly in between. The first decade of the twenty-first century was the hottest on record – and quite possibly warmer than any others in the past millennium (see p.244).

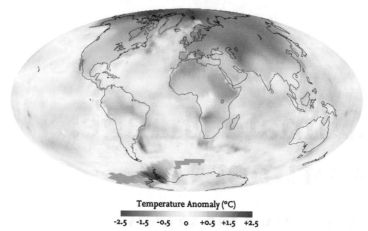

Temperature Anomaly (°C)

-2.5 -1.5 -0.5   0   +0.5 +1.5 +2.5

The change in surface temperature for the period 2000–2009 relative to 1951–1980, was calculated by NASA's Goddard Institute for Space Studies using data from surface weather stations for land areas and satellite-derived temperature estimates for the oceans. Grey areas denote places where temperatures were not recorded. The warmth was especially prominent at high northern latitudes, as predicted by computer models.

Apart from what temperatures tell us, there's also a wealth of circumstantial evidence to bolster the case that Earth as a whole is warming up.

▶ **Ice** on land and at sea is melting dramatically in many areas outside of interior Antarctica and Greenland. Montana's Glacier National Park could lose its glaciers as soon as 2020. Arctic sea ice lost nearly half its average summer thickness from 1980 to 2008, and by mid-century the ice may disappear completely each summer, perhaps for the first time in more than a million years. The warmth is already heating up international face-offs over shipping, fishing and oil-drilling rights in parts of the Arctic once written off as inaccessible.

▶ **The growing season** has lengthened across much of the Northern Hemisphere, with a boost from urban heat islands in some areas. The average blooming date of Japan's famed sakura (cherry blossoms) is now earlier than at any time in more than a thousand years of record-keeping, pushing the traditional festivals from April to March. According to the UK Climate Impacts Programme, the British spring is arriving two to six days earlier, and autumn two days later, each decade.

▶ **Mosquitoes, birds and other creatures** are being pushed into new territories, driven to higher altitudes and latitudes by increasing warmth. The range of twelve bird species in Britain shifted north in the 1980s and 1990s by an average of 19km (12 miles). And Inuits in the Canadian Arctic report the arrival over the last few years of barn swallows, robins, black flies and other previously unseen species. (As we'll see later, however, not all fauna will migrate so successfully.)

But don't many experts claim that the science is uncertain?

There is plenty of uncertainty about details in the global-warming picture: exactly how much it will warm, the locations where rainfall will increase or decrease, and so forth. Some of this uncertainty is due to the complexity of the processes involved, and some of it is simply because we don't know how individuals, corporations and governments will change their greenhouse emissions over time. But there's near-unanimous agreement that global climate is already changing and that fossil fuels are at least partly to blame (see p.272).

The uncertainty that does exist has been played both ways in the political realm. Sceptics use it to argue for postponing action, while others point out that many facets of life require acting in the face of uncertainty (buying insurance against health or fire risks, for example).

Is a small temperature rise such a big deal?

While a degree or so of warming may not sound like such a big deal, the rise has been steeper in certain locations, including the Arctic, where small changes can become amplified into bigger ones. And the planet as a whole could warm by 6°C (11°F) or more by the end of this century, which is hardly a small rise. Any warming also serves as a base from which heat waves become that much worse – especially in big cities, where the **heat-island effect** comes into play. Like a thermodynamic echo chamber, the concrete canyons and oceans of pavement in a large urban area heat up more readily than a field or forest, and they keep cities warmer at night. During the most intense hot spells of summer, cities can be downright deadly, as evidenced by the hundreds who perished in Chicago in 1995 and the thousands who died in Paris in 2003 and Moscow in 2010 (see p.55).

Climate change or global warming?

The phrases that describe climate in transition have a history of their own. Early in the twentieth century, researchers preferred **climatic change** or **climate change** when writing about events such as ice ages. Both terms are nicely open-ended and still used often. They can describe past, present or future shifts – both natural and human-produced – on global, regional or local scales.

Once scientists began to recognize the specific global risk from human-produced greenhouse gases, they needed a term to describe it. In 1975 Wallace Broecker, of New York's Lamont-Doherty Earth Observatory, published a breakthrough paper in the journal *Science* entitled, "Climatic Change: Are We on the Brink of a Pronounced Global Warming?" By the early 1980s the phrase **global warming** – without the "a" in front – was gaining currency among scientists. Meanwhile, the term **global change** emerged as a way to embrace all modes of large-scale human tampering with the planet. When 1988's watershed events arrived (see p.274), the global-warming label broke into headlines worldwide and became standard shorthand among media and the public.

Of course, the planet as a whole *is* warming, but many scientists avoid that term, preferring "global change" or more specifically **global climate change**. One of their concerns is that global warming could be interpreted as a uniform effect – an equal warming everywhere on the planet – whereas in fact a few regions may cool slightly, even as Earth, on average, warms up.

Politicians hoping to downplay the reality of global warming gravitate towards "climate change" for entirely different reasons. US political pollster and consultant Frank Luntz reportedly advised clients that "climate change" sounds less frightening to the lay ear than "global warming". Scary or not, a number of other surveys support the idea that "global warming" gets people's attention more quickly than the less ominous (though more comprehensive) "climate change". And a few activists and scientists, including the Gaia theorist James Lovelock, now favour **global heating** – a phrase that implies humans are involved in what's happening.

How could humans change the whole world's climate?

By adding enormous quantities of carbon dioxide (CO_2) and other **greenhouse gases** to the atmosphere over the last 150 years. As their name implies, these gases warm the atmosphere, though not literally in the same way a greenhouse does. The gases absorb heat that's radiated by Earth, but they release only part of that heat to space, which results in a warmer atmosphere (see p.23).

The amount of greenhouse gas we add is staggering – in carbon dioxide alone, the total is more than thirty billion metric tonnes per year, which is more than four metric tonnes per person per year. And that gas goes into an atmosphere that's remarkably shallow. If you picture Earth as a soccer ball, the bulk of the atmosphere would be no thicker than a sheet of paper wrapped around that ball.

Even with these facts in mind, there's something inherently astounding about the idea that a few gases in the air could wreak havoc around the world. However, consider this: the eruption of a single major volcano – such as Krakatoa in 1883 – can throw enough material into the atmosphere to cool global climate by more than 1°C (1.8°F) for over a year. From that perspective, it's not so hard to understand how the millions of engines and furnaces spewing out greenhouse gases each day across the planet, year after year, could have a significant effect on climate. (If automobiles spat out chunks of charcoal every few blocks in proportion to the invisible carbon dioxide they emit, the impact would be more obvious.) Yet many people respond to the threat of global warming with an intuitive, almost instinctive denial.

The densest part of Earth's atmosphere extends only a few kilometres above the surface, as shown in this photo taken from the International Space Station (ISS). Dark clouds and an orange sky near the bottom represent the troposphere, or "weather layer" where temperatures have been warming. Above this layer is the stratosphere, where temperatures have been cooling and ozone depletion has been a concern over recent decades.

When did we discover the issue?

Early in the twentieth century, the prevailing notion was that people could alter climates locally (for instance, by cutting down forests and ploughing virgin fields) but not globally. Of course, the ice ages and other wrenching climate shifts of the past were topics of research. But few considered them an immediate threat, and hardly anyone thought humans could trigger worldwide climate change. A few pioneering thinkers saw the potential global impact of fossil-fuel use (see p.30), but their views were typically dismissed by colleagues.

Starting in 1958, precise measurements of carbon dioxide confirmed its steady increase in the atmosphere. The first computer models of global climate in the 1960s, and more complex ones thereafter, supported the idea floated by mavericks earlier in the century: that the addition of greenhouse gases would indeed warm the climate. Finally, global temperature itself began to rise sharply in the 1980s, which helped raise the issue's profile among the media and the public as well as scientists.

Couldn't the changes have natural causes?

The dramatic changes in climate we've seen in the past hundred years are not proof in themselves that humans are involved. As sceptics are fond of pointing out, Earth's atmosphere has gone through countless temperature swings in its 4.5 billion years. These are the results of everything from cataclysmic volcanic eruptions to changes in solar output and cyclic variations in Earth's orbit (see pp.212–17). The existence of climate upheavals in the past raises the question asked by naysayers as well as many people on the street: how can we be sure that the current warming isn't "natural" – i.e. caused by something other than burning fossil fuels?

That query has been tackled directly over the last couple of decades by an increasing body of research, much of it through the **Intergovernmental Panel on Climate Change (IPCC)**, a unique team that draws on the work of more than one thousand scientists. We'll refer often throughout this book to the IPCC's work (see p.308 for more on the panel itself). Back in 1995, the IPCC's Second Assessment Report included a sentence that made news worldwide:

> "The balance of evidence suggests a discernible human influence on global climate."

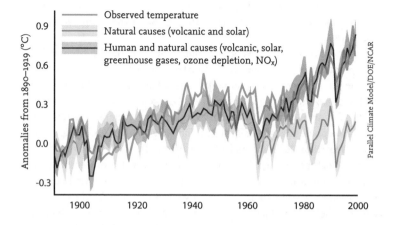

The red and green shaded areas show the departure in global temperature from the 1890–1919 average, as produced by four computer model simulations. The red and green lines show averages of the four models. The green runs factor in natural agents of climate change only; the red runs include both human and natural factors. The blue line shows the temperature measured in the real world.

By 2001, when the IPCC issued its third major report, the picture had sharpened further:

> "There is new and stronger evidence that most of the warming observed over the last 50 years is attributable to human activities."

And in its fourth report (2007), the IPCC spoke even more strongly:

> "Human-induced warming of the climate system is widespread."

To support claims like these, scientists call on results from two critical types of work: **detection** and **attribution** studies. Detection research is meant to establish only that an unusual change in climate has occurred. Attribution studies try to find the likelihood that humans are involved.

One way to attribute climate change to greenhouse gases is by looking at the signature of that change and comparing it to what you'd expect from non-greenhouse causes. For example, over the past several decades, Earth's surface air temperature has warmed most strongly near the poles and at night. That pattern is consistent with the projections of computer models that incorporate rises in greenhouse gases. However, the pattern agrees less well with the warming that might be produced by other causes, including natural variations in Earth's temperature and solar activity.

As computer models have grown more complex, they've been able to incorporate more components of climate. This allows scientists to tease out the ways in which individual processes helped shape the course of the last century's warm-up. One such study, conducted at the US National Center for Atmospheric Research, examined five different factors: volcanoes, sulphate aerosol pollution, solar activity, greenhouse gases and ozone depletion. Each factor had a distinct influence. The eruption of Mount Pinatubo in 1991 helped cool global climate for several years. Sulphate pollution (see p.207) peaked in the middle of the twentieth century, between World War II and the advent of environmentalism, and it may have helped produce the mid-century cool-down already discussed. Small ups and downs in solar output probably shaped the early-century warming and perhaps the mid-century cooling as well. However, the Sun can't account for the pronounced warming evident since the 1970s. The bottom line is that the model couldn't reproduce the most recent warming trend unless it included greenhouse gases.

Couldn't some undiscovered phenomenon be to blame?

Although many people would love to find a "natural" phenomenon to blame for our warming planet, such as the relationship between clouds and cosmic rays (see p.275), it's growing extremely unlikely that a suitable candidate will emerge. Even if it did, it would beg a difficult question: if some newly discovered factor can account for the climate change we've observed, then why aren't carbon dioxide and the other greenhouse gases producing the warming that basic physics tells us they should be?

And there's another catch. Any mystery process could just as easily be a cooling as a warming agent, and if it were to suddenly wane, it could leave us with even greater warming than we imagined possible. Trusting in the unknown, then, is a double-edged sword. As such, most scientists in the trenches trust in Occam's razor, the durable rule credited to the medieval English logician and friar William of Occam: "One should not increase, beyond what is necessary, the number of entities required to explain anything."

How do rainforests fit into the picture?

As a whole, Earth's land-based ecosystems absorb more CO_2 than they release, but when deforestation is considered on its own, it's a major source of carbon dioxide. Though deforestation has eased somewhat in the last few years, it still accounts for more than 10% of recent human-produced CO_2 emissions.

Tropical forests hold nearly half of the carbon present in vegetation around the world. When they're burned to clear land, the trees, soils and undergrowth release CO_2. Even if the land is eventually abandoned, allowing the forest to regrow, it would take decades for nature to reconcile the balance sheet through the growth of replacement trees that pull carbon dioxide out of the air. In addition to the CO_2 from the fires, bacteria in the newly exposed soil may release more than twice the usual amount of another greenhouse gas, nitrous oxide, for at least two years. Brazil's National Institute for Amazon Research estimates that deforestation puts four times more carbon into the atmosphere than the nation's fossil-fuel burning does.

Rainforests also cool the climate on a more local level, their canopy helping to trap moisture and allow it to slowly evaporate, providing a natural air-conditioning effect. When the rainforest has been slashed and burned over large areas, hotter and dryer conditions often set in, although the exact strength of this relationship is difficult to quantify. Across eastern Brazil, where nearly 20% of the Amazonian rainforest has been destroyed, 2005 saw the region's worst drought in a century, perhaps related to changes in the nearby Atlantic and to rain-suppressing smoke from fires as well as to the deforestation itself.

By contrast, in mid-latitude and polar regions, it appears that forests actually tend to warm the climate (see p.385).

Are hurricanes getting worse because of global warming?

There's tremendous variation in hurricane activity over time and from place to place. A number of studies since 2005 indicate that the number and/or strength of hurricanes have increased in various regions, especially since the 1970s. However, it's likely that some hurricanes at sea went unnoticed in the days before satellites and hurricane-hunter aircraft, and that complicates the assessment. There's no doubt that hurricane activity has stepped up since the mid-1990s in the North Atlantic; here, ocean temperatures have risen through long-term warming and an apparent multidecadal cycle in Atlantic currents. The tropics are part of a global trend toward ocean warming that goes hand in hand with atmospheric warming, and warm oceans provide the energy to drive hurricanes. As for the future, computer models tend to point toward fewer hurricanes overall (for reasons that aren't yet firmed up) but a general strengthening of winds and rainfall in the hurricanes that do form.

Trends aside, a catastrophic storm can strike in any year, and it's impossible to tie any single hurricane or other weather event directly to global warming. Take Hurricane Katrina, which ravaged New Orleans in 2005. Several hurricanes of comparable strength have been observed across the Atlantic over the last century. And the horrific damage caused to the city was the result not only of Katrina's strength but also the storm's track, the weakness of levees and many other factors. That said, the waters of the Gulf of Mexico that fuelled Katrina were at near-record warmth at the time. See p.141 for more on Katrina and other hurricanes and p.116 for details on oceanic changes.

Haven't we had "global cooling" lately?

The planet did cool slightly from the 1940s to the 1970s, mainly in the Northern Hemisphere and most likely a result of the post-war boom in industrial aerosol pollutants that bounce sunlight away from the Earth. Despite a flurry of 1970s media reports on an imminent ice age (see p.270), there was never anything approaching a scientific consensus on the likelihood of further cooling, and it appears that greenhouse warming has long since eclipsed the mid-century cool spell.

After temperatures reached a new global high in 1998, the following decade saw smaller ups and downs. The absence of another show-stopping record led many sceptics and pundits to claim that "global warming stopped in 1998". In truth, nobody expects the global average temperature to rise smoothly from one year to the next. Just as any April in London, New York or Beijing will see a few cold snaps, we can expect long-term warming to be punctuated by periods of little rise in global temperature, or even slight cooling, that could last for a few years. As for record warm years, these will likely occur in association with El Niño events, as noted above.

Recent temperature trends have sometimes been ascribed to "global dimming". While a slight decrease in the amount of sunlight reaching Earth has been measured over the last few decades, it's not been enough to counteract the overall warming (see p.275).

And the ozone hole?

There are a few links between ozone depletion and global warming, but for the most part they're two separate issues. The world community has already taken steps to address the Antarctic ozone hole, which is expected to disappear by the end of this century (see p.32).

The outlook

How hot will it get?

According to the 2007 IPCC report, the global average temperature is likely to rise anywhere from 1.1°C to 6.4°C (2.0–11.5°F) by 2080–99, relative to 1980–99. This range reflects uncertainty about the quantities of greenhouse gases we'll add to the atmosphere in coming decades and also about how the global system will respond to those gases. Some parts of the planet, such as higher latitudes, will heat up more than others. The warming will also lead to a host of other concerns – from intensified rains to melting ice – that are liable to cause more havoc than the temperature rise in itself.

Is global warming necessarily a bad thing?

Whether climate change is bad, good or neutral depends on your perspective. Some regions – and some species – may benefit, but many more will suffer intense problems and upheavals. And some of the potential impacts, such as a major sea-level rise, increased flooding and droughts, intensified hurricanes and many species being consigned to extinction, are bad news from almost any perspective. So while it may be a bit of a reach to think in terms of "saving the planet" from global warming, it's perfectly valid to think about preserving a climate that's sustaining to as many of Earth's residents as possible.

Perhaps the more pertinent question is whether the people and institutions responsible for producing greenhouse gases will bear the impacts of their choices, or whether others will – including those who had no say in the matter. Indeed, people in the poorest parts of the world – such as Africa – will generally be least equipped to deal with climate change, even if the changes are no worse there than elsewhere. Yet those regions have released only a small fraction of the gases that are causing the changes.

How many people might be hurt or killed?

Quantifying the human cost of climate change is exceedingly difficult. The World Health Organization has estimated that in 2000 alone, more than 150,000 people died as a result of direct and indirect climate-change impacts. A more comprehensive study, released in 2009 by the Global Humanitarian Forum (a think tank headed by former UN chief Kofi Annan) found that disasters related to climate change kill some 300,000 people each year. The medical journal *Lancet*, building on IPCC findings, warned in a 2009 report that "climate change is potentially the biggest global health threat in the twenty-first century".

It's important to keep in mind that many of the projected impacts of global warming on society are the combined effects of climate change, development patterns and population growth (some claim the last factor is the most important). It's thus hard to separate out how much of the potential human suffering is due to each factor. Moreover, weather-related disasters kill thousands of people each year regardless of long-term changes in the climate.

In the decades to come, the warming of the planet and the resulting rise in sea level will likely begin to force people away from some coastlines. Low-lying islands are already vulnerable, and entire cities could eventually be at risk. The implications are especially sobering for countries such as **Bangladesh**, where millions of people live on land that may be inundated before the century is out. Residents may not have the luxury of retreating from a slow rise in water level, as deadly surges from hurricanes and other oceanic storms will strike on top of the ever-rising sea.

Another concern is moisture – both too much and too little. In many areas rain is falling in shorter but heavier deluges conducive to flooding, as evidenced by Pakistan's unprecedented monsoon rains and the resulting flood catastrophe in 2010 (see p.66). However, **drought** also seems to be becoming more prevalent. (See p.65 for more on this seeming paradox.) Changes in the timing of rainfall and runoff could complicate efforts to ensure clean water for growing populations, especially in the developing world.

Warming temperatures also stand to facilitate the spread of vector-borne diseases such as **malaria** and dengue fever (see p.172), although other factors are critical and the science on this continues to evolve.

The present coastline of Bangladesh and West Bengal, India, (left) compared with a visualization (produced using tools developed by Dr Mark Mulligan and Google) of the vast inundation that would result (below left) from a 4-metre rise in sea level.

Will agriculture suffer?

That depends on where the farming and ranching is done. Global agricultural productivity is predicted to go *up* over much of the next century, thanks both to the fertilizing effect of the extra CO_2 in the atmosphere and to now-barren regions becoming warm enough to bear crops. However, the rich world looks set to reap the benefits: crop yields in the tropics, home to hundreds of millions of subsistence farmers, are likely to drop. And once global temperatures rise beyond a certain point – perhaps 3°C – then global food production as a whole may begin to decline. Long before then, food shortages could be triggered if large swathes of cropland are used to grow biofuels (see p.180).

And wildlife?

Because climate is expected to change quite rapidly from an evolutionary point of view, we can expect major shocks to some ecosystems – especially in the Arctic – and possibly a wholesale loss of species. According to a 2004 study led by Chris Thomas of the University of Leeds and published in the journal *Nature*, climate change between now and 2050 may commit as many as 37% of all species to eventual extinction – a greater impact than that from global habitat loss due to human land use. Similar figures emerged from the 2007 IPCC report, which pegs the percentage of plant and animal species that are at risk from a temperature rise of 1.5–2.5°C (2.7–4.5°F) at 20–30%.

How will the economy be affected?

There is a divergence of views regarding the impact of climate change – and indeed emissions cuts – on national and international economies in the coming century. The most widely cited research into this question was carried out on behalf of the UK government by the economist Nicholas Stern. The Stern Review, released in 2006, concluded that global warming could reduce global GDP by 20% within the current century; it recommended immediately spending 1% of global GDP each year on reducing emissions. Stern later concluded that his report underestimated the speed, scale and likelihood of some serious climate impacts, and he now advocates spending 2% of GDP on mitigation. Other economists have questioned Stern's methodology, however, and have estimated either higher or lower figures for the costs of avoiding and ignoring climate change.

Will rising seas really put cities such as New York and London under water?

Not right away, but it may be only a matter of time. In its 2007 report, the IPCC projects that sea level will rise anywhere from 180mm to 590mm (7–23in) by 2090–2100. This range is smaller than in the IPCC's 2001 report, but it excludes some key uncertainties about how quickly warming will melt land-based ice. The last few years have seen glaciers accelerating their seaward flow in many spots along the margins of Greenland and West Antarctica. Computer models don't depict the dynamics behind this speed-up very well, so it's not explicitly included in the IPCC projection, but the report does note the added risk at hand. After considering new data on such melting, the US National Research Council estimated in 2010 that sea-level rise from all sources may total 500–1000mm (20–40in) by 2100.

Unfortunately, sea level won't stop rising then, which means coastal areas will have to contend with an ever-worsening threat. If emissions continue to rise unabated through this century, the Greenland and/or West Antarctica ice sheets could be thrown into an unstoppable melting cycle that would raise sea level by more than 7m (23ft) each. This process would take some time to unfold – probably a few centuries, although nobody can pin it down at this point – but should it come to pass, many of the world's most beloved and historic cities would be hard-pressed to survive.

Will the Gulf Stream pack up, freezing the UK and northern Europe?

The Gulf Stream and North Atlantic Drift bring warm water (and with it warm air) from the tropical Atlantic to northern Europe. This helps keep the UK several degrees warmer than it would otherwise be. Although this system is unlikely to pack up entirely, there is a possibility that it could be diminished by climate change. The reason is that increasing rainfall and snow-melt across the Arctic and nearby land areas could send more freshwater into the North Atlantic, pinching off part of the warm current. The best estimate is that the flow might weaken by 10–50% over the next century or so. That's probably not enough to offset global warming completely for the UK or northwest Europe, although it could certainly put a dent in it. In any case, the impacts would be much smaller – and would take much longer to play out – than the scenario dramatized in the film *The Day After Tomorrow* (see p.131).

Will we reach a "tipping point"?

Some aspects of climate change are expected to unfold in a roughly linear fashion. Computer models, and the changes we've seen to date, suggest that many of the regional and seasonal fingerprints of our future climate will be roughly twice as strong for 2°C of global warming as for 1°C.

On top of these incremental changes, there's a real risk of even bigger trouble because of positive-feedback processes. These tend to amplify change and make it worse, pushing conditions well beyond what one would otherwise expect. The challenge is to identify the points at which the most dangerous positive feedbacks will kick in. For instance, scientists consider it likely that the Greenland ice sheet will begin melting uncontrollably if global temperatures climb much more than 2°C (3.6°F). Because of the implications for coastal areas, as noted above, this is a particularly worrisome threshold.

Since each positive feedback has its own triggering mechanism, some would no doubt occur sooner than others. Thus, there is no single temperature agreed upon as a tipping point for Earth as a whole. However, scientists, governments and activists have worked to identify useful targets that might focus attention and action. One goal adopted by the European Union, as well as many environmental groups, is to limit global temperature rise to 2°C (3.6°F) over pre-industrial levels. But that ceiling looks increasingly unrealistic – we're already close to 40% of the way there, and only the lower fringes of the latest IPCC projections keep us below the 2°C threshold by century's end.

Another approach is to set a stabilization level of greenhouse gases – a maximum concentration to be allowed in the atmosphere – such as 500 parts per million of CO_2 as compared to 270–280ppm in pre-industrial times and about 390ppm today. Still another concept is to limit the total amount of carbon added to the air over time. You can learn more about these and other goals in "The predicament", p.298.

What can we do about it?

What would it take to stop climate change?

It isn't possible to simply "stop" climate change. The century-plus lifespan of atmospheric CO_2 means that the planet is already committed to a substantial amount of greenhouse warming. Even if we turned off every fuel-burning machine on Earth tomorrow, climate modellers tell us that the world would warm at least another 0.5°C (0.9°F) as the climate adjusts to greenhouse gases we've already emitted.

The best the world can do, therefore, is to try and limit the amount of warming that will occur. And this will only be possible if and when changes in technology and lifestyle enable us to pull back far beyond our current emission levels, or unless we find some safe method to remove enormous amounts of carbon from the atmosphere, or both. That's a tall order – but if we're determined to reduce the risk of a wide range of climate impacts, we have no choice but to fulfil it.

Given the global nature of the climate problem, slashing emissions on the scale needed will almost certainly only be possible as part of an ambitious political agreement between the world's nations.

Which countries are emitting the most greenhouse gases?

For many years the United States was in first place, with 30% of all of the human-produced greenhouse emissions to date and about 20% of yearly totals – despite having only a 5% share of global population. However, China has now taken the lead. Its emissions are considerably lower per capita, but due to its growing population and affluence, China leapfrogged the US as the world's leading greenhouse emitter around 2006. As shown in the diagram on p.314, the world's industrialized countries vary widely in how much they have increased or decreased their total emissions since 1990. Some of the decreases were due to efficiency gains, while others were due to struggling economies.

Has the Kyoto Protocol made a difference?

This United Nations-sponsored agreement among nations was the first to mandate country-by-country reductions in greenhouse-gas emissions. Kyoto emerged from the UN Framework Convention on Climate Change, which was signed by nearly all nations at the 1992 mega-meeting popularly known as the Earth Summit. The framework pledges to stabilize greenhouse-gas concentrations "at a level that would prevent dangerous anthropogenic interference with the climate system". To put teeth into that pledge, a new treaty was needed, one with binding targets for greenhouse-gas reductions. That treaty was finalized in Kyoto, Japan, in 1997, after years of negotiations, and it went into force in 2005. Nearly all nations have now ratified the treaty, with the notable exception of the United States (see map overleaf). Developing countries, including China and India, weren't mandated to reduce emissions, given that they'd contributed a relatively small share of the current century-plus build-up of CO_2.

Under Kyoto, industrialized nations pledged to cut their yearly emissions of carbon, as measured in six greenhouse gases, by varying amounts, averaging 5.2%, by 2012 as compared to 1990. That equates to a 29% cut in the values that would have otherwise occurred. However, the protocol didn't become international law until more than halfway through the

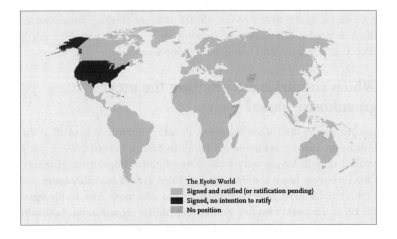

The Kyoto World
- Signed and ratified (or ratification pending)
- Signed, no intention to ratify
- No position

1990–2012 period. By that point, global emission amounts had risen substantially. Some countries and regions, including the European Union, were on track by 2011 to meet or exceed their Kyoto goals, but other large nations were falling woefully short. And the two biggest emitters of all – the United States and China – churned out more than enough extra greenhouse gas to erase all the reductions made by other countries during the Kyoto period. Worldwide, emissions soared by nearly 40% from 1990 to 2009, according to the Netherlands Environmental Assessment Agency.

What will the next global deal look like?

Even if every nation lived up to its Kyoto goals, that would only make a tiny dent in the world's ever-increasing output of greenhouse gases. Reducing greenhouse gas emissions by a few percent over time is akin to overspending your household budget by a decreasing amount each year: your debt still piles up, if only at a slower pace. Hence most people who are concerned about climate change are pinning their hopes on world leaders agreeing to much more ambitious emissions cuts.

At one point, hopes were high for a strong post-Kyoto deal, especially following the election of US president Barack Obama in 2009. However, the momentum faded during the run-up to a massive UN meeting that December in Copenhagen, where the outlines of a new global agreement were to be set. That meeting descended into near-chaos, with countries far apart on a variety of issues and global economic turmoil overshadowing it all. The final outcome – mainly a set of voluntary national goals – struck many observers and participants as a huge disappointment. On top of that, a series of controversies involving climate scientists and the IPCC (see p.294) likely weakened the hand of politicians in gaining public support for a far-reaching new deal. With all this in mind, it could take years for the next global agreement on emission cuts to emerge.

Does the growth of China and India make a solution impossible?

Not necessarily. Although its growth in coal production is hugely worrisome, China is already making progress on vehicle fuel efficiency and other key standards. And because so much of the development in China and India is yet to come, there's a window of opportunity for those nations to adopt the most efficient technologies possible. At the same time, the sheer numbers in population and economic growth for these two countries are daunting indeed – all the more reason for prompt international collaboration on technology sharing and post-Kyoto diplomacy.

If oil runs out, does that solve the problem?

Hardly. It's true that if oil resources do "peak" in the next few years, as some experts believe, we're likely to see economic downswings, and those could reduce oil-related emissions, at first over periods of a few years and eventually for good. The same applies to natural gas, although that peak could arrive decades later. Then, the question becomes what fuel sources the world will turn to: coal, nuclear, renewables or some combination of the three. If the big winner is coal – or some other, less-proven fossil source such as shale or methane hydrates – it raises the potential for global warming far beyond anything in current projections.

Even if renewables win the day later in this century, we're still left with the emissions from today's stocks of oil, gas and coal, many of which would likely get burned between now and that eco-friendly transition. With this in mind, research has intensified on **sequestration** – how carbon might be safely stored underground. The idea appears promising, but big questions remain (see p.335). There's also geoengineering – a set of grand schemes to pull carbon directly from the atmosphere or block the Sun through mirrors, soot or artificially boosted clouds. All of these geoengineering ideas have huge unknowns, but scientists and governments are taking a closer look at them, if only as last-ditch alternatives to unstoppable warming (see p.359).

Won't nature take care of global warming in the long run?

Only in the *very* long run. The human enhancements to the greenhouse effect could last the better part of this millennium. Assuming that it takes a century or more for humanity to burn through whatever fossil fuels it's destined to emit, it would take hundreds of years before most of the CO_2 emissions were absorbed by Earth's oceans and ecosystems, and some of the added CO_2 would remain in the air for thousands of years.

There are few analogies in the geological past for such a drastic change in global climate over such a short period, so it's impossible to know what will happen after the human-induced greenhouse effect wanes. All else being equal, cyclical changes in Earth's orbit around the Sun can be expected to trigger an ice age sometime within the next 50,000 years, and other warmings and coolings are sure to follow, as discussed in the box on p.244. In the meantime, we'll have our hands full dealing with the next century and the serious climate changes that our way of life may help bring about.

The greenhouse effect

How global warming works

Imagine our planet suddenly stripped of its atmosphere – a barren hunk of rock floating in space. If this were the case, then Earth's near-ground temperature would soar by day but plummet by night. The average would be something close to a bone-chilling -18°C (0°F). In reality, though, Earth's surface temperature now averages a much more pleasant 14.4°C (57.9°F). Clearly, there's something in the air that keeps things tolerably warm for humans and other living things. But what?

One of the first people to contemplate Earth's energy balance was the French mathematician and physicist Joseph Fourier. His calculations in the 1820s were the first to show the stark temperature contrast between an airless Earth and the one we actually enjoy. Fourier knew that the energy reaching Earth as sunlight must be balanced by energy returning to space, some of it in a different form. And though he couldn't pin down the exact process, Fourier suspected that some of this outgoing energy is continually intercepted by the atmosphere, keeping us warmer than we'd otherwise be.

Harking back to experiments by others on how a glass box traps heat, Fourier likened the atmosphere to a hothouse (or greenhouse). *Voilà* – the concept of the **greenhouse effect** was

"**Remove for a single summer-night the aqueous vapour from the air which overspreads this country, and you would assuredly destroy every plant capable of being destroyed by a freezing temperature.**"
John Tyndall, 1863

born. It's been with us ever since, even though it's a flawed analogy. The atmosphere doesn't imprison the air the way a glass box does. Instead, it absorbs **infrared radiation** rising from Earth's Sun-warmed surface. All else being equal, the more greenhouse gas there is, the less radiation can escape from Earth to space, and the warmer we get (but there are a few twists along the way, as we'll see).

The diagram opposite shows what happens to the sunlight that reaches our planet:

▶ **About 30%** gets reflected or scattered back to space by clouds, dust or the ground, especially from bright surfaces like ice.

▶ **More than 20%** is absorbed in the atmosphere, mainly by clouds and water vapour.

▶ **Almost 50%** gets absorbed by Earth's surface – land, forests, pavement, oceans and the rest.

The incoming radiation from the intensely hot Sun is mostly in the visible part of the spectrum – which is why you shouldn't stare at the Sun. Earth, being much cooler than the Sun, emits far less energy, most of it at infrared wavelengths we can't see.

Some of Earth's outgoing radiation escapes through the atmosphere directly to space. Most of it, though, is absorbed en route by clouds and greenhouse gases (including water vapour) – which in turn radiate some back to the surface and some out to space. Thus, Earth's energy budget is maintained in a happy balance between incoming radiation from the Sun and a blend of outgoing radiation from a warm surface and a cooler atmosphere (an important temperature distinction, as we'll see below).

The air's two main components, nitrogen (78%) and oxygen (20%), are both ill suited for absorbing radiation from Earth, in part because of their linear, two-atom (diatomic) structure. But some other gases have three or more atoms, and these branched molecules capture energy far out of proportion to their scant presence. These are the **greenhouse gases**, the ones that keep Earth inhabitable but appear to be making it hotter.

Most greenhouse gases are well mixed throughout the **troposphere**, the lowest 8–16km (5–10 miles) of the atmosphere. (Water vapour is the big exception; it's much more concentrated near ground level.) As mountain climbers well know, the troposphere gets colder as you go up, and so these greenhouse gases are cooler than Earth's surface is. Thus, they radiate

Sunlight (shortwave radiation)

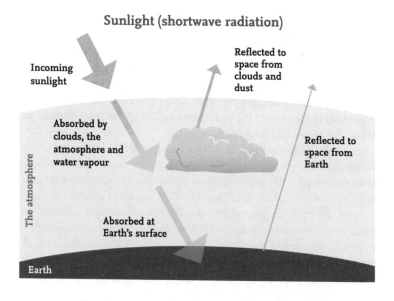

Incoming
sunlight

Reflected to
space from
clouds and
dust

Absorbed by
clouds, the
atmosphere and
water vapour

Reflected to
space from
Earth

The atmosphere

Absorbed at
Earth's surface

Earth

Incoming and outgoing radiation: the diagram above shows sunlight entering
the atmosphere. The light absorbed (as opposed to reflected) is given out as
infrared radiation, as shown in the diagram below. The widths of the arrows in
both diagrams reflect the warming power of each process, relative to the total
sunlight coming in.

Infrared (longwave radiation)

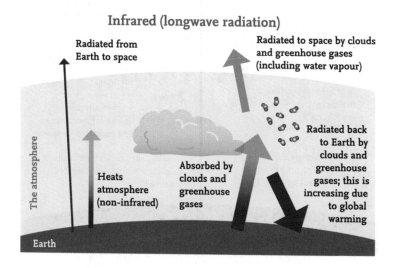

Radiated from
Earth to space

Radiated to space by clouds
and greenhouse gases
(including water vapour)

The atmosphere

Heats
atmosphere
(non-infrared)

Absorbed by
clouds and
greenhouse
gases

Radiated back
to Earth by
clouds and
greenhouse
gases; this is
increasing due
to global
warming

Earth

less energy to space than Earth itself would. That keeps more heat in the atmosphere and thus helps keep our planet livable.

All well and good – but the more greenhouse gas we add, the more our planet warms. As carbon dioxide and other greenhouse gases accumulate, the average elevation from which they radiate energy to space ascends to even higher, and colder, heights. This means less energy leaving the system, which in turn causes the atmosphere to heat up further. The diagram below shows how this works in more detail.

Once the added gases get the ball rolling through this process, a series of atmospheric readjustments follows. These are largely **positive feedbacks** that amplify the warming. More water evaporates from oceans and lakes, for instance, which roughly doubles the impact of a carbon-dioxide increase. Melting sea ice reduces the amount of sunlight reflected to space. Some feedbacks are less certain: we don't know whether cloud patterns will change to enhance or diminish the overall warming. And, of course, we're not talking about a one-time shock to the system. The whole planet is constantly readjusting to the greenhouse gases we're adding.

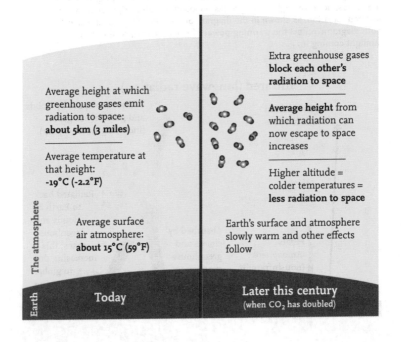

The atmosphere

Average height at which greenhouse gases emit radiation to space: **about 5km (3 miles)**

Average temperature at that height: -19°C (-2.2°F)

Average surface air atmosphere: **about 15°C (59°F)**

Extra greenhouse gases **block each other's radiation to space**

Average height from which radiation can now escape to space increases

Higher altitude = colder temperatures = **less radiation to space**

Earth's surface and atmosphere slowly warm and other effects follow

Earth

Today

Later this century (when CO$_2$ has doubled)

A rogue's gallery of gases

In the 1860s, eminent Irish scientist **John Tyndall** became the first to explore and document the remarkable power of greenhouse gases. Intrigued by the shape-shifting behaviour of light as it passes through various substances, Tyndall put a number of gases to the test in his lab, throwing different wavelengths of light at each one to see what it absorbed. Almost as an afterthought, he tried coal gas and found it was a virtual sponge for infrared energy. Tyndall went on to explore carbon dioxide and water vapour, which are both highly absorbent in certain parts of the infrared spectrum. The broader the absorption profile of a gas – the more wavelengths it can absorb – the more powerful it is from a greenhouse perspective.

Here are the main greenhouse gases of concern. The diagram shows the prevalence of each gas and how much of an impact it's had in enhancing the overall greenhouse effect.

▶ **Carbon dioxide (CO$_2$)**, the chief offender, accounts for about 390 of every million molecules in the air, or 390 **parts per million (ppm)**. That number has been climbing by 1–3ppm, or about 0.25 to 0.75 percent, per year. The worldwide **emissions** of CO$_2$ are increasing at several percent per year on average, but that annual ramp-up becomes a smaller percentage when it joins the large amount of CO$_2$ already in the air.

Nitrous oxide (0.32ppm) 6%

CFCs, HFCs (about 1ppb) 12%

Carbon dioxide (380ppm) 63%

Methane (1.77ppm) 18%

The prevalence of human-produced greenhouse gases and their relative importance in affecting Earth's radiative balance in 2005, according to the 2007 IPCC report. The percentages depend on each gas's prevalence and its molecular ability to trap energy. The numbers total less than 100% due to rounding and other trace gases. Not shown are tropospheric ozone (produced indirectly) and water vapour.

Both a pollutant and a natural part of the atmosphere, carbon dioxide is produced when fossil fuels are burned as well as when people and animals breathe and when plants decompose. Plants and the ocean soak up huge amounts of carbon dioxide, which is helping to keep CO_2 levels from increasing even more rapidly (see p.39). Because the give-and-take among these processes is small compared to the atmospheric reservoir of CO_2, a typical molecule of carbon dioxide stays airborne for over a century.

Atmospheric levels of CO_2 held fairly steady for centuries – at around 270–280ppm – until the Industrial Revolution took off. In past geological eras, the amount of CO_2 has risen and fallen in sync with major climate changes, although there's chicken-or-egg uncertainty in whether CO_2 led or lagged some of these transitions (see p.226).

▶ **Methane** emerges from rice paddies, peat bogs and belching cows as well as from vehicles, homes and factories. It's a greenhouse powerhouse: though it stays in the air for less than a decade on average, its effect on global warming is roughly 25 times that of carbon dioxide per molecule when both are considered over a century-long span. Methane's total annual contribution to the greenhouse effect is thought to be roughly a third as big as carbon dioxide's, even though it makes up only around 2ppm of the atmosphere (see pie chart, p.27).

After soaring in the last few decades, the amount of atmospheric methane rose more fitfully in the 1990s and through most of the following decade. A 2006 study by the US National Oceanic and Atmospheric Administration pinned the slowdown on widespread drought in Northern Hemisphere wetlands, together with sagging economies across eastern Europe and Russia in the 1990s. Starting in 2007, methane concentrations began to rise once more, apparently due to a combination of heavy tropical rains, Arctic warmth and biomass burning.

▶ **Ozone** Like a versatile actor, **ozone** takes on multiple roles, but in the greenhouse drama, it's only a supporting player. Instead of being emitted when fossil fuels are burned, ozone forms when sunlight hits other pollutants and triggers ozone-making reactions. Its presence can spike to unhealthy levels when stagnant air takes hold for days near ground level, where people, animals and plants live and breathe. Fortunately, ozone

Venus: a cautionary tale

Scientists don't have the luxury of an extra Earth for experimentation. However, our nearest planetary neighbour serves as an example of what can happen when the greenhouse effect runs amok. Many scientists once saw Venus as a mysterious sibling, one whose climate might be mild enough to support human or humanlike life. But although the planet is named for the Roman goddess of love and beauty, it's actually a rather harsh place. That's because of carbon dioxide, which makes up less than 0.04% of Earth's atmosphere but more than 96% of Venusian air. Factor in the planet's location – about 25% closer to the Sun than Earth is – and you end up with a surface air temperature in the neighbourhood of 460°C (860°F), hot enough to melt

lead. (All that CO_2 is invisible, by the way; it's a dash of sulphuric acid that produces Venus's legendary cloak of haze.)

Carl Sagan, the US astronomer and science popularizer, built his early career on Venus, as it were, studying its sizzling atmosphere. In the early 1960s Sagan drew on radio observations and simple mathematical modelling to explain how the planet's dense, superheated atmosphere could be produced by what he called a "runaway greenhouse effect". Sagan's findings led to his concern about the fate of Earth's own atmosphere: in 1984 he was the first witness in then-senator Al Gore's landmark congressional hearings on global warming.

survives only a few days in the **troposphere**. This makes it hard to assess its global concentration, though the limited data that exist indicate an average of around 34 parts per billion. There's apparently been little change in tropospheric ozone amounts since the 1980s, but models hint at a global increase of about 30% since the Industrial Revolution.

Higher up, a protective natural layer of ozone warms the stratosphere while helping to shield Earth from ultraviolet light. The

human-induced hole that appears each year in this layer of "good"
ozone has led to record-low temperatures in the lower stratosphere.

▶ **Water vapour** isn't a very strong greenhouse gas, but
it makes up for that weakness in sheer numbers. If
you're on a warm tropical island with 100% relative
humidity, that means the balmy breezes are wafting
as much water vapour as they can – perhaps 3 out
of every 100 molecules in the air as a whole. When
averaged globally, the concentration of water vapour is much less, and
it varies greatly by location and time of year. It appears to be increasing
by perhaps 1% a decade, far more slowly than carbon dioxide.

Water vapour works to propagate itself and boost global warming
through an interesting twist. As global temperatures rise, oceans
and lakes release more water vapour, obeying a well-known law of
thermodynamics. The extra water vapour, in turn, adds to the warming
cycle. This is one of several positive feedbacks critical to the unfolding
of future climate change.

▶ **A few other gases** – extremely scant but extremely powerful – make up
the rest of the greenhouse palette. **Chlorofluorocarbons** (CFCs) and
related compounds increased rapidly until they were identified as a
key player in stratospheric ozone depletion (see box, p.32). Under the
Montreal Protocol, they've begun to level off. Along with helping to
destroy "good" ozone, they are also powerful, long-lived greenhouse
gases – another good reason we're phasing them out, though some
replacements are just as risky for climate. **Nitrous oxide** is also an
industrial byproduct, showing up at only about 300 parts per billion
(ppb), but with about 300 times the effect of CO_2, molecule for molecule,
over its century-long lifespan in the atmosphere. Though its immediate
effect on radiative balance is less than that of CFCs (as shown in the
chart on p.27), its long lifetime makes it far more important than CFCs
when the effects are measured over a 100-year time frame.

Greenhouse pioneers

Virtually no one at the peak of the Victorian age reckoned that burning
coal or oil would tamper with our climate. But it was becoming clear that
great changes in climate had occurred before. Chief among those were the
ice ages, which coated North America and Eurasia with kilometre-thick

sheets of ice as far south as modern-day Germany and Illinois. Fossils proved that this ice cover had persisted well into the era of early human civilization, some 12,000 years ago. What made the climate plummet into such frigidity and then recover? And could it happen again?

In the mid-1890s, the Swedish chemist **Svante Arrhenius** took a look at how carbon dioxide might be implicated in ice ages. His unexpected results provided our first real glimpse of a greenhouse future. Scientists already knew that volcanoes could spew vast amounts

Svante Arrhenius

of greenhouse gas, so Arrhenius wondered if a long period of volcanic quiet might allow carbon dioxide levels to draw down and perhaps help plunge Earth into an ice age. He set out to calculate how much cooling might result from a halving of CO_2. When he included the role of water vapour feedback, Arrhenius came up with a global temperature drop of around 5°C (9°F).

Soon enough, a colleague inspired him to turn the question around: what if industrial emissions grew enough to some day double the amount of carbon dioxide in the air? Remote as that possibility seemed, Arrhenius crunched the numbers again and came up with a similar amount, this time with a plus sign: a warming of about 5°C (9°F).

Amazingly, this century-old calculation isn't too far off the mark. Scientists now use a **doubling of CO_2** as a reference point for comparing computer models and a benchmark for where our atmosphere might stand by around 2050 or a little afterwards, if present trends continue. Most models peg the likely warmup from doubled CO_2 at somewhere between 1.5°C and 4.5°C (2.7–8.1°F). All in all, Arrhenius's initial forecast was quite impressive for the pre-computer era.

In his own time, though, Arrhenius was a lone voice. He himself far underestimated the pace of global development, figuring it would take until the year 4000 or so for his projected doubling to occur. Even then, he figured, a little

> **"By the influence of the increasing percentage of carbonic acid [CO_2] in the atmosphere, we may hope to enjoy ages with more equable and better climates, especially as regards the colder regions of the Earth..."**
> Svante Arrhenius, 1908

Is the ozone hole linked to global warming?

The saga of ozone depletion in the stratosphere is conflated with global warming in the minds of many. Both topics came into public view during the 1980s, often lumped together under the heading of **global change**. There are several links between the two – just enough to cause confusion – but at heart they're two distinct issues.

Ozone, a greenhouse gas, is a pollutant at ground level, harmful when we breathe it. However, the ozone layer that sits within the **lower stratosphere** (especially at about 25–40km/15–25 miles high) is a godsend. Even though the ozone makes up only a tiny fraction of the stratospheric air, it intercepts much of the **ultraviolet light** that can produce sunburns and skin cancer, damage our eyes, and cause other kinds of trouble for people and ecosystems.

A dramatic seasonal depletion in this layer of ozone was found over Antarctica in 1985. Shortly thereafter, scientists identified the three factors that conspire to form the ozone hole, which waxes and wanes during each Southern Hemisphere spring. The first ingredient is a special type of cloud, a **polar stratospheric cloud**, that only forms when winter temperatures fall below about -80°C/-112°F at high altitudes and latitudes. Also needed are **chlorofluorocarbons (CFCs)**, used since the 1920s in spray cans, air conditioners and many other places. CFCs are heavier than clean air, but they mix easily through the atmosphere; once lofted into the stratosphere, they can remain there long enough to do damage. (CFCs are also greenhouse gases themselves; they account for roughly 10% of the human-enhanced greenhouse effect.)

The final protagonist is **sunlight**. As the six-month Antarctic night comes to an end each September, round-the-clock sunshine helps break down the CFCs. This releases **chlorine**, and the chlorine uses the surface of the polar stratospheric cloud to break down ozone into **oxygen**. A single molecule of chlorine can destroy many ozone molecules over a few weeks. During that time, about half of all the ozone through the depth of the Antarctic atmosphere typically vanishes, with near-

global warming might not be such a bad thing. Europe was just emerging from the hardships of the Little Ice Age, which had much of the continent shivering and crops withering from about 1300 to 1850. Arrhenius figured a little extra CO_2 might help prevent another such cold spell.

Another greenhouse pioneer came along in the 1930s, by which time the globe was noticeably heating up. **Guy Stewart Callendar**, a British engineer who dabbled in climate science, was the first to point to human-produced greenhouse gases as a possible cause of already observed warming. Callendar estimated that global temperatures might rise around 1.0°C (1.8°F) by 2200. Like Arrhenius, he wasn't too perturbed about this prospect.

complete ozone loss in parts of the lower stratosphere. By December of each year, though, the stratosphere has warmed up, the clouds disappear, and the ozone hole fills in.

Fortunately, the ozone hole has never extended much beyond Antarctica, although it has encroached on southern Chile. Southern Australia and New Zealand, while outside the hole per se, have seen ozone reductions of more than 10% at times. As far as the Arctic goes, its wintertime vortex is less stable than its Antarctic counterpart, which limits the growth of polar stratospheric clouds and helps keep a bona fide ozone hole from forming. Still, the springtime depletion over the Arctic can be as high as 60% in some years. A weaker but broader and more persistent **ozone depletion** of some 5–10% extends across much of the globe. Some of this worldwide depletion is likely due to the yearly dispersal of the Antarctic ozone hole and the mixing of that ozone-depleted air around the globe.

Since ozone absorbs sunlight, its depletion can have a **cooling** effect. This helps explain why, even as Earth's surface air temperatures reach record highs, record lows are being notched up in the stratosphere (see p.202). The resulting changes in air circulation over Antarctica sometimes extend to the surface, where the frozen interior has warmed far less than coastal areas (see p.99).

Unlike global warming, there's an end to ozone depletion in sight, at least on paper. The 1987 **Montreal Protocol**, orchestrated by the United Nations and ratified with amazing speed, called for CFCs to be replaced by substitutes such as **halo-chlorofluorocarbons**, which have shorter lifetimes and are far less likely to break down in a way that damages ozone. Chlorine concentrations have stabilized in the stratosphere and may already be going down. According to a special IPCC report in 2005, the ozone layer should be rebuilding over the next several decades, although we can expect a few ups and downs along the road to recovery. Indeed, while the yearly extent of the ozone hole has generally levelled off after the year 2000, the most extensive ozone hole on record occurred in 2006.

In any case, many scientists discounted the message of Arrhenius and Callendar because of the fact that CO_2 and water vapour absorbed energy at overlapping wavelengths. Laboratory tests seemed to show that the two components of the atmosphere were already doing all the absorbing of infrared energy that they could: enlarging the atmospheric sponge, as it were, could have only a minuscule effect. Only after World War II did it become clear that the old lab tests were grievously flawed because they were carried out at sea level. In fact, carbon dioxide behaved differently in the cold, thin air miles above Earth, where it could absorb much more infrared radiation than previously thought. And CO_2's long lifetime meant that it could easily reach these altitudes. Thus, the earthbound absorption tests proved fatefully wrong, one of many dead ends that kept us from seeing the power of the greenhouse effect until industrialization was running at full tilt.

The tale told by a curve

If there's one set of data that bears out the inklings of Arrhenius and Callendar, it's the record of CO_2 collected atop Hawaii's Mauna Loa Observatory since 1958 (see graph below). **Charles Keeling** convinced the Scripps Institution of Oceanography to fund the observing site as part of the International Geophysical Year. Because of CO_2's stability and longevity, Keeling knew that the gas should be well mixed throughout Earth's atmosphere, and thus data taken from the pristine air in the centre of the Pacific could serve as an index of CO_2 valid for the entire globe. After only a few years, Keeling's sawtoothed curve began to speak for itself (see graph below). It showed a steady long-term rise in CO_2, along with a sharp rise and fall produced each year by the wintering and greening of the Northern Hemisphere (whose land area and plant mass far outstrip those of the Southern Hemisphere).

With Keeling's curve as a backdrop, a growing number of scientists began to wonder whether human-induced warming might take shape much sooner than Callendar or Arrhenius predicted, as population grew

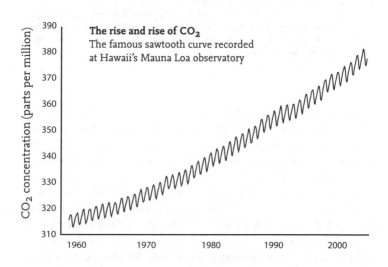

The rise and rise of CO₂
The famous sawtooth curve recorded at Hawaii's Mauna Loa observatory

-20 -16 -13 -11 -9 -7 -5 -3.6 -2.8 -2 -1.2 -0.4 0.4 1.2 2 2.8 3.6 5 7 9 11 13 16 20°F

Surface Air Temperature Change [°F]

Computer models of the global atmosphere confirm the long-held understanding that rising concentrations of greenhouse gases lead to a warmer climate. These simulations from the NOAA Geophysical Fluid Dynamics Laboratory show the projected changes in average seasonal temperatures (June through August on the left, December through February on the right) for the period 2051–2060 compared to 1971–2000. The model incorporated increases in greenhouse gases and aerosols based on a "middle of the road" estimate of future emissions.

and industrialization proceeded. In the 1960s, climate scientists devised their first primitive renditions of Earth's atmosphere in **computer models**. Like the first crude motion pictures, even these early models had a compelling quality, a whiff of the plausible. One model created by Syukuro Manabe and Richard Wetherald at the US Geophysical Fluid Dynamics Laboratory showed in 1967 that a doubling of CO_2 could produce warming of 2.0°C (3.6°F).

The increasingly complex models developed since the 1960s have refined the picture, but they haven't changed it all that much (see p.249). The best-guess warming for a doubling of CO_2 is in the ballpark of 3.0°C (5.4°F). The Mauna Loa data continue to show carbon dioxide concentrations rising inexorably. And despite ratification of the Kyoto Protocol, the world remains perilously far from even beginning to stabilize CO_2 levels in the atmosphere. Although Arrhenius foresaw a century ago that humans could accentuate the natural greenhouse effect, he would doubtless be amazed at how quickly the process is unfolding and the far-reaching effects it's having on society.

Who's responsible?

Which countries, regions and activities are warming the world?

It took an awful lot of coal, gas and oil passing through engines and stoves to boost the atmosphere's level of carbon dioxide by 35% over the last century and a half. There's been able assistance from other heat-trapping greenhouse gases, such as methane, as well as from deforestation and other changes in land use. Still, most of the human-produced global warming to date appears to be due to CO_2.

Where did it all come from? And what are the main sources today? Asking these questions isn't merely an exercise in finger-pointing. In order to get a handle on where global emissions are headed, it's critical to know what parts of the world and what economic sectors are contributing the most to the current situation.

When it comes to the greenhouse effect, one nation's emissions are everyone's problem. Ordinary pollution disperses in a few days and tends to focus its worst effects in the region where it was generated, plus some areas downwind. Carbon dioxide is much longer lived: once added to the air, it stays there for more than a century. That gives the natural mixing processes of the atmosphere time to shake and stir CO_2 evenly around the globe. Some of the countries producing minimal amounts of greenhouse gases, such as tiny island nations, are among the most vulnerable to the climatic fallout from emissions produced largely by a few big countries.

Scientists can't yet draw a straight line from emissions to a specific climate impact, but in some cases they can now characterize the extent to which greenhouse warming hiked the odds of a particular climate event, such as the 2003 European heat wave (see p.61). It's not a stretch to imagine courts in the future using such findings, along with the tally of who emits what and where, in order to assign partial responsibility for major climate shocks.

Most of the data below are drawn from analyses by the **Intergovernmental Panel on Climate Change** (IPCC) and the **Pew Center on Global Climate Change** (working with the **World Resources Institute**), as well as the **US Carbon Dioxide Information Center**.

How much greenhouse gas is in the air right now?

The global atmosphere currently carries about **3000 gigatonnes** of carbon dioxide, which includes about 800 gigatonnes of carbon. (A gigatonne is a billion metric tonnes.) That's by far the largest presence of any human-produced greenhouse gas. The atmosphere also holds about four giga-tonnes of carbon in the form of methane, which is a much stronger but shorter-lived greenhouse gas.

How much are we adding each year?

As of 2009, humans were directly adding more than 31 gigatonnes of carbon dioxide into the atmosphere per year – a stout increase from around 15 gigatonnes in 1970 and about 26 gigatonnes as recently as 2002. This is mainly from burning fossil fuels, though about 4% of the total comes from cement production. Almost half of this yearly output remains in the atmosphere, while the rest is absorbed (more on this later). Deforestation adds even more carbon dioxide to the air – an amount equal to more than 10% of the total when considered together with fossil fuels.

The global **emission rate** of CO_2 doesn't go up every year. During economic slowdowns, the rate can level off, as it did in 2009, or even decline slightly, as it did in the early 1990s. Even in those years, though, we're still adding lots of carbon dioxide to the atmosphere. Indeed, the **total amount** of CO_2 in the air – as seen in the graph on p.34 – has risen every year since measurements began in the late 1950s.

To put these figures in perspective, consider **per-person emissions**. Each human on Earth is responsible, on average, for putting about 4.7 metric tonnes of carbon dioxide into the air each year. That's 4700kg, or about 10,300lb. In other words, someone who weighs 68kg (150lb) puts the equivalent of her weight in carbon dioxide into the atmosphere every five to six days. Of course, the real figures vary enormously based on where and how you live. For most people in Europe and North America, the average is much higher. For more on the "carbon footprints" of individuals, see p.365.

Comparing greenhouse gases

Because greenhouse gases vary so widely in the power of their climatic effects (see p.27), researchers often rely on a unit called **global warming potential**. The global warming potential of a gas is a measure of its contribution per unit mass to greenhouse warming in the atmosphere over a given time span as compared to that of carbon dioxide. Methane, for example, is shorter-lived than CO_2 but much more powerful in its ability to trap heat in the atmosphere. Thus, over a century's time, methane's global warming potential is estimated to be somewhere around 25 (a number that's not yet set in stone), compared to a value of 1 for carbon dioxide. These figures can be multiplied by the prevalence of each gas to produce a **carbon equivalent** that enables all emissions to be considered as a group. By using carbon equivalents, researchers get a better sense of the impact of the atmosphere's total greenhouse-gas burden. Some studies use **carbon dioxide equivalent,** a number obtained by multiplying the carbon equivalent by 44/12 (the ratio of the molecular weight of carbon dioxide to that of carbon).

Human-produced methane emissions are harder to quantify, but they could represent another 0.3 to 0.45 gigatonnes per year, with natural processes adding almost that much methane each year. Interestingly, the total amount of methane in the atmosphere hasn't changed much at all in the new millennium, though it did rise by roughly 0.5% over the period 2007–08.

What happens to the greenhouse gas we put into the air?

It's easy to assess how much CO_2 is burned each year, at least in principle. And because CO_2 is mixed throughout Earth's atmosphere over its long lifetime, it's also fairly straightforward to measure the global amount of CO_2 in the air by taking measurements at a few well-chosen observation points. From these two factors, we know that about 45% of the carbon that enters the atmosphere each year due to human activity stays there, adding to airborne CO_2 for a century or more (though individual molecules cycle through the system more quickly, like water flowing through a drain as more water enters). The other 55% is either absorbed by the ocean or taken up by land-based ecosystems – trees, crops, soils and the like. Plants take up carbon dioxide when they photosynthesize and return it to the soil and the atmosphere when they die and decompose.

This simplified picture gets a bit more complex when you consider that Earth's vegetation evolves from year to year and decade to decade. El Niño and other atmospheric cycles can spawn drought over huge areas. The dry

Will Earth's carbon sink get stopped up?

One of the critical issues in climate-change science is why and how Earth's land areas are able to soak up roughly 25% of our carbon emissions. Several possibilities are on the table. One is that the CO_2 we're adding to the atmosphere may be stimulating plants to grow more vigorously. Many laboratory tests support this notion; after all, CO_2 is a fertilizer. Even so, we can't count on this process indefinitely, because plant growth may run up against other constraints, such as the supply of other nutrients or changes in land use. Indeed, climate change itself could turn the tide by reducing water supply, raising temperatures beyond an optimal level, or increasing the risk of fires – which themselves would release even more CO_2.

Reforestation is also likely to have an important effect. Many parts of eastern North America that were deforested and then farmed for decades are now tree-covered once more. That growing forest is pulling additional CO_2 from the air, but this effect will slow down as the trees reach maturity. On the other side of the equation, the massive deforestation of Earth's tropics (see p.11) is releasing vast amounts of CO_2, which only adds to the carbon puzzle. A 2007 study led by Britton Stephens of the US National Center for Atmospheric Research found that northern land areas were taking up less CO_2 than earlier believed. This and subsequent studies imply that the tropics might be taking up more CO_2 than previously thought – perhaps even enough to compensate for tropical deforestation. On top of these regional differences, the normal respiration of Earth's plant life goes through large multiyear changes related to climate patterns such as El Niño. Researchers are still working on ways to measure the year-to-year and decade-to-decade variability.

The bottom line is that, one way or another, Earth's land areas as a whole appear to be serving as a sink for carbon right now. Happily, this takes a substantial edge off the impact of our greenhouse emissions. The land sink may have even strengthened over the last couple of decades, according to a 2010 analysis by Jorge Sarmiento (Princeton University). However, this may not be enough to counter the ever-increasing pace of fossil-fuel burning. A 2009 study led by Corinne le Quére (University of East Anglia) found that, while the land sink's absorption of CO_2 has been increasing in recent years, the percentage staying in the atmosphere has been increasing even more quickly.

conditions make for struggling plants that absorb less CO_2, and they allow for more deforestation and large-scale fires that add CO_2 to the atmosphere. And recent work hints that the absorption of CO_2 by plants and oceans could decrease in the long term as the atmospheric load increases. It's very hard to measure and separate out all these effects; for one thing, what we add is folded into a much larger exchange of carbon that takes place naturally. Over the last few years, the balance has played out as follows:

▶ **The atmosphere** retains about 45% of each year's added CO_2 when averaged across several years, though in a given year the percentage can range anywhere from 30% to 80% as land absorption rises and falls.

▶ **Oceans** absorb close to 25%.

▶ **Land-based ecosystems** take up the rest of the atmosphere's carbon – around 30% on average. There are large variations from year to year due to climate cycles and changes in land use, and scientists aren't yet sure which regions account for most of the absorption (see p.39).

Because most of Earth's plant life is north of the Equator, the northern spring produces a "breathing in" effect (a dip in CO_2 levels), with a "breathing out" (a rise in CO_2) each northern autumn. These ups and downs show up as the sawtooths atop the steady multiyear rise in CO_2 visible on the graph on p.34.

How much do different activities contribute?

The chart opposite shows two different ways to analyse the sources of human-produced greenhouse gas: where the emissions come from (left) and the activities that generate them (right). As our emissions climb at roughly 1–2% per year, the balance among various sources is slowly shifting.

By and large, the world's most technologically advanced nations have become more energy efficient in recent years, with **industrial emissions** actually declining a few percent in some developed countries since 1990. Counterbalancing this progress is the explosion of industry in the world's up-and-coming economies, such as China's and India's. In sum, the global CO_2 emissions from industry are climbing a bit more slowly (less than 1% a year) than the average emissions for all sectors combined. Another area of progress is deforestation, which accounted for more than 25% of emissions in the 1990s but only about 10–15% in recent years.

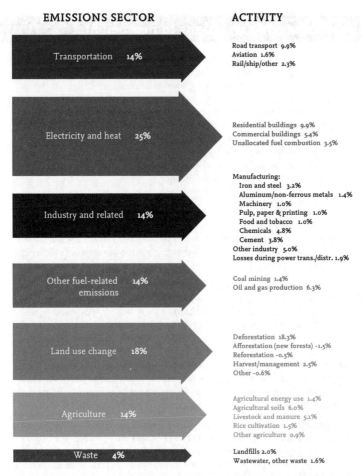

EMISSIONS SECTOR

ACTIVITY

Transportation **14%**

Road transport 9.9%
Aviation 1.6%
Rail/ship/other 2.3%

Electricity and heat **25%**

Residential buildings 9.9%
Commercial buildings 5.4%
Unallocated fuel combustion 3.5%

Industry and related **14%**

Manufacturing:
  Iron and steel 3.2%
  Aluminum/non-ferrous metals 1.4%
  Machinery 1.0%
  Pulp, paper & printing 1.0%
  Food and tobacco 1.0%
  Chemicals 4.8%
  Cement 3.8%
Other industry 5.0%
Losses during power trans./distr. 1.9%

Other fuel-related emissions **14%**

Coal mining 1.4%
Oil and gas production 6.3%

Land use change **18%**

Deforestation 18.3%
Afforestation (new forests) -1.5%
Reforestation -0.5%
Harvest/management 2.5%
Other -0.6%

Agriculture **14%**

Agricultural energy use 1.4%
Agricultural soils 6.0%
Livestock and manure 5.1%
Rice cultivation 1.5%
Other agriculture 0.9%

Waste **4%**

Landfills 2.0%
Wastewater, other waste 1.6%

Breakdown of total global greenhouse gas emissions (CO_2 equivalent) by sector (left) and end use (right). Because of rounding, columns do not total 100%, and some emissions sectors at left contribute to more than one end use at right. (Adapted from World GHG Emissions Flow Chart (2000), World Resources Institute, wri.org.)

The emissions produced by heating, cooling and powering **homes** have been rising by close to 2% a year, thanks in large part to bigger homes stuffed with more energy-hungry devices. In many parts of the world, there are ambitious plans to reverse the growth in household emissions, but it remains to be seen whether such targets will be met. Meanwhile, **commercial buildings** are tending to become more

Aviation: taking emissions to new heights

The advent of low-cost flying has sent millions more people hopping from country to country over the last decade, especially across Europe. In many developed countries, aviation is the fastest-growing transportation segment, and it threatens to counteract progress in reducing greenhouse-gas emissions on other fronts. The Netherlands Environmental Assessment Agency estimates that CO_2 emitted by international transport now totals more than 1 billion metric tonnes per year. That's about 3% of all global emissions, but in developed countries where air travel is most common the percentage is rising quickly. In fact, though huge homes and hulking SUVs are familiar symbols of emissions excess, frequent flyers are among the people with the very biggest carbon footprints.

Part of the reason that aviation is such a hot topic for climate change is the sheer rise in pure volume. In countries such as the UK, emissions from aviation have more than doubled since 1990. To make matters worse, it appears that contrail clouds generated by aircraft also contribute to warming, perhaps boosting aviation's greenhouse impact several times over (though the numbers are still being researched – see p.208).

International airplane and ship traffic aren't considered in the Kyoto Protocol, so their emissions won't play a role in whether the UK or other nations meet their Kyoto requirements. However, they're likely to enter into any post-Kyoto agreement, and the EU has already taken action to control emissions in the sky. As of 2012, all international flights to or from member nations – including flights operated by US carriers – fall under the EU's emissions trading system. Airlines are mandated to keep their greenhouse emissions 3% below 2004–06 levels, with a 5% decrease beginning in 2013. Some activists consider these targets weak ones, as they allow aircraft to continue sending far more greenhouse gas into the atmosphere than they did in 1990, the benchmark year for the Kyoto Protocol.

It's a good thing that aviation is a fairly small piece of the greenhouse emissions picture for the time being, because it may be one of the most troublesome ones to address in the long term. Electric cars are already taking to the streets, and fuel-cell vehicles could become viable later on, but there's no obvious way to wean aircraft off fossil fuels in the next several decades. Biofuels are being tested, but these are unlikely to make up more than a sliver of airlines' carbon footprint. Some progress is being made in how airplanes fly, though. SAS began experimenting with a go-slower strategy in 2008. Test flights between Oslo and Bergen showed that cutting

efficient more quickly than homes, with companies motivated by the potential for long-term savings.

Transport is where the most trouble appears to be brewing. Emissions from this sector are climbing at well over 2% a year, stoked by the global growth in motor vehicle use and the ever-longer distances many people are

A grim future ahead if we can't rein in our appetite for flight.

in-flight speeds by about 10% reduced emissions for the entire flight by a few percent per passenger. Landings – the most fuel-intensive legs of any flight – can become more than 30% more efficient (and quieter to boot) when planes descend in a smooth, continuous manner, rather than the traditional stairstep technique. Though "green landings" require more airspace, they're growing in popularity in Europe, with tests now under way at many US airports.

Rough Guides travel books encourage you to "fly less and stay longer", and to make smart choices when you do fly. See p.365 for details on how you can minimize the greenhouse impact of your own flights.

driving and flying. Transport-related emissions are growing most quickly in the developing world, as millions of people in newly prospering nations take to the roads for the first time ever in their own cars.

For more on how personal choices in home life and transportation feed into the global picture, see pp.365–79.

Which countries are most responsible?

Establishing which countries are most responsible for climate change is more complicated than simply adding up the amount of fossil fuels that each nation burns each year. First, there's the fact that a populous country's total greenhouse-gas emissions may be relatively large even if its per-capita emissions – i.e. its emissions per person – are relatively small, as is the case with China. Then there's the issue of greenhouse gases emitted in the past (some of them before global warming was a known problem), many of which still remain in the atmosphere.

Carbon intensity: an easy way out?

George W. Bush's administration in the US found itself under intense pressure in 2002 to ratify the Kyoto Protocol, which was then making its way through the world's legislatures. Instead, Bush steered the US away from Kyoto and towards a different way of assessing progress on climate change. His plan emphasized greenhouse gas intensity, aka **carbon intensity**. This is a measure of how much fossil fuel it takes to produce a certain amount of economic output. Thus, carbon intensity is not the actual amount of carbon emitted, but a number pro-rated by the Gross Domestic Product (GDP) or purchasing power parity (PPP). For example, if GDP and emissions both climbed 3% in a given year, the carbon intensity would remain unchanged even though the actual emissions had risen.

The Bush administration called for an "ambitious but achievable" reduction in carbon intensity of 18% by the year 2012. Environmentalists pointed out that the US carbon intensity dropped 17.4% from 1990 to 2000 without any special attempt to reduce it. Thus, they claimed, the plan offered little more than business as usual. The picture is similar elsewhere. Globally, carbon intensity (looking only at CO_2) dropped by 13% from 1990 to 2000, even as total emissions grew. In China, a booming economy helped reduce carbon intensity by 47% at the same time that CO_2 emissions climbed by 39%.

Despite its shortcomings, the concept of carbon intensity is still widely used. As part of their 2007 Sydney Declaration, the 21-nation Asian–Pacific Economic Cooperation group held back from any direct targets for emission reduction, but agreed on an "aspirational" goal of reducing energy intensity by at least 25% by 2030, as compared to 2005 values. Similarly, China put forth carbon-intensity goals at the 2009 Copenhagen summit rather than pledging any specific level of emission cuts.

In the long run, then, intensity is a useful way of gauging the impact of greenhouse-gas reductions on the economy. But when it comes down to effects on the physical world, a molecule of gas is still a molecule of gas.

Outsourcing emissions

The nation-by-nation figures cited in this section look only at the direct emissions of greenhouse gases within those countries. But there's an important side effect of globalization to be considered: the shift it produces in the balance of greenhouse emissions. When a country imports consumer goods, should the emissions produced by the manufacture of those goods be assigned to the destination country rather than the supplier? If they were, the United States would leap back into its longtime role as the world's leading greenhouse emitter, because so many of its household products are made in other countries, particularly China. In recent years, as much as half of the increase in China's greenhouse emissions has arisen from the manufacture of exports. On the receiving end, one preliminary study led by Shui Bin (US Pacific Northwest National Laboratory) and Robert Harriss (Houston Advanced Research Center) estimated that US emissions would have been 6% higher in 2003 if the products imported by Americans had been made in the US. Extending this concept worldwide, a 2010 study led by Ken Caldeira (Carnegie Institution) found that US imports from all countries together accounted for nearly 11% of the nation's carbon footprint in 2004, the most recent year for which international trade figures were available (see graphic below). Shifting emissions isn't the main purpose of US trade with China, of course, because the US is not a party to the Kyoto Protocol. But what if a country whose emissions are limited by Kyoto did decide to transfer its greenhouse-intensive industry to a nation unfettered by the protocol? The risk of this so-called **carbon leakage** has been studied in some depth. Economists are still tussling over how big a concern it is. A 2005 study by Mustafa Babiker of the Massachusetts Institute of Technology argues that carbon leakage could be as much as 30% in some countries – more than enough to counteract any domestic emissions reductions made in those nations under Kyoto. Confirming the point, the Caldeira study found that some European nations were already outsourcing more than 30% of emissions as of 2004. For Switzerland, the percentage was more than 50%.

This schematic from Ken Caldeira and colleagues shows the largest international flows of emissions embodied in trade between major partners, with nine western European nations considered as a group.

Suburban Chicago's grid, ablaze at dusk, stretches to the horizon.

The question of how countries are using – or abusing – their landscapes also needs to be considered. Countries undergoing major deforestation, such as Brazil and Indonesia, would rank significantly higher in the list shown on p.47 of nations with the highest total emissions if the greenhouse-boosting effects of forest destruction were taken into account. To add another level of complexity, there's the claim that emissions figures are currently rigged in favour of nations that tend to import, rather than export, goods (see box on previous page).

When it comes to overall greenhouse gas output, however, two countries stand above the rest. The United States is responsible for around 30% of the **cumulative** CO_2 emissions to date (that is, all emissions since the industrial era began) and continues to generate almost 20% of annual greenhouse-gas emissions – roughly four times its share, considering that the US represents less than 5% of world population. Some of the US's outsized emission rate is clearly due to a lack of emphasis on energy efficiency and a focus on economic growth as opposed to environmental virtue. There are also historical factors in the mix that are difficult to change: a car-loving and car-dependent culture, an economy built on once-vast reserves of fossil fuels, and the simple fact that the US is large in both population and land area. The country now has more than fifty years' worth of suburban development that virtually forces millions of Americans to drive to work, school and just about anywhere else outside the home.

Quickly outpacing the US in terms of current emissions is China. With its billion-plus citizens, the country's emissions per person are still comparatively low. But as the nation continues to industrialize, its total emissions have been climbing at a staggering pace – close to 10% annually. Only a few years ago, it was thought that China might surpass the US

Percentage of global CO_2 emissions (fossil fuels, cement and gas flaring only)	Emissions per capita (tonnes of CO_2 from fossil fuel use)	Carbon intensity (tonnes of CO_2 emitted per thousands of US dollars in GDP/PPP, using 2000 prices)
China 21.3	Qatar 51.7	Iraq 3.21
United States 19.0	Kuwait 34.1	Uzbekistan 2.01
India 5.3	Netherland Antilles 32.3	Qatar 1.67
Russia 5.0	UAE 30.8	Mongolia 1.63
Japan 4.1	Bahrain 29.7	North Korea 1.56
Germany 2.6	Trinidad/Tobago 27.9	Neth. Antilles 1.53
Canada 1.8	Aruba 23.1	Kazakhstan 1.49
United Kingdom 1.8	Luxembourg 22.7	Trinidad/Tobago 1.43
South Korea 1.6	Brunei 19.4	Bahrain 1.32
Iran 1.6	Falkland Islands 19.4	Jamaica 1.32
Mexico 1.5	United States 19.1	Turkmenistan 1.19
Italy 1.5	Australia 17.6	UAE 1.15
South Africa 1.4	Montserrat 16.9	Yemen 1.06
Saudi Arabia 1.3	Saudi Arabia 16.9	Serbia 1.03
Indonesia 1.3	Canada 16.9	Russia 0.99
Australia 1.2	Estonia 15.4	Saudi Arabia 0.99
France 1.2	Kazakhstan 14.7	Ukraine 0.95
Brazil 1.2	Faeroe Islands 14.3	Kuwait 0.94
Spain 1.2	Gibraltar 13.9	Brunei 0.97
Ukraine 1.0	Oman 13.6	Moldova 0.87

Columns 1 & 2 show 2007 data from the US Carbon Dioxide Information Analysis Center (cdiac.ornl.gov), with the data converted from carbon to CO_2 for column 2. Column 3 shows 2007 data from the International Energy Agency (www.iea.org/co2highlights).

in total emissions by the year 2025. But that point was actually reached in 2006, according to the Netherlands Environmental Assessment Agency. In recent years the annual *increase* in China's emissions has often run higher than Britain's *total* emissions.

The **United Kingdom**, for its part, has produced 6% of cumulative emissions, thanks to its head start as the birthplace of the Industrial Revolution. In the course of a major industrial transformation, the UK has cleaned up its act significantly. It now generates only 2% of the yearly global total of greenhouse gases. Still, that's twice its rightful share, considering that the UK houses less than 1% of Earth's population.

On the previous page is a list of the world's top twenty greenhouse emitters, measured in three different ways: percentage of all global emissions, emissions per capita, and carbon intensity (see p.47). For **total emissions**, the list is a mixed bag, with the contributions split almost evenly between developing and developed nations. (If considered as a unit, the European Union would rank third in total emissions behind the US and China.) The **emissions per capita** column tells a different story, with the list topped by the tiny oil-producing nations of **Qatar**, **Kuwait**, **Netherland Antilles**, **United Arab Emirates** and **Bahrain**. These countries have so few residents that their national contributions to the global greenhouse effect remain small, but because they are heavy producers and consumers of oil, they have a high per-capita emissions rate. Otherwise, industrialized nations lead the way, with the sprawling, car-dependent trio of **Australia**, the **US** and **Canada** clustered behind the oil producers. For more about **carbon intensity** – the data presented in the third column – see p.44.

PART 2

THE
SYMPTOMS

What's happening now, and what might happen in the future

Dry riverbed, Xiangjiang River in Changsha, capital of Hunan Province, China, 2009.

Extreme heat

Too hot to handle

To those who think of Russia as a perpetual icebox, the summer of 2010 came as a sizzling shock. The nation roasted in unprecedented heat from July into early August. Moscow, whose summer afternoons are typically close to room temperature, soared above 38°C (100°F) for the first time on record and hit 30°C (86°F) every day for more than four weeks. Smoke from peat and forest fires blocked out skyscrapers and clogged lungs. Even before the heat wave subsided, it was clear that thousands of lives had been lost, and Russian leaders – long sceptical on global warming – openly expressed concern about the nation's climatic future.

The horror in Russia triggered memories of the super-hot European summer of 2003 – a relentless barrage of heat that produced an even greater slow-motion catastrophe, killing more than 50,000 people. In at least one study, a group of climate scientists confirmed the suspicions held by many: the odds of getting the great heat wave of 2003 were boosted severalfold by the helping hand of fossil fuels.

Heat waves themselves are nothing new, of course. Many parts of the US Great Plains have yet to top the all-time records they set during the 1930s Dust Bowl, including a reading of 49.5°C (121°F) as far north as North Dakota. However, climate-modelling studies and our understanding of the greenhouse effect both indicate that the next few decades could bring hot spells that topple many longstanding records across mid-latitude locations. In 2010 alone, at least nineteen nations – an unprecedented number, most of them in Eurasia and Africa – set all-time temperature peaks. The roll call included 37.2°C (99.0°F) in Finland, 48.2°C (118.8°F) in Niger, and a blistering 52.0°C (125.6°F) in Saudi Arabia.

Many Muscovites wore masks in August 2010 when a heat wave – accompanied by smog and smoke from nearby wildfires – gripped the Russian capital. The city's mortality rate doubled as the concentration of atmospheric carbon monoxide soared to more than double acceptable safety norms.

Striking as these statistics are, a heat wave doesn't have to bring the warmest temperatures ever observed to have catastrophic effects. All you need is a long string of hot days combined with unusually steamy nights. Stir in a complacent government, cities that were built for cooler times, the lung-clogging effects of air pollution, and a population that can't or won't respond to the urgency, and you have a recipe for the kind of disaster that 2003 brought to Europe and 2010 delivered to Russia.

Off the charts

Extremes in summertime heat are one of the long-anticipated outgrowths of a warming planet. Part of this is due to simple mathematics. If the average temperature goes up, the deviations from that average should rise as well, so the most intense spikes in a hot summer ought to climb accordingly. Computer models and recent data bear out this intuitive concept.

Could the spikes themselves become larger? It's too soon to know whether temperatures will become more variable as the climate warms. Researchers are finding that the increased intensity of heat waves generated in the latest simulations can be explained largely by the rise in average summer temperatures – the base on which any heat wave would build. Some simulations of greenhouse-warmed climate do show an increase in the strength of warm extremes, especially across interior areas. In other

Pollution: heat's hidden partner in crime

Some victims of heat waves die not because the air is so warm but because it's so dirty. The sunny, stagnant conditions prevalent during heat waves make an ideal platform for the sunlight-driven processes that create **ozone** – which is a blessing in the stratosphere (see p.28) but a dangerous pollutant at ground level. Ozone irritates the lungs and makes them more vulnerable to other chemicals. Moreover, the relative calm of a heat wave allows tiny bits of heavy metals, sulphates, nitrates and other substances to accumulate in the air. These are often grouped into the classes **PM10** (particulate matter smaller than 10 microns or 0.0004 inches) and **PM2.5** (particles smaller than 2.5 microns). The smallest of these particulates can sneak past the body's natural respiratory filters, causing a variety of lung problems and raising the risk of heart attacks.

After focusing on other pollutants for decades, scientists have only recently learned how deadly ozone and fine particulates can be. The World Health Organization (WHO) has estimated that mortality goes up by 0.3% during low-level ozone episodes. The WHO's Global Burden of Disease project estimates that 100,000 deaths a year in Europe may be related to fine particulates, with US fatalities estimated in the tens of thousands annually. The growing megacities of the developing world are at particular risk. One study led by US scientists Mario and Luisa Molina showed that a 10% reduction in fine particulates in Mexico City's air could save roughly 1000 lives a year.

Several studies have linked a substantial fraction of 2003's heat-wave deaths to ozone and particulates. A group of British epidemiologists and atmospheric scientists concluded that 21–38% of the UK deaths classified as heat-related could instead be attributed to ozone and PM10. Writing in the *Swiss Medical Weekly*, analysts at the universities of Basel and Bern connected ozone to 13–30% of Switzerland's 2003 heat fatalities. And three scientists with the Dutch government suggest that around 40% of that country's heat-attributed deaths were triggered in roughly equal numbers by ozone and PM10. If there's an upside to these numbers, it's that a concerted effort to reduce ozone and particulate pollution might help save many of the people who die in the worst heat waves.

words, even as the typical summertime temperatures rise, warm departures from that norm (both night and day) could become even larger.

However high it might get, the peak temperature on any given day isn't the ideal measure of a heat wave's ferocity. When it comes to the toll on people, animals and plants, the **duration** of a heat wave and the warmth of the **night-time lows** are the real killers. A weeklong stretch of severe heat can be far more deadly than just two or three days of it. If a heat wave is defined at a certain location by a certain number of days with readings above, say, 35°C (95°F), then a warmer climate would likely bump more days above such a threshold.

There's a different connection between global warming and steamy nights. As noted in the Floods & droughts chapter (see p.65), higher temperatures have boosted **atmospheric water vapour** on a global scale. When there's more water vapour in the air, nights tend to stay warmer. And cool nights act as a critical safety valve, giving people a physiological and psychological break from the intense daytime stress of a heat wave. If the nights warm up too much, that safety valve goes away.

The 2003 disaster in Europe brought home these points vividly. During the first two weeks of August, large swathes of the continent topped 35°C (95°F) and found themselves hard-pressed to cool below 20°C (68°F). Such temperatures might not sound like a big deal to an American from the Sunbelt. But most of Europe's urban areas lack the ubiquitous air conditioning of the hottest US cities, and certainly most Europeans aren't accustomed to dealing with weather like this. Among the history-making statistics from 2003:

▶ 10 August was **London's** hottest day on record, with highs of 37.9°C (100.2°F) at **Heathrow Airport** and 38.1°C (100.6°F) at the Royal Botanic Gardens. Not to be outdone, Faversham, Kent, climbed to 38.5°C (101.3°F) that same afternoon. It was the first day in nearly 300 years of record-keeping that any place in Britain had topped the 100°F mark.

▶ All-time national records were also broken with 40.4°C (104.7°F) in **Roth, Germany;** 41.5°C (106.7°F) in **Grono, Switzerland**; and a scorching 47.3°C (117.1°F) in **Amareleja, Portugal**.

▶ Germany notched its warmest night on record, with the weather station at **Weinbiet**, a mountain near the Rhine, recording nothing lower than 27.6°C (81.7°F) on 13 August.

Numbers can't do full justice to the impact of this atmospheric broadside on society. In **England**, railway tracks buckled, tube stops resembled ovens, and the London Eye and its glassed-in Ferris-wheel pods had to shut down. Schools and offices closed for "heat days" in parts of Germany. But perhaps no place suffered so acutely as did **Paris**. With little air conditioning and block after block of tall, closely packed stone buildings, the French capital – ill prepared for weather that was more suited to Paris, Texas – became a literal heat-trapping machine. The city ended up falling a whisker short of its all-time high of 40.4°C (104.7°F), which was set in 1947, but it did endure nine consecutive days that topped 35°C (95°F).

The areas hit hardest by the 2003 heat wave (relative to July averages for 2000, 2001, 2002 and 2004).

Even worse for many were the nights of surreal heat. On 12 August, the temperature never dropped below 25.5°C (77.9°F), making it Paris's warmest night on record. The early-August heat was so sustained that it kept the temperature within the intensive-care unit of one suburban Paris hospital, which lacked air conditioning, at or above a stifling 29°C (84°F) for more than a week.

Russia's heat wave of 2010 brought similar misery. A vast dome of high pressure encased Moscow and surrounding areas from mid-July to early August, setting the stage for a vicious brew of heat, smog and smoke from nearby fires. During the worst bouts of pollution, visibility dropped to a few city blocks. On these days, Muscovites faced the no-win choice of opening windows for a smoky breeze or keeping their homes sealed and sweltering. Wind shifts pushed the smoke away from Moscow at times, but the heat was virtually constant, with high temperatures running more than 10°C (18°F) above average for weeks on end. Alexander Frolov, the head of Russia's weather service, dubbed the heat wave the worst in the nation's thousand-year history. Relief wasn't easy to come by, as smoke and heat also plagued the thousands of summer homes (dachas) where Muscovites traditionally flee from urban swelter. Many people took to the beach after drinking, which – combined with the disorientation that extreme heat can produce – helped lead to more than 1200 drownings in June alone. The heat also decimated wheat crops across a huge swath of Russia's breadbasket, which sent world grain prices soaring.

As with any weather event, climate scientists couldn't immediately quantify the role of climate change in Russia's heat wave. However, the event was in sync with longstanding predictions of high-latitude warming. It also galvanized Russian leaders who'd long viewed such warnings with suspicion. Early in 2010, President Dmitry Medvedev had dubbed climate change "some kind of tricky campaign made up by some commercial structures to promote their business projects". Only a few months later, the heat had produced a remarkable about-face. In late July, Medvedev said, "What's happening with the planet's climate right now needs to be a wake-up call to all of us."

The human cost of heat

It can take a surprisingly long time to gauge how many people die in a heat wave, but the broad strokes of the disaster often show up quickly. Moscow's death rate in July 2010 doubled that of the preceding July, and the number of funerals soared. At the height of the heat, one Moscow crematorium operated around the clock for days. (In a gruesome but telling bit of irony, its furnaces were overheating.)

Likewise, though it was clear in early August 2003 that people were dying in large numbers across Europe, it took several weeks to start getting a handle on the heat's human cost. By September more than 20,000 Europeans were listed as casualties of the summer of 2003, and the summer's toll continued to mount over time. It hardly made the news when, in 2005, authorities in Italy abruptly raised their nation's toll from 8000 to 20,000. Simply adding the various national counts would imply that more than 50,000 people died as a result of the heat wave, though the exact number is difficult to discern (see below). No heat wave in global history has produced so many documented deaths. Even the horrific heat-driven fires across Greece in summer 2007, which killed at least 67 people, fell far short of 2003's human toll.

How can heat victims go untallied for a year or more? The answer lies partly in the way mortality statistics are compiled and partly in the way heat waves kill. Many people die indirectly in heat waves – from **pollu-**

The nights Chicago fried

A little more wind in the Windy City might have been appreciated in the second week of July 1995. During four days of searing conditions, more than 700 people died from heat-related causes. The days were hot indeed – ranging from 37°C to 41°C (98°F to 106°F) at the in-town Midway Airport – but in this case, it truly was the humidity as much as the heat. Tropical amounts of moisture in the air, coupled with the heat-island effect (see p.194), helped produce overnight lows of 27°C and 29°C (81° and 84°F) on two consecutive nights. As in Paris, many of the city's older multistorey buildings lacked air conditioning. When children and teens sought relief by opening fire hydrants, police put a stop to it.

Chicago's heat-emergency plan sat on the shelf, and the city's mayor and other key officials remained on vacation until the disaster became dire. Then things really heated up – at least politically, according to New York University sociologist Eric Klinenberg, author of *Heat Wave: A Social Autopsy of Disaster in Chicago*. Mayor Richard Daley acknowledged the heat but asked people not to blow it out of proportion. The city's commissioner of human services blamed victims for not taking care of themselves. "They were often interpreted as individual failure", says Klinenberg of the Chicago heat deaths. "In Europe, the heat wave was immediately framed as a political event."

Chicago did learn from its calamity. In 1999, when another heat wave struck, the city's action plan included not only the usual warnings but also free bus rides to "cooling centers". Crews of city workers phoned and checked in person on elderly people living alone. Those steps helped keep the death toll to 110, a number that Klinenberg still finds too high. As global warming unfolds, he says, "we know that more heat waves are coming." And, he adds, dying from heat is not a nice way to go. "If you look closely at the police reports, or the medical autopsies, they're just horrific. These are isolated, lonely, painful deaths."

In Chicago, densely settled neighbourhoods with busy streets and public spaces, such as the heavily Latino Little Village, fared much better than areas where people were more disconnected from neighbours and had few places to gather, such as the African-American neighbourhood of North Lawndale. As Klinenberg sees it, heat waves are "invisible disasters that kill largely invisible people. Perhaps that's the reason we don't care enough about them."

tion, for instance (see box, p.53), or from pre-existing conditions that are exacerbated by the heat – so it's not always apparent at first that weather is the culprit. And unlike a tornado or hurricane, a hot spell doesn't leave a trail of photogenic carnage in its wake. People tend to die alone, often in urban areas behind locked doors and closed windows. The piecemeal nature of heat deaths over days or weeks makes it hard to grasp the scope of an unfolding disaster until the morgue suddenly fills up. Furthermore, it often takes months or even years for countries to collect and finalize data from the most far-flung rural areas.

On top of this, heat is a selective killer. It targets the very young and especially the very old, people whose metabolisms can't adjust readily to temperature extremes. In Rome, more than half of the 700 deaths attributed to 2003's heat were among people older than 85. Because the elderly die so disproportionately in heat waves, it's tempting to assume that weather is simply claiming those fated to die shortly anyhow. If that were the case, however, then you'd expect mortality to dip below average in the months following a heat wave. Careful studies of major heat disasters have, in fact, shown that such dips typically account for only about 20% to 30% of the spikes above average, or what epidemiologists call the **excess deaths**, that were observed in the preceding heat. It thus seems that most victims of heat waves, even the elderly, die well before their time.

In France, the situation was exacerbated by local culture. The worst of the heat struck during the August vacations, when much of professional Paris heads out of town. Left behind to fend for themselves were many thousands of elderly Parisians, the ones most susceptible to heat-related health problems and least able to seek refuge in a cool location. As many as half of France's fatalities occurred at rest homes, few of which had air conditioning.

Could a better **warning system** have saved lives in Europe? That's been the case in the United States, where most parts of the country can easily top 35°C (95°F) in a typical summer. After years of perfunctory heat advisories, issued by the US National Weather Service (NWS) but seldom heeded, the country was shocked into awareness by a 1995 heat wave in Chicago that killed more than 700 people (see box, previous page). That same year, Laurence Kalkstein of the University of Delaware teamed up with the NWS to launch a new watch-warning system for extreme heat. The scheme also included intensive **neighbourhood interventions**, such as block captains assigned to check on neighbours. It's been estimated that Philadelphia's heat plan saved more than one hundred lives in four years.

As it happened, Kalkstein and Italian colleagues had just launched a US-style warning system in Rome when the 2003 heat wave struck. The city ended up declaring heat emergencies on eighteen days that summer. Even with its new plan in place, hundreds more Romans died than the scientists expected. In a post-mortem analysis, the team found that the heat was worse than forecast and, indeed, transcended anything they'd considered in planning their warning system.

Chicago, 1995: a heat-wave victim's body is moved into a refrigerated truck.

The future of summer sizzle

On a broader scale, the European disaster of 2003 and Russia's 2010 meltdown were merely blips – albeit spectacular ones – in a long-term warming trend that spans most of the world, particularly large parts of North America and Eurasia. While the planet as a whole has warmed nearly 0.8°C (1.4°F) since 1900, the rise has generally been greater over the Northern Hemisphere continents (especially at higher latitudes) than over the tropics. That simplified picture doesn't hold quite as well, though, when you go beyond the average temperature and look at various types of heat extremes, some of which have increased in the tropics as well as mid-latitudes.

It's taken a long while for scientists to verify how heat waves have changed globally over the last few decades, not to mention how they might evolve in the future. Why is this so? After all, the average worldwide temperature has been tracked for decades (see p.191), and individual stations keep close tabs on daily weather. Between the local and the global, though, it's surprisingly difficult to assess regional trends in temperature and precipitation extremes.

Nations have traditionally shared more general data, such as the average high and low temperatures for each month at major reporting sites. But knowing that the average August high in a given city rose, say, 1°C in fifty years doesn't tell you whether stretches of the most intense heat are

What counts as a heat wave?

Half the battle in assessing climate extremes is simply deciding what to measure. What defines an extreme spell of heat, aside from a sticky shirt or a wilting garden? Experts have tried out a variety of indices to capture climate extremes. In the realm of heat, these include the following. Caution: these may take a minute or two to digest.

▶ **Absolute thresholds** The number of days that exceed a given temperature. While this is nicely concrete, one threshold doesn't always fit all. A week of July afternoons at 35°C (95°F) might feel miserable in London but normal in air-conditioned Houston. Thus, each location might need a different threshold, which makes it hard to compare heat intensity among locations.

▶ **Monthly maximums and minimums** Changes over time in the highest and lowest single temperature observed during a given month of the year. These can provide useful, easy-to-grasp illustrations of a shifting climate, like the all-time highs recorded during the 2003 Euro heat wave. However, a single day or night of record warmth doesn't necessarily correspond to the kind of sustained multi-day heat that causes major problems.

▶ **Threshold departures** The number of days when temperatures climb above average by a fixed amount, such as 5°C (9°F). This gives a more location-appropriate sense of how unusual a hot stretch might be. However, it doesn't acknowledge that one city might normally have more variability than the next. For example, a 5°C jump in Denver's dry climate, where temperatures can gyrate wildly from day to night or across a few days, would be less obvious than the same leap in Miami's sultry summers, where temperatures often change little from day to day.

▶ **Percentile departures** The number of days that land among the hottest of all days in that month's long-term record, based on percentage (the hottest 10%, 5%, 1%, etc). This index provides a tailored-to-fit measure of a heat wave's intensity, based on each city's unique characteristics. No matter where you live, a day that's among the warmest 1% observed in the past decade or century means something (especially if such heat were to start occurring 5% or 10% of the time rather than 1%). Some researchers combine this with a measure of duration – for instance, the number of consecutive days on which the temperature reaches a given percentile ranking of heat.

becoming more frequent, intense or longer-lived. To find that out, you need a consistent way to measure heat extremes, and you need day-by-day data extending over decades. That's been hard for researchers to obtain. Some nations balk at releasing such data – considering it valuable intellectual property – and some aren't set up to provide the data easily.

Climate scientists made big headway on this problem from 2001 to 2005 through a series of regional workshops. These meetings provided a venue in which countries could freely share the daily-scale data needed to understand changes in heat waves. Also, scientists agreed on a few key indices of heat, cold and precipitation that were used in preparing the 2007 IPCC report. Emerging from this work is the most detailed picture yet of how heat waves and other climate extremes are evolving across the world. Key findings include the following:

▶ On average, the level of heat that occurred only one day in ten during the 1950s is now observed about 15% more often. The strongest such trends are across Saharan Africa, Russia, central Europe, southern Australia, western Canada and Alaska.

▶ The average range in daily temperature shrank significantly across nearly half of the globe from the 1950s to the 1980s, especially in eastern Asia and the central United States. This is largely because the rise in low temperatures was stronger and more widespread (about 1°C or 1.8°F globally since the 1950s) than the rise in high temperatures (closer to 0.75°C). In other words, nights have warmed even more quickly than days. Interestingly, for reasons unknown, there's been little change in the daily temperature range since the 1980s.

How much of 2003's heat can be pinned on greenhouse gases? The ever-growing heat islands of modern European cities no doubt played a role, and these are a form of climate change in themselves (see p.194). Yet the continent-wide, rural-plus-urban scope of the heat points to something more going on. Many campaigners trumpeted the heat wave as a classic example of global warming at work, but it was a year later before scientists had marshalled a statistical case for this accusation. The resulting study, published in *Nature* in 2004 by Peter Stott (University of Reading) and colleagues at the University of Oxford, is a landmark – one of the first to lay odds on whether a specific weather or climate event was rendered more likely to occur by global warming.

Using a climate model from the UK Hadley Centre, Stott's team simulated European summers since 1900 a number of times, sometimes including the century's ramp-up in greenhouse gases and sometimes omitting it. The authors found that human-induced greenhouse emissions made it roughly four times more likely that a summer like 2003 might occur. In the US, a team led by NOAA's Martin Hoerling used similar techniques to find that human-produced emissions have led to a fifteen-fold jump in the risk of the type of record-setting heat the US saw in 2006. Such efforts – often dubbed attribution studies – are expected to increase through this decade, providing much-needed perspective on the role of climate change vis-à-vis extreme weather events such as the Russian heat wave.

Looking further ahead, Stott and colleagues extended their Europe study to the year 2100 using the IPCC's high-emission A2 scenario. If the model pans out, the *typical* European summer of the 2040s would be even warmer than was the extreme summer of 2003. But even if this outlook turns out to be true, do warmer summers necessarily mean more, or longer, heat waves? In 2004, Gerald Meehl and Claudia Tebaldi of the US National Center for Atmospheric Research examined the worst three-day stretches of heat by location. Their modelling shows these periods intensifying across all of the United States and Europe over the next century, especially across the US South and West and the Mediterranean. The model also showed typical heat waves in Chicago and Paris lengthening by a few days.

Other studies have added fuel to the fire, as it were. A compilation of data from eight global models that fed into the 2007 IPCC report found the average length of heat waves (defined as a stretch of at least five days with highs 5°C above average) increasing by the year 2100 across nearly all of the globe's interior land areas. If the models apply the IPCC's lower-emission B2 scenario, which assumes a good deal of progress towards energy alternatives, then heat waves still lengthen substantially until at least 2050. The IPCC concluded that heat waves are likely to become longer, more intense and more frequent across most continents, especially in central Europe, the western US and eastern Asia.

A massive European Union project called **PRUDENCE** has produced more specific projections of future European temperature. PRUDENCE employed nine regional climate models nested within two global models, producing 50 simulations for the IPCC's A2 scenario. Among the most harrowing findings in the model consensus:

▶ The number of days per year that reach 30°C (86°F) in the Paris region jumps from the current 6–9 days to 50 days by the end of this century. The region's longest consecutive streak of such days in a typical year leaps from three to nineteen, a value more typical of present-day Sicily or Spain.

▶ Unlike the global trend found in the past century, the hottest 10% of days across Europe warms even more than the coolest 10%, especially across the heart of Europe (from France to Hungary), where drier summers take hold.

Many other heat-prone regions of the world have yet to be studied in the detail shown above. However, Europe is an unusually instructive case in point. Long tempered by its proximity to water, Europe now appears to be losing some of that protection, as huge domes of hot, dry air build over the continent more frequently and gain more influence in summer.

A wild card in the Atlantic

There's one character waiting in the wings that could change this picture dramatically: a slowdown or shutdown of the **Atlantic thermohaline circulation**, which helps keep Europe unusually warm for its latitude. This is the scenario made famous in the 2004 film *The Day After Tomorrow* (see p.293). Although evidence shows that the thermohaline circulation has ground to a halt more than once in climate history, it's believed that this process takes at least a few years to play out, and sometimes many decades, rather than the few days portrayed in the movie.

In any case, a complete shutdown of the thermohaline circulation isn't expected any time soon, but the IPCC deems a slowdown "very likely" over the next century. The best guess from the most sophisticated computer models is that the circulation might slow by 10% to 50% over the next century, assuming greenhouse gas emissions continue unabated, with the model average around 25%. If this happens, the expected climate warming might be nearly erased across the United Kingdom and diminished across many other parts of Europe. However, summers could still be warmer and more drought-prone across the UK and Europe than they are now. For a full discussion of the thermohaline shutdown hypothesis, see p.131.

Handling the heat

In the long run, even if the world tackles climate change wholeheartedly over the next few years, Europe will clearly need to adapt to the risk of heat waves like the one it endured in 2003. Better warning systems will help; more air conditioning, and the associated cost, seems inevitable. The latter may put a dent in Europe's goal of reducing greenhouse emissions, though it's possible that some of the energy spent to cool the continent will be counterbalanced by a drop in the need for wintertime heating fuel. However, nobody knows how many of the poorest and most vulnerable of Europeans will simply be left to suffer through future hot spells in un–air-conditioned misery.

Poverty is certainly a major co-factor in heat deaths across the **developing world**. The people of India are long accustomed to spells of intense heat during the late spring, just before the monsoon arrives, when temperatures can soar well above 40°C (104°F) across wide areas. As is the case elsewhere, it's the extremes on top of that already scorching norm that cause the most suffering. Several pre-monsoonal heat waves in recent years have each killed more than one thousand people, many of them landless workers forced by circumstance to toil in the elements. The heat was especially fierce in 2010, as India endured its hottest pre-monsoon weather on record. On 26 May, MohenjuDaro, Pakistan, reported an astounding 53.5°C (128.3°F) – apparently the highest temperature ever reliably observed in Asia.

As for the United States, much of its Sunbelt, including the vast majority of homes and businesses in places like Atlanta and Dallas, is already equipped with air conditioning. Even as temperatures soar further, these regions may prove fairly resilient – as long as the power stays on. Older cities in the Midwest and Northeast appear to be more at risk of occasional heat crises, even though their average summer readings fall short of those in the South, where air conditioning is the accepted standard. Indeed, fatalities appear to be more common in places where intense heat is only an occasional visitor. Another factor is diabetes, a global epidemic hitting the US especially hard. Because diabetes inhibits sweating and destabilizes blood sugar levels, it could put millions of Americans at an increased risk of heat-related illness, according to a 2010 study by the Mayo Clinic and NOAA.

Floods and droughts

Two sides of a catastrophic coin

Although it's natural to think of temperature first when we think of global warming, the impact of climate change on precipitation may be even more important in the long run for many places and many people. Too much rain at once can cause disastrous floods, while too little can make an area unproductive or even uninhabitable.

Anyone who was caught in the epic 24-hour rainfall of 944mm (37.2in) in Mumbai, India, in July 2005, or who experienced Britain's destructive summer rains in 2007, might wonder if the planet's waterworks are starting to go into overdrive. That thought must have crossed the minds of people in Pakistan during the summer of 2010, as torrential monsoon rains inundated an area the size of England. At least four million people were left homeless – one of the worst such displacements in modern times.

Full-globe observations of rainfall are difficult to obtain, in no small part because 70% of the planet is covered by oceans (scientists are currently hard at work trying to improve the data). Land areas *are* now believed to be getting about 1% more precipitation than they did a hundred years ago – an increase of about 1cm (0.4in) on average – but this small boost (which isn't statistically significant) hides much larger variations in space and time. For the United States, the increase is closer to 5%, and the range across most mid- and high latitudes of the Northern Hemisphere is anywhere from 5% to 20%, according to the 2007 IPCC assessment. Meanwhile, rainfall across many parts of the tropics and northern subtropics (from 10°S to 30°N) is on the decrease, although recent satellite studies hint that some tropical locations are getting up to 5% more rain than they did in the late 1970s.

Scientists expect these trends to continue over the next century, with a further increase in rain and snow poleward of about 50°N (including most of Canada, northern Europe and Russia) and in tropical areas that

A more potent monsoon for Asia?

The swollen Swat River flows past a collapsed road bridge in August 2010. The worst flooding in Pakistan's history devastated the country's agricultural heartland and left millions of people homeless or injured. At the height of the deluge, approximately a third of the country's total land area was underwater.

There's no better example of seasonality in rainfall than the Asian monsoon, which delivers life-giving water – and flooding that is sometimes deadly – to more than two billion people from Pakistan to China. The monsoon normally kicks into gear as the Indian subcontinent cooks in the intense heat of late spring. Updraughts rising from the sun-baked ground help pull in moist air from the nearby ocean, just as a hot day by the seashore stimulates a sea breeze. While the basic process is the same from year to year, the strength of a given year's monsoon can vary greatly. For instance, heavy winter snows over the Tibetan plateau can slow the process by which land heats up in the spring and pulls in summer monsoon moisture. Also, El Niño events tend to dampen the Indian monsoon's strength, while most La Niña years are wetter than usual. Seasonal forecasters have gained skill at incorporating these and other factors into outlooks for each year's monsoon, although most Indians still wait on tenterhooks until the first raindrops arrive.

Since the last ice age, the monsoon has apparently strengthened and weakened in tandem with Earth's temperature. About 10,000 years ago, summer sunlight across the Northern Hemisphere was a few percent stronger than today, due to a cyclic variation in Earth's orbit (see p.216). The extra sunlight not only helped melt the ice sheets but also intensified the Asian monsoon. After that effect started to wane about 6000 years ago, the trend has been towards weaker monsoons during cooler periods and stronger monsoons when Earth is relatively warm. Oddly enough, the link is especially strong with the North Atlantic: when those surface waters are warm, it's a sign that the global conveyor belt of ocean circulation is vigorous (see p.131), and this appears to help keep the Indian monsoon humming as well.

Consistent with this picture, most of the 21 global models analysed in the 2007 IPCC report show an increase in summer monsoonal rains across south Asia over the next century, averaging 11% and ranging as high as 23%. However, average monsoon rainfall across India has actually dropped by 5% to 10% since the 1950s, even as global temperatures have warmed, and some models hint that the drying could continue. In 2009, a fine-scale regional analysis led by Purdue University – one of the most detailed simulations to date – projected an eastward shift in the summer rains over the coming century, which would push the core of the monsoon from India toward Bangladesh and Burma.

India's recent drying may be due to the rapid growth of sulphate pollution, according to a 2006 study by Chul Eddy Chung and V. Ramanathan of the Scripps Institution of Oceanography. The soot and other pollutants blowing offshore appear to be blocking enough sunlight to shift heating patterns across the Indian Ocean and in turn weaken the circulations that drive India's monsoon. If these pollutants persist or increase, the monsoons may tend to stay relatively weak; if not, then the monsoons could gain strength as global warming proceeds, which could bring an increased chance of summers with excessive rains, such as Mumbai's 2005 inundation (which was more than 50% beyond the city's previous record).

Another possible culprit behind weaker Indian monsoons is the very irrigation that helped fuel the much-touted Green Revolution. A 2010 study led by Purdue's Dev Niyogi used satellite data to reveal that the average amount of soil moisture during the hot weeks before the monsoon's arrival jumped threefold during the period 1988–2002. Yet rainfall over the same region has plummeted as much as 40% since the 1950s. It's possible that, even as parched conditions prod farmers to irrigate, the moistened soil might be making things worse by weakening the heat-driven updraughts that help pull monsoon flow inland.

No matter how the average monsoon evolves, it's the extremes that cause havoc. Monsoon failures of the past have led to massive famines, and excessive rains can put enormous stretches of land underwater. A classic case in point is 2010's monsoon, which came on the heels of some of the hottest weather in Asia's history (see p.64). Instead of India's usual wet-to-the-east/dry-to-the-west pattern, which has intensified in recent years, 2010 saw the tables turned. Some of India's lush eastern states received as little as half of their usual rainfall, while desert areas in the far west got close to double their normal quotient. But 2010 also brought far too much of a good thing to some areas. In India's far northwest, the normally bone-dry town of Leh was raked by a vicious flash flood on 6 August that killed more than 150 people and injured thousands more. Further south, along the Indus River valley in Pakistan, monsoonal torrents led to a mind-boggling round of flooding that took more than a thousand lives and displaced millions of residents across a huge area.

benefit from monsoons. Meanwhile, rainfall is projected to decrease across the subtropics, including much of Australia, southern Africa, the Mediterranean and Caribbean, and the southwest US and Mexico. While confidence in this general picture has become more robust, there's still uncertainty about just where the transition zones between increasing and decreasing precipitation will lie.

When it comes to floods, though, it's the character of the precipitation that really counts. How often is rain concentrated in short, intense bursts that produce flash floods, or in multi-day torrents that can cause near-biblical floods across entire regions? And how often is it falling in more gentle, non-destructive ways? Here, the global-change signal is worrisome. Data show a clear ramp-up in **precipitation intensity** for the US, Europe and several other areas over the last century, especially since the 1970s. When it rains or snows in these places, it now tends to rain or snow harder, over periods ranging from a few hours to several days. Projections from computer models suggest that this trend will intensify in the decades to come, although the details will vary from place to place and not every location will follow the trend.

There's a cruel twist to this story. You might expect droughts to diminish on a global basis as rainfall goes up. But higher temperatures not only allow more rain-producing moisture to enter the atmosphere – they also suck more water out of the parched terrain where it hasn't been raining. Thus, in addition to triggering more rainfall, global warming could also increase the occurrence of drought, a seeming paradox that already appears to be taking shape. One study in 2005 by the US National Center for Atmospheric Research found that the percentage of Earth's land area undergoing serious drought has more than doubled since the 1970s. Where and how the worst droughts manifest hinges on a number of factors, including a tangled web of influence through which the world's tropical oceans can help trigger multiyear drought in places as far-flung as the southwest United States and southern Africa. Even in a single location, both drought and flood can become more common if the timing of rainfall shifts in not-so-fortunate ways.

A wetter world (but not everywhere)

The stakes are high when it comes to assessing the future of floods and droughts, because both are notoriously deadly and destructive. According to the World Meteorological Organization, flooding affected 1.5 billion people between 1990 and 2001. China's Yangtze River flows over its banks so reliably during monsoon season (its floods and related crop failures have claimed more than a million lives since 1900) that it's prompted the world's largest flood-control project, the colossal and controversial Three Gorges Dam. Many other flood victims die on a smaller scale, unlucky enough to be caught solo or in a small group along a rampaging creek during a lightning-quick flash flood.

Though drought is a less dramatic killer than flooding, its toll is even more staggering. Monsoon failures across India reportedly killed more than a million people each in 1900 and 1965–67. More than thirty million Chinese died in the first half of the twentieth century as a result of drought and related famine, according to the US National Climatic Data Center. The World Meteorological Organization estimates that more than a million people died throughout Africa's Sahel during 25 years of poor rainfall that peaked in the devastating droughts of 1972–75 and 1984–85 (see box on p.74). Southern and eastern Africa have also been prone to recurring drought over the last several decades. Looking further back, the history of civilization is chequered with cultures believed to have met their downfall due to drought (see "The long view", p.239).

The roots of both flooding and drought lie in the physical process known as evaporation. As global warming heats the world's oceans, the water molecules near the sea surface become more energetic and tend to evaporate into the atmosphere more readily. Thus, the air gains water vapour (or, as it's often put, the air holds more water vapour). This is a nonlinear relationship, meaning the effect gets stronger for each additional degree of warming.

Because the connection between warmth and evaporation is so well established, scientists have considered it a fairly safe bet that a warmer atmosphere would carry more water vapour around the globe. That's been confirmed by satellite observations since 1988 over the ocean, where the amount of water vapour has been rising at just over 1% a decade. The air over some land areas has shown even more of a jump: roughly 5% per decade from 1973 to 1993, according to one study that examined the United States, the Caribbean and Hawaii.

Rain and snow tend to develop where air is converging and rising. If the air is warmer and has a bit more water vapour, it ought to rain or snow more intensely, all else being equal. Based on the physics involved, Kevin Trenberth of the National Center for Atmospheric Research estimates that, overall, precipitation intensity should rise by about 7% for every 1°C (1.8°F) of warming. However, rainfall and snowfall often vary greatly over small distances. Thus, a small increase in global precipitation could mask regional and local trends that are more dramatic, downwards as well as

England and the rains of autumn

You'd have to have been a contemporary of King Charles I in order to have seen Yorkshire's River Ouse higher (in 1625) than it was in autumn 2000. A series of punishing rains and floods swept much of the United Kingdom during October and November 2000, flowing into over 10,000 homes and prompting more than 1400 flood warnings. The total cost was of the order of £1 billion. It was the UK's most widespread flooding since March 1947, when many areas were inundated by the melting of a six-week snowpack. The floods in 2000 came during the wettest autumn in the UK's 330-plus years of record-keeping. Moreover, the November damage could have been far worse, according to the consulting firm Risk Management Services: "It is clear that a flood event of even a slightly greater magnitude would have overtopped and/or breached many defences, causing a significantly greater loss."

By themselves, the floods of autumn 2000 – or summer 2007 (see opposite) – can't be laid at the doorstep of climate change, but the character of rainfall in the UK is clearly changing. So says Hayley Fowler of the University of Newcastle upon Tyne. As part of a European Commission-funded project, she and Chris Kilsby analysed rainfall extremes at 204 UK stations in nine regions for the period 1961 to 2000. She focused on "return periods" – the amount of time one would expect to go between experiencing extreme rainfall or flooding of a given magnitude.

The project produced a cloudburst of attention-grabbing statistics. Across eastern Scotland, for instance, a ten-day rainfall of 150mm (5.9in) could be expected every 50 years during the mid-twentieth century. But since 1990, it's now moved closer to every eight years. Other parts of the UK have also seen substantial changes, with many return periods in the north less than half what they were before 1990. Such dramatic shifts have come unexpectedly early, says Fowler: "This pattern of change is the same as that projected by climate models under global warming for the end of the twenty-first century."

In the past, the northern and western UK would tend to experience its worst floods in autumn, while the southern UK would get them in wintertime. But now, the worst floods in the south – including London and environs – seem to be occurring in the September-to-November time period. That's likely a reflection of the recent shift in extreme rainfall events towards autumn, says Fowler. Not only does

upwards. To complicate the picture further, rain or snow totals at some locations can rise or fall sharply for a year or more due to the climate cycles such as El Niño discussed below.

As with temperature, it's the extremes that matter most in rainfall and the lack of it. The UK's soggy summer of 2007 vaulted into the record books largely on the strength of two extremely wet days: 25 June, which produced 103.1mm (4.06in) of rain in Fylingdales, North Yorkshire, and 20 July, when 120.8mm (4.76in) fell at Pershore College in Worcestershire. In

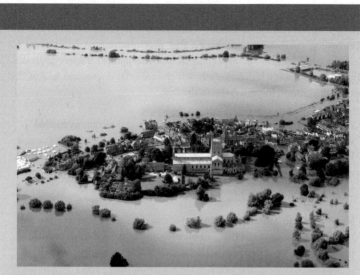

Flood waters surround Tewkesbury Abbey in central England, July 2007. Torrential rain over five days caused flooding that isolated the centre of the Cotswolds town and left some 1800 properties underwater.

this confirm the perceptions of many weather-savvy Brits, it also bodes ill for the future. "The change in timing appears to be contributing to the recent increase in flooding across the south of the UK", says Fowler. "This has huge economic and social implications, especially if these trends continue under global warming, as climate models suggest."

If the trends do continue, they may dovetail with projections from the UK Climate Impacts Programme, whose twenty-first-century outlook predicts wetter winters and drier springs, summers and autumns across the UK. However, UKCIP also expects autumn-to-autumn variability in rainfall to increase markedly across southeast England. Thus, it's possible that tranquillity could alternate with torrents from one October to another.

both cases, the amount of rain that fell over 24 hours was roughly half of what one would expect during an entire summer. The resulting floods put large sections of northern and western England under water. Hundreds of thousands of residents lost power and access to drinking water, and damages were estimated at more than £2 billion.

To help quantify the links between downpours and climate change, the scientists who came up with a set of criteria for measuring temperature extremes (see p.60) have done the same for precipitation. Those indices are a key part of research that fed into the IPCC's 2007 assessment. In one of the first related studies, Povl Frich and colleagues from the Danish Meteorological Institute confirmed that precipitation extremes seem to be increasing in general, especially across the planet's higher latitudes. Highlights from Frich's 2002 study, which looked at the second half of the twentieth century, included the following:

▶ **The number of days that see at least 10mm (0.4in) of rain or melted snow** rose by 5–15% across large stretches of the United States, Europe, Russia, South Africa and Australia. These locations also tended to see upward trends in the peak five-day totals of rain or snow in a given year.

▶ **Much of North America and Europe** showed jumps in the fraction of rainfall and snowfall that fell on the soggiest 5% of all days with precipitation. In other words, the wettest days became wetter still.

As noted in the 2007 IPCC report, a number of studies zeroing in on regional trends have confirmed this general picture. They've shown intense rainfall events on the increase in a variety of places, including the Caribbean, the western United States, the United Kingdom and Italy. However, as Frich found, there's a lot of hidden variability: even a country where precipitation is on the increase can have a few locations that buck the overall trend.

Are floods increasing?

It's surprisingly hard to compile a global picture of whether floods are becoming more frequent or intense due to climate change. In part, that's because the chain of events that leads from an unusually heavy rain to a flood involves many factors other than immediate weather: for example, how wet the region's soils already are, how high rivers and reservoirs are running, what kind of flood-control devices are in place and – perhaps most critically – how much the landscape has been altered by development. One of the few attempts at a global flood census was published in the science journal *Nature* in 2002. Led by Christopher Milly of the US Geological Survey, the study examined 29 of the world's largest river basins. It found that 16 of their 21 biggest floods of the last century occurred in the century's second half (after 1953). However, as noted above, a warming climate makes both downpour and drought more intense, so an increase in floods doesn't necessarily mean rivers are running consistently higher around the globe. A 2009 study from the National Center for Atmospheric Research estimated that about a quarter of the world's largest rivers saw a reduction in streamflow between 1948 and 2004, while only about 10% of the rivers showed an increase.

Land-use changes play a huge role in flooding potential. Deforestation appears to exacerbate the risk of flooding and landslides in most cases, since the water that falls can flow more quickly when it's unimpeded by trees and undergrowth. Water flows particularly easily across the acres of pavement that are laid down as cities expand into the countryside. Faster-flowing water is especially likely to feed into small-scale flash floods, which are even harder to monitor and analyse than larger river floods.

Timing is another important factor when looking at how global precipitation has evolved. For instance, are the heaviest one-hour deluges getting heavier? This is virtually impossible to calculate across the globe, because many national meteorological centres don't compile such fine-scale data or won't release it without collecting a hefty fee. Also, the heaviest of heavy downpours tend to be so localized that they can be measured only with the help of the dense observing networks found mainly in highly industrialized countries. This leaves us in the dark as to changes in flash floods across poorer nations.

Seasonal timing is critical when it comes to both flooding and drought. For example, much of northern and central Europe as well as the United Kingdom is expected to see a pair of coexisting trends in the coming century: heavier wintertime rains as well as drier, more drought-prone summers. The high-summer months of July and August have been getting drier across the UK since the 1970s, though this doesn't prohibit downpours (as the flooding rains of 2007 showed). There are also intriguing changes elsewhere in the year across the UK (see box on p.70).

Rising carbon dioxide levels may also influence both flooding and drought through their impact on plants. In a 2007 *Nature* paper, a team led by Richard Betts of the UK Met Office's Hadley Centre found that the average worldwide runoff from rain and snow increased a further 6% in a model with doubled carbon dioxide affecting plants, above and beyond the effects of carbon dioxide on global warming. The additional increase is due to plants' natural tendency to close their pores in conditions of enhanced CO_2, thus allowing more of a given round of precipitation to remain in the soil and drain into rivers instead of being taken up by the plants. All else being equal, this could make flooding more likely while ameliorating some aspects of drought.

Which way will the Sahel go?

Only a few years after the promising 1960s, when many African nations were shaking off their colonial past, weather stepped into the role of oppressor. A prolonged period of low rainfall – which hit hardest in 1972–75 and 1984–85 – caused drought-driven despair across much of the continent, with the north-central Sahel region at the epicentre. The region's name is derived from the Arabic *sahil*, or shore, and it's an apt label. Each summer, like the tide coming in, monsoon moisture sweeps north from the Gulf of Guinea. It brings the only substantial rain of the year to the semi-arid belt of land that extends from Senegal to Sudan across the great breadth of Africa. Stretching more than 7000km (4400 miles), the Sahel separates the wet tropics to the south from the parched Sahara to the north. There's a lot of climatic contrast packed into this thin ribbon, which spans only about 800km (500 miles) from north to south. Across the width of this narrow ribbon from bottom to top, the average rainfall drops from about 750mm (30in) – more than in London – to about 250mm (10in), less than in Los Angeles.

It doesn't take much of a shift in the monsoon to put the region in jeopardy. Major droughts struck the Sahel in the 1910s and 1940s, but the dry spell of the early 1970s was particularly intense, and the world responded with a mass outpouring of relief. Another widespread drought struck in the mid-1980s, this one extending east to Ethiopia. This time the world's response – though unconscionably delayed until the BBC captured the misery on camera – was even more resounding. Bob Geldof organized the supergroup charity single "Do They Know It's Christmas?", followed by the US counterpart, "We Are the World", and the first Live Aid concert.

Despite the loss of dozens of rain-measuring stations as the tragedy unfolded, the drought's signature is clear enough. Analysing the years 1920 to 2003, a team of US and British climate scientists led by Aiguo Dai of the US National Center for Atmospheric Research (NCAR) found that rainfall during the worst of the drought, in 1984–85, fell nearly 30% short of the long-term average of about 500mm (19.7in). In more recent years, the Sahel's rainfall figures have been both unsettling and encouraging. Though the region has moistened gradually, some years (such as 1994, 1999 and 2003) have produced more rain than the pre-1970 average, while others have been notably dry (occasionally triggered by strong El Niños, as in 1997–98).

The outlook for Sahelian rain in the twenty-first century offers similarly mixed signals. In part, geography remains the Sahel's destiny: the region will stay poised between desert and wet tropics, leading to whipsaws in year-to-year rainfall. If the variability goes up further, as is forecast for many parts of the world, then Sahelians could face even more challenges. The region's population is booming – it's projected to grow from fifty to a hundred million by 2050 – so the resulting stress on wood-based fuel sources and on food stocks will make the region even more vulnerable to climatic swings. The story of Lake Chad shows what can happen when climate and land use conspire. Once the sixth largest freshwater lake in the world, Lake Chad has shrivelled by 95% over the last forty years, a victim of the 1970s/80s droughts as well as intensive irrigation.

Lake Chad as seen by satellite in 1966 (left), when water covered an area of 25,400 square kilometres (9800 square miles), and in 1990, when the lake had shrunk to cover only 2200 sq km (850 sq mi) due to drought and water use.

Despite these grim signs, there's some ground for optimism. Although many farmers and pastoralists have pushed into ever more marginal lands across the Sahel, there's also growing diversity in the ways land is used – a sign of resilience in a place where adaptation is essential – and more local involvement in development decisions. As for the climate outlook, when the 21 global models in the 2007 IPCC report are pooled, the average shows a 10–20% increase in Sahelian summer rains by 2080–99 as compared to 1980–99. However, there's a good deal of spread in the models, with everything from strong drying to a markedly wetter climate showing up in the mix.

Some of the most encouraging projections come from James Hurrell (NCAR) and Martin Hoerling (US National Oceanic and Atmospheric Administration). In 2005 the two ran a series of eighty simulations using five different computer models. What emerged was a link between Sahel rains and the Atlantic. During the worst of the Sahel droughts, the North Atlantic surface waters ran cool relative to the South Atlantic's. That state of affairs reversed in the 1990s, as the North Atlantic began to warm more strongly. Hurrell and Hoerling believe the North Atlantic's warm-up helped pull monsoon moisture across the Sahel, bringing back the long-sought rains. Their simulations show further warming across the North Atlantic through at least 2050, with the Sahel moistening by 20–30% compared to the 1950–99 average. On the down side, a set of NASA simulations found that if aerosol pollution increases this century across the Northern Hemisphere, it could keep the North Atlantic from warming as quickly as its southern counterpart – which might help push the Sahel into a progressively drier state.

On top of this uncertainty, multi-decadal swings in rainfall can strike the Sahel entirely apart from human-enhanced greenhouse gases. In fact, paleoclimate data shows that even greater shifts in rainfall have occurred across this climatically precarious land in the distant past. From that angle, says Hurrell, "the recent African dryings appear to be neither unusual or extreme."

Defining drought

Anybody can see drought at work: wilting crops, bleached lawns, a receding lakeshore. Yet scholars have debated for many years how best to define this strange condition, the result not of the presence but of the absence of something.

Meteorological drought is the most obvious type: the rains stop, or at least diminish, for a given length of time. In places where it normally rains throughout the year, such as London or Fiji, you might see regular rain and still find yourself in a meteorological drought if the amounts don't measure up. One way to gauge a drought in such places is to add up the cumulative rainfall deficit over some period of time (for example 30, 60 or 90 days). Another would be to calculate how long it's been since a single day produced a given amount of rain. In semi-arid or desert locations, such as Melbourne or Phoenix, you could measure how many days it's been since it's rained at all. No matter where you are, it's the departure from average that makes the difference, since flora and fauna adapt to the long-term norms at a given locale. A year with 500mm (20in) of rain would be disastrously dry for Tokyo but a deluge in Cairo.

Since drought affects us by drying up water supplies and damaging crops, experts now classify drought not only by rainfall deficits but also in terms of **hydrological** and **agricultural drought**. How low is the water level in the reservoir that your city relies on? Are the lands upstream getting enough rain or snow to help recharge the system? These factors, and others, help determine if a hydrological drought is under way. Especially when a dry spell first takes hold, it may take a while for rivers and lakes to show the effects. Likewise, long-parched ground may soak up enough of a drought-breaking rain to keep the hydrologic system from responding right away. Thus, hydrological drought often lags behind meteorological drought, a process that may unfold over a few weeks in a fairly moist climate or across months or even years in a semi-arid regime.

Crops can be affected in even more complex ways by drought, making a declaration of agricultural drought highly dependent on what's normally planted in a given location. If a dry spell is just starting, the topsoil may be parched but the deeper subsoil still moist enough to keep an established crop going. Conversely, a long-standing drought may break just in time to help germinating plants get started, even if the subsoil remains dry.

What do oceans have to do with drought?

Oceans might seem to be the last thing that would control the onset or departure of a drought. But the more scientists learn about the 70% of Earth's surface that's covered by water, the more they learn about how oceans can dictate the amount of water falling on the other 30%. And as greenhouse gases heat them up, the oceans may start issuing their droughty decrees in hard-to-predict ways.

Some parts of the world are dry by virtue of their location. Most of the planet's major deserts are clustered around 30°N and 30°S, where sinking air predominates, while rising air keeps most of the deep tropics between 15°N and 15°S quite rainy on a yearly average. The fuel for those showers and thunderstorms is the warm tropical water below. Should a large patch of ocean warm or cool substantially, it can take months or years for it to return to more typical temperatures. During that time, the anomalous water can reshuffle the climatic cards, bringing persistent wetness or dryness to various parts of the globe.

The supreme examples of such an ocean-driven shift are El Niño and La Niña (see p.130), which bring warmer- or cooler-than-normal surface waters, respectively, to the eastern tropical Pacific. The region of warming or cooling can be as large as the continental United States, and the resulting shifts in rainfall patterns can affect much of the globe. Typically, El Niño hikes the odds of drought across Indonesia, Australia, India, southeast Africa and northern South America, and it can cause winter-long dryness across parts of Canada and the northern United States. Most El Niño droughts only last a year or two, but that's enough to cause major suffering where societies can't easily adapt (by changing crops or shifting their planting schedules, for instance).

Although El Niño and La Niña are the undisputed champions in drought-making, other oceanic cycles (see p.132) can also have an influence. There's some evidence, for instance, that the warm phase of the Atlantic Multidecadal Oscillation (see p.134) is linked to a greater chance of drought in the US Southwest. Similar cyclic warmings and coolings across the Pacific and Indian Oceans may play a role in sustained droughts that can parch regions such as the US Southwest and south Asia for years on end.

On top of these natural waxings and wanings, there's climate change. Warmer temperatures should help promote more evaporation from dry areas and thus create a tendency for droughts to be somewhat more severe where they do occur. Beyond that, the picture gets fuzzy, because

The drying of southern Australia

Water – and the lack of it – are key elements in the psyche of Australia. The unofficial national poem, Dorothea Mackellar's *My Country*, extols a land "of drought, and flooding rains." As a severe dry spell raged in 1888, Australian Henry Lawson bemoaned his fate in verse: "Beaten back in sad dejection, after years of weary toil / On that burning hot selection where the drought has gorged his spoil."

With much of the continent at the mercy of wild swings in precipitation, many Australians have viewed the southwestern coastal belt, roughly from Perth to Cape Leeuwin, as a corner of relative climatic sanity. One of the world's most biodiverse areas, it features a Mediterranean climate with hot, dry summers and mild, dependably damp winters. A 1920 book called *Australian Meteorology* noted the region's reliable moisture: "Here the rains rarely vary 10% from their average amount, and the lot of the farmer should be a happy one."

The farmers – and the urban water managers – haven't been so happy lately. Rainfall across far southwest and eastern Australia has declined notably over the last half-century. (It's also become wetter across the sparsely populated northwest, though the cause of that isn't entirely clear.) Since the mid-1970s, wet-season rainfall around Perth has consistently run some 10–15% lower than before, most noticeably in the late autumn and early winter. These days a wet year in Perth is one that merely reaches the long-term average of 869mm (34.2in). As of 2009, the city had gone more than four decades without mustering a yearly total of 1000mm (39in), a mark once hit every few years. For a booming city with 1.5 million thirsty people, those are scary statistics.

The story is little better across the southeast, Australia's most populous corner. Places like Sydney – unlike Perth – do get a substantial amount of summer rain in the form of showers and thunderstorms. However, new analyses for the last fifty years show that even warm-season rains have slackened slightly across far southeast Australia. Intense drought has dogged the region for most of the past decade, and relentless heat has helped deplete water for cities and agriculture even more than the rainfall deficit might suggest. Water storage systems for Melbourne and Sydney each dipped below 30% of their capacities between 2007 and 2009, and by 2010 they had recovered only modestly to the 40–60% range.

How much of Australia's drying is related to greenhouse gases? And will it continue? Long-term computer simulations hint that natural variability can push the region into and out of multi-decade dry spells. Like much of Australia, the southwest corner tends towards dryness during El Niño years, so the increase in El Niño's frequency and strength since the 1970s may be a factor. Furthermore, much of the countryside was deforested after European settlement, and some models show that the resulting landscape evaporates less moisture into the air, perhaps exacerbating drought. Even with all that in mind, however, the last few decades of drying do bear some of the fingerprints of global warming, especially for southwest Australia. Scientists have verified a poleward shift in the storm track that girdles the Southern

Annual precipitation (mm) at Perth Airport, 1950–2009

Average before downturn

Average after downturn

Ocean, encircling Antarctica. The low-pressure centres that race eastward along this track are the source of nearly all of Perth's rainfall, and they also provide important cool-season rains in Melbourne and Sydney.

Global computer models tend to agree that the storm track's winter position is likely to shift even further south as the century unfolds (see the Southern Annular Mode discussion, p.128). This could mean real trouble. Australia's Bureau of Meteorology and the nation's Commonwealth Scientific and Industrial Research Organization employed an ensemble of 23 computer models for a major set of climate projections released in 2007. The predictions showed an increased risk of further warming and drying across all but northernmost Australia in the coming century. Some models indicated winter and spring drying of up to 10–15% across southern Australia as soon as 2030. Warming temperatures will only add to the water woes, helping evaporate what rain does fall before it can recharge water supplies.

In Perth, the dry shift has already wrought major impacts on water systems. Between 1997 and 2005, area stream-flows dropped 30% below those observed in the previous 23 years, which were themselves on the dry side. Faced with a deepening crisis, the city responded in 2005–06 by launching Australia's first major desalination plant in the industrial town of Kwinana, about 25km (16 miles) south of Perth. The largest such plant in the Southern Hemisphere, it uses reverse osmosis to render water from the Indian Ocean drinkable. The plant now supplies an estimated 17% of Perth's water needs, and construction on a second unit began in 2009. Meanwhile, Sydney gained its own desalination plant – nearly twice the size of Perth's – in 2010. Both facilities are powered by nearby wind farms, making them about as climate friendly as they come. More controversial is a mammoth desalination plant being built more than 100km (60 miles) southeast of Melbourne. It's expected to run on fossil fuels, with offsets at a separate wind farm. Managed by a private consortium called AquaSure, with an eye toward selling water on a statewide open market, the project was lambasted by activists and the nation's Green Party.

it's exceedingly difficult for climate models to replicate the interwoven network of ocean cycles and how they might evolve in a warmer climate. One of the most dramatic signals of the last few years has occurred over the Indian Ocean, where NASA satellite data show average surface water temperatures rose by around 0.25°C (0.45°F) from 1992 to 2000 – a substantial warming for this large a body of water. The NASA study shows that trade winds across the south Indian Ocean weakened enough to reduce up-welling and cooling currents by up to 70%. Should the Indian Ocean continue warming over the next century, as it does in some climate simulations, it may trigger persistent drought in some adjoining areas, particularly southern Africa. To the east, there's also some evidence that a quasi-permanent El Niño-like state could set up in the central Pacific, shifting the current zones of recurrent drought somewhat. All in all, it seems there's much yet to learn about how ocean-modulated droughts will unfold as the world warms.

The plough and its followers: farming and rainfall

Perhaps no other part of weather and climate humbles human ambition as does rainfall and the lack of it. For centuries we've flattered ourselves by thinking we can control the atmospheric tap, whether it be through rainmaking rituals or cloud seeding. Indeed, one of the first ways in which Western thinkers conceptualized human-induced climate change was the notion that "rains follow the plough" – an interesting spin-off from the manifest-destiny mindset that led to US and Australian expansion in the 1800s.

As Europeans moved into progressively drier parts of North America and Australia, they believed that cultivating the prairies and bushland – watering the crops and thus moistening the air above them – would help bring rainfall. Occasional setbacks, like Australia's vicious drought of 1888, didn't quash the belief at first. It wasn't until the 1930s, when unprecedented multiyear drought created the Dust Bowl across the US heartland, that farmers and land managers began to rethink their relationship to climate.

The flipside of this faith that humans could produce a rainy regime via agriculture was the belief that improper use could dry out the land permanently. The notion of **desertification** got its start in Africa's Sahel (see p.74), where it was postulated by colonial explorers as far back as the 1920s and long supported by actual events. During moist years, government policy in some African nations had encouraged farmers to cultivate northward towards the desert's edge. In turn, that pushed nomadic farmers and their

Dust storm approaching Stratford, in the northern Texas panhandle, 1935.

herds into even more marginal territory further north. When the drought hit, both grazing and farming were too intensive for their land's now-fragile state. Wind and erosion took hold, and the zone of infertile land grew southward. At the time, it was accepted by many that the disaster was being produced, or at least perpetuated, by its victims.

Now it seems, researchers are coming around to the idea that neither drought nor rain may follow the plough in a predictable way. It stands to reason that both dry and wet conditions have a certain self-sustaining quality. All else being equal, when the ground is soaked, the air above it is more likely to produce further rain than when the ground is dusty. But the big question is, how powerful is this process against the vast energy reserves of the oceans and the natural variations of weather?

Thus far, there's no consensus on exactly how much of a difference land use can make. In any event, drought-prone areas can take simple, useful steps to reduce their vulnerability. Some of the techniques that proved helpful in the US Great Plains, such as wind-breaking belts of trees, were successfully adapted to the Sahel. Prudent land management can also help people get the most benefit out of the rain that does fall. At the same time, a bit of modesty is in order: oceanic shifts and other natural processes appear to be able to produce drought in some cases no matter how much we do to keep it away.

The big melt

Climate change in overdrive

If there's any place where global warming leaps from the abstract to the instantly tangible, it's in the frigid vastness of the Arctic. Here, the ice is melting, the ground is heaving, plants and animals are moving, and people are finding themselves bewildered by the changes unfolding year by year. Due to a set of mutually reinforcing processes, climate change appears to be progressing in the Arctic more quickly than in any other region on Earth.

Not that some Arctic residents haven't had to adapt to climate change before. Signs of Stone Age life dating back nearly 40,000 years – well before the last ice age ended – have been found near the Arctic Circle in northwest Russia. What appears to be different this time is the pace of the change and the sheer warmth now manifesting. Long-time residents of western Canada and Alaska have seen wintertime temperatures climb as much as 4°C (7°F) since the 1950s. That's several times beyond the average global pace. Across parts of the Arctic, recent summers have been the balmiest in at least 400 years.

Some of the permafrost that's undergirded Alaska for centuries appears to be thawing. Sea levels are slowly rising across the Arctic, as they are elsewhere around the globe, and longer ice-free stretches are increasing the risk of damaging, coast-pounding waves. And experts believe the twenty-first century will likely bring a summer when the Arctic Ocean is entirely free of ice for the first time in nearly a million years. As we'll see, these changes threaten to reorder Arctic ecologies in dangerous and unpredictable ways, with potentially vast long-term consequences for other parts of the planet.

The big melt isn't limited to the poles, however. Many glaciers in the highest mountains of the tropics and mid-latitudes are receding at a startling pace. It's tempting to chalk this up to a warmer atmosphere, but in fact drying may be a bigger culprit than warming in the erosion of glaciers on some tropical peaks, including the one best known to many: Africa's Kilimanjaro (see pp.105–6). Meanwhile, the biggest ice repositories of all – the vast sheets that cover most of Antarctica and Greenland – are sending frozen and melted water to the ocean at accelerating rates, an ominous development that will touch every corner of the world's oceans and coastlines.

On thin ice

The North Pole sits at the heart of an ocean ringed by the north fringes of eight countries: **Canada**, **Greenland**, **Finland**, **Iceland**, **Norway**, **Russia**, **Sweden** and the **United States**. All this adjoining land provides plenty of toeholds for **sea ice**, which overspreads the entire Arctic Ocean and portions of nearby waters each winter. This ice takes on many roles: a home base for polar bears, a roadbed for people and a shield for coastal towns and underwater creatures.

In the summer, as warm continental air sweeps poleward from North America and Eurasia, the ice releases its grip on land and contracts into the Arctic Ocean, with only the oldest, most resistant sections clinging to Greenland and the islands of the Canadian Arctic. Much of the rest disintegrates into a messy patchwork that allows icebreaker cruise ships to penetrate all the way to the North Pole (see box on p.85). While some patches of ice erode in place, winds and waves push other fragments south to meet their doom, primarily through the Fram Strait between Greenland and Spitsbergen, Norway. Eventually, winter returns, and by February the Arctic Ocean is once again encased in ice.

Minimum Arctic sea ice in 1979 (left) ... and in 2007 (right).

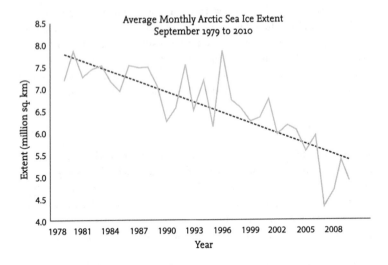

Nobody expects the Arctic to completely lose its wintertime ice cover, but summer is another matter. As of 2009, the average coverage in September (when the ice usually reaches its minimum) was only about 70% of what it was in the late 1970s, according to the US National Snow and Ice Data Center (see diagram above). Even the wintertime ice extent is shrinking by close to 4% per decade, with the depletion becoming especially rapid since 2003. Satellite images paint a stark portrait of these trends, which have reached truly dramatic proportions (see illustration on previous page).

The summer of 2007 smashed all prior records, with even experienced ice watchers surprised by the pace of melt. By that September, the ice extent was barely over four million square kilometres, more than a million square kilometres less than its previous low, set in 2005. Only two thirds of the usual late-summer pack remained, and ice-free waters stretched from Alaska and Siberia nearly to the North Pole. For the first time in memory, the fabled Northwest Passage above Canada was clear enough of ice so that a typical ocean-going vessel could have sailed directly from Europe to Asia unimpeded. Scientists later connected the melt of 2007 to unusually sunny skies in early summer, along with an ice-threatening pattern of atmospheric pressure called the Arctic Dipole. Summer winds normally tend to encircle the Arctic, which helps preserve ice locked at high latitudes. But when the dipole sets up, warm air shuttles across the Arctic from North America to Eurasia, and ocean currents push ice toward the North Atlantic, where it's more likely to melt. In a 2008 study, a team led

Where did that North Pole go?

A group of Arctic tourists got less than they bargained for in the summer of 2000 when they headed for the North Pole aboard an icebreaker ship. Planning to set foot on the ice at 90°N, they made it to the pole – and found only a few scattered floes strewn among the cold water. The group's unnerving experience soon made it onto the front page of *The New York Times*. A paleontologist and a zoologist aboard the ship were quoted beneath the headline "Ages-Old Icecap at North Pole is Now Liquid, Scientists Find". Climate experts hastened to point out that winds and summer warmth can easily push open sections of ice as far north as the pole in a typical summer. But the image of an unnaturally soggy North Pole persisted, a symbol of the very real Arctic melting that's been gaining momentum over the last few years. Extreme swimmer Lewis Gordon Pugh called on the archetype in 2007 when he swam a kilometre amid chunks of ice at the pole to draw attention to climate change.

by Xiangdong Zhang (University of Alaska Fairbanks) targeted the dipole pattern – only recently recognized, but increasingly common – as evidence of a drastic change in Arctic climate.

Summer ice didn't retreat quite as dramatically in the Arctic during the three summers that followed 2007, a weak recovery that was trumpeted by climate-change contrarians. Yet each of those years saw a lower September ice extent than was observed any time before the 2007 record-breaker. Moreover, in each year (2008, 2009 and 2010), the Northwest Passage opened again. So did the Northeast Passage above Siberia, marking three of the first four times in recorded history this route has opened. Dozens of adventurous sailors now trek to the region each year, and in 2010 two yachts – *Northern Passage* and *Peter 1* – became the first to complete both routes in a single summer, circumnavigating the Arctic. "It is, unfortunately, the dramatic changes in Arctic sea ice conditions in recent years that have made this trip possible", said Thorlief Thorliefson and Børge Ousland, who led the *Northern Passage* crew.

> **"The rules are starting to change and what's changing the rules is the input of greenhouse gases."**
>
> Mark Serreze, US National Snow and Ice Data Center

The Arctic's ice isn't just contracting horizontally; it's getting shallower as well. Widespread summer melt and shifting ocean currents are eliminating much of the ocean's thick multiyear ice, which can extend 2m (6.6ft) or more below the surface. The thinner coat that forms in place of multiyear ice is easier to erode when summer returns. Data on ice thickness collected by US nuclear submarines from the late 1950s to 2000 show that sea ice thinned by up to 40% over the

What makes the Arctic so vulnerable?

Trouble seems to beget trouble when it comes to far-northern climate. As a team of scientists noted in the journal *Eos* in 2005, "The Arctic system balances on the freezing point of water. Each summer, the system swings toward the liquid phase; each winter, it returns to the solid phase." While a tropical city might warm a degree or two without obvious effects, the same amount of warming could transform parts of the Arctic, were it to bring the system from just below to just above the freezing point. If the tropics are the lumbering long-haul trucks of the climate world, responding slowly but powerfully to greenhouse gases, the Arctic is a finely tuned sports car, reacting sensitively to even small changes.

Much of this responsiveness is because of **positive feedbacks**: self-reinforcing processes that tend to amplify change. In its definitive 2005 study, the Arctic Climate Impacts Assessment (ACIA) noted several positive feedbacks that allow the Arctic to warm more quickly than other parts of the globe in response to a given increase in greenhouse gases. These include:

▶ **A darkening surface** Several kinds of change in the Arctic all help to produce a darker surface that absorbs more sunlight and, in turn, warms the air above it more strongly. Open ocean typically absorbs more than 90% of the solar energy reaching it, while snow and ice absorb as little as 10%, reflecting the rest of the sunlight off their bright white surface. Thus, a patch of ocean that remains unfrozen can soak up as much as nine times more solar energy. Likewise, when snow cover over land disappears for longer periods, the exposed tundra absorbs more of the sunlight than the snow did. Shrubs and forests, in turn, absorb more solar energy than the lighter tundra does, so as these larger plants migrate northwards (see p.94), they act to further warm the region. (This is in contrast to the tropics, where trees tend to cool the climate as they catch rainfall and evaporate it back into the air.) And pollution wafting into the Arctic from more populated areas helps darken some of the snow and ice, increasing its absorptivity. About a third of the soot in the Arctic atmosphere arrives from South Asia, according to a 2005 study by James Hansen (NASA), and a 2007 study led by Charles Zender (University of California, Irvine) found that the effects of regional soot may explain a third or more of the Arctic's warming since pre-industrial times. Mark Jacobson (Stanford University) found in a 2010 modelling study that reducing soot and related gases could be the quickest way to limit Arctic ice loss.

▶ **A more direct route to warming** As the Sun passes over the tropics, much of its energy goes into evaporating water, leading to the sultry air we associate with the Caribbean or Indonesia. But the evaporation rate in the frigid Arctic is far less. Thus, of the sunlight that reaches the Arctic, a higher percentage goes directly into warming the air.

▶ **A thinner atmosphere** The Arctic's troposphere, or "weather layer", typically extends about 8–10km (5–6 miles) up, compared to 16–18km (10–11 miles) in the tropics. In the shallow troposphere of the poles, it takes less energy to produce a given amount of warming. There's another wrinkle as well. Especially during clear, calm, cold spells, Arctic landscapes often experience a thin, ground-hugging layer of air (an inversion) that's much colder than the air just a few hundred metres higher up. When the inversion breaks up, the surface air can warm dramatically. Thus, anything that punctures inversions or makes them form less often – such as a change in atmospheric circulation, or a loss of snow cover – could produce major ground-level warming.

Potential changes in **heat-trapping clouds** aren't a feedback effect per se, but they do represent a huge uncertainty in polar regions – and elsewhere across the globe (see p.204). US satellites show that, between 1982 and 1999, Arctic cloud cover increased in spring and summer by 5–10% and decreased in winter by about 15%. Overall, these changes point to an increased cooling effect over time, because the low clouds and fog of Arctic summertime screen out the midnight Sun, whereas clear skies in winter allow heat to easily escape to space. If these trends continue, they might help counter the positive feedbacks above. But for now, the Arctic's headlong warm-up seems to be crashing through any cloud-erected barriers.

With all this in mind, we might hope to find a few **negative feedbacks** – processes that tend to dampen change over time. One possibility is the vast flow of heat from the tropics towards the poles. If the Arctic continues to warm at a faster pace than lower latitudes, as expected, then the pole-to-Equator contrast would be weakened and the flow of heat ought to slacken. However, those weakened winds will carry more moisture than before – and that's a warming influence, all else being equal.

All of these feedbacks, plus the tendency of glacial melting to accelerate (see p.97), point towards a potential cascade of effects – a so-called "tipping point", as it's been characterized by some scientists and journalists. In fact, there may be a number of tipping points, each with its own setting. If so, each point we pass makes it that much harder to avoid the others. Conversely, the sooner that greenhouse warming is stabilized, the less likely the Arctic will enter a period of runaway change.

period. The submarines traversed only a few select routes below the ice, though. Estimating ice thickness over a broader area from space isn't easy, but NASA started doing so in 2003 using a laser-based sensor aboard a satellite called ICEsat. (Similar data are on the way from a European satellite launched in 2010, Cryosat-2.) By linking submarine and satellite records, scientists at the California Institute of Technology and the University of Washington found in 2009 that the thinning had worsened. The average depth of midwinter ice had plummeted by nearly 50% since the 1970s, and late-summer ice had withered to less than half its previous depth.

Even the stubborn remnants of multiyear ice hanging on for dear life aren't what they used to be. A team led by David Barber (University of Manitoba) trekked deep into the Arctic Ocean aboard an icebreaker in September 2009. They expected to grind their way through multiyear or thick first-year ice along most of their route. Instead, they zipped along at cruising speeds through a transformed Arctic. Ponds of meltwater had apparently cut through the multiyear ice, slicing it into chunks that lay beneath a coating of fresh ice only about 5cm (2 in) thick. When scanned by NASA's satellite-based sensor, this new icescape resembled a more typical, multiyear ice field – a masquerade hiding a far more serious decline than scientists had realized.

How long will the Arctic Ocean hang onto its summertime ice? Until the last few years, the best consensus among top climate modellers was that the Arctic would retain some ice throughout each summer until at least 2080, perhaps into the next century. But as new data and model results come in, that consensus is fragmenting even more quickly than the ice itself. The breakneck pace of recent summertime loss has raised concern that the iceless-summer benchmark could be upon the Arctic waters far more quickly than scientists once thought. A number of experts now project the milestone to arrive by 2030 or even sooner. This is no minor development: as far as anyone can tell, the Arctic hasn't been free of major ice cover at any point in at least the last 800,000 years. Of course, it's still possible that the intense melting of the last few years is a short-term variation superimposed on a slower long-term trend. However, there's no obvious natural cause for the recent speed-up.

People, animals and ice

So what's wrong with a few extra weeks of open ocean in a place that's usually locked tight in ice? Indeed, the prospect of ice-free summers has many industries and entrepreneurs salivating, especially those hoping to cash in on trans-Arctic shipping (see p.328). Along the Arctic coast, however, the increasingly tenuous state of the ice during autumn and spring complicates life for indigenous residents. Subsistence whalers and seal hunters use the ice sheets to venture beyond the coastline during the transition seasons. The timing of those seasons, and the quality of the ice, has now become less dependable. Also, the near-shore ice normally helps keep wind-driven waves well offshore during the violent weather of autumn and spring. But storms now smash into coastal towns unimpeded by ice well into autumn. The resulting erosion threatens the very survival of some Arctic settlements (see p.91).

The loss of sea ice has especially grave implications for **polar bears**. These white giants retreat to land-based dens by winter and then bring their hungry young across the thinning springtime ice to find seals. After the ice is gone, the bears spend their summers fasting on land, which has become a more stressful affair for both mothers and gestating cubs now that ice-free spells are lengthening. At Hudson's Bay, near the southern end of the polar bears' range, the summer fast is now a month longer than it was decades ago, and the average birthweight of cubs dropped 15% from 1981 to 1998.

Greater extremes in precipitation (see p.65) are also taking their toll. Heavier and more frequent spring rains have collapsed some of the bears' dens, and fires that rage through drought-stricken boreal forests destroy permafrost and the dens lodged within it. Then there's the spectre of interbreeding between polar bears and grizzlies now taking shape in Canada, as a warming climate shifts the grizzlies' range toward coastal lands where polar bears roam in between shorter sea-ice seasons. One bear shot in 2006 in northern Canada, was later found through DNA testing to be a grizzly/polar hybrid: a "pizzly".

Despite all of these concerns, the polar bear is hardly disappearing just yet. Since the bears' territory is so remote, it's tough to obtain a firm count, but the most common estimate is that there are now between 20,000 and 25,000 of them. That's more than double a 1960s count, though census techniques have improved notably since then. Ultimately, the fate of polar bears comes down to sea ice. Beyond a certain point, there simply won't be enough of it over a large enough area to sustain polar bears as they now live. In a 2007 study, the US Geological Survey estimated that

> **"You don't have to be a polar scientist to see that if you take away all the sea ice, you don't have polar bears any more."**
>
> Andrew Derocher,
> University of Alberta

two thirds of Earth's polar bears will be gone by 2050. Only a few months after the Arctic's dramatic melt of 2007, the US government listed polar bears as a threatened species. Unless these iconic creatures can quickly adapt to land-based life like their brown-bear cousins and predecessors – or unless Arctic warming can be slowed down – polar bears appear to be headed for extinction in the wild before the century is out.

Several species of **seal** are also at serious risk from ice loss. **Ringed seals**, the Arctic's most common type, are the main food source for many Inuit communities as well as for polar bears. In spring, the seals nurse their pups in lairs atop the snow-coated ice sheet, so the stability of springtime ice is critical for them. A recent survey of ringed seal populations near the northwest coast of Hudson's Bay showed a seesawing pattern, with the numbers dropping in the 1970s and 1990s but rising in the 1980s, in sync with the snowfall that the seals use to build their winter dens. As the ice retreats, ringed seals may get a short-term boost from new access to Arctic cod and other prey that become easier to snatch. However, pups will find themselves exposed to the cold ocean – and to predators – at earlier points in the spring. There's no evidence that ringed seals will take to land, which goes for several other seal species as well, so in the long run the loss of summertime ice may prove devastating. In contrast, grey seals and other temperate species already accustomed to land have a better shot at dealing with reduced Arctic ice successfully.

Birds – made for mobility – appear to have a head start in adapting to changes across the Arctic. The task won't be easy for all species, though. White gulls are scavengers that feed mainly on the carcasses of polar bears and ringed seals, so a decline in those species will at least force the gulls to find other meal tickets. Fish and other marine creatures may also find their diets shaken up, as schools shift with the warming oceans and as industrial fishing moves northward into formerly ice-encased areas.

It's hard enough for some forms of Arctic wildlife to deal with climate change, but **pollution** may be making matters worse. A witch's brew of contaminants, including mercury and other heavy metals, has drifted into the Arctic from industrial nations over the years. Many of these toxins are currently locked in snow and ice. A warming climate could melt this snow and ice, releasing decades' worth of pollutants into rivers, ponds and oceans. Some of these contaminants can work their way up the food chain, and they've already been found in the fat reserves of some creatures near the top of the chain – notably polar bears.

Move or drown: one town's tough choice

Poised at a geographic crossroads, Shishmaref, Alaska, also lies in the crosshairs of a warming climate, earning the village of 600 people a modest measure of fame. Shishmaref is perched on a slender island the length of New York's Central Park but barely half as wide. It lies off the north coast of the Seward Peninsula, not far from Russia (in fact, Shishmaref is closer to Moscow than to Washington DC). Inupiat peoples, with close ethnic ties to their counterparts across the Bering Strait, have lived on the island for centuries. Today, like many indigenous communities, their town is a hybrid of the modern (a school, contemporary homes) and the frontier (no television, no running water).

Each winter the Chukchi Sea surrounds Shishmaref with a thick layer of ice that's integral to the subsistence-based, hunting-and-fishing culture. The ice also keeps out seawater that would otherwise smash into the island's cliffs during violent winter storms. Residents have long counted on the freeze-up to occur in October, but lately there have been patches of open water near shore as late as December, leaving the town at the mercy of autumn elements. Recent October storms have flung winds of 144kph (90mph) and waves of nearly 4m (13ft) at the tiny island, whose top elevation is only 6.7m (22ft).

In an earlier time, the people of Shishmaref might have simply packed up and moved on, but abandoning a modern township with ancient roots isn't so easy. Though it goes against their cultural grain, locals pleaded for state and federal officials to recognize their plight. An Erosion and Relocation Commission was formed in 2001, and in 2002 the townspeople voted by a count of 161 to 20 to move.

> "It's not our fault that the permafrost is melting, or that there's global warming that's causing us to go farther away from our home in Shishmaref. But we'll survive."
>
> Luci Eningowuk,
> Shishmaref resident

But as of 2010 it still wasn't certain where, or even whether, Shishmaref would relocate. One option is Tin Creek, a mainland site about 8km (5 miles) away – close enough that many homes could be moved there and townspeople could maintain connections to the land and sea. The US Congress had considered moving the residents to the larger towns of Kotzebue or Nome, but the idea of their community's identity being swallowed up didn't sit well with many residents.

Wherever they end up, the move will be a costly process for Shishmaref, with the tab estimated at up to $200 million and no ready source of funding available. The commission established a website in 2007 (shishmarefrelocation.com) to publicize its plight and seek supporters. Other settlements along the Arctic coastlines of Alaska, Canada and Russia may also be forced to move, a prospect that serves as eloquent – and expensive – testimony to a climate in flux.

A softening landscape

As the Arctic Ocean slowly sheds its mantle of year-round ice, terrestrial life nearby is taking on new forms. Perhaps the most emblematic is the "drunken forest" – larch and spruce trees that lean at crazy, random angles as the soil beneath them thaws out.

Permafrost – land that's been frozen for at least two years – holds the key to this and many other transitions across the Arctic landscape. Permafrost covers about a quarter of the Northern Hemisphere's land area, including half of Canada, most of Alaska and much of northern Russia, as well as a small part of Scandinavia and several mountainous regions closer to the Equator. On the fringes of the Arctic, the permafrost is **discontinuous**, forming in a mosaic of favoured locations. In the coldest areas, **continuous permafrost** extends throughout the landscape.

Permafrost is topped by an **active layer** that typically thaws and refreezes each year. The permafrost itself – which starts anywhere from a few centimetres to a few tens of metres below ground – remains below freezing year-round. It can extend hundreds of metres further down, until the soil temperature again rises above freezing en route to the Earth's molten centre. The thicknesses of the active layer and the permafrost depend on both recent and past climate, making them convenient diaries of global change. As part of the International Polar Year (which ran, despite its name, from 2007 to 2009), scientists drilled more than 500 boreholes across the frozen north to gauge the health of its permafrost. Most areas showed a continuation of the rising soil temperatures documented over the last few decades, with the coldest permafrost warming the most quickly.

It's the deepening of the active layer, and the thawing of permafrost just below it, that's causing trouble in certain parts of the Arctic. Tucked within the permafrost is water, some of it in crystals bonded to the soil structure and some in much larger ice beds or wedges. If a thaw descends below the active layer, it can cause pockets of ice in the topmost permafrost to liquefy and create an underground lake (a **thermokarst**). Should this lake then drain off, the resulting cavity may fill with surface soil slumping inward. Such is the process that has put many trees and buildings in a slow-motion fall, leaning at ever-sharper angles over time, especially across the fast-warming lower Arctic between about 55°N and 65°N.

Because of the variability of year-to-year warming and the random locations of ice pockets, it's almost impossible to predict which house or tree in a melt zone will start leaning next. In **Yakutsk**, the major city of eastern Siberia, many buildings are on stilts a few metres deep. Even so, over 300

As the permafrost melts beneath its foundations, an Alaskan barn collapses into the ground. The area near Fairbanks is currently one of the worst hit by rising ground temperatures.

structures and an airport runway have been compromised by unstable permafrost (perhaps coupled with less-than-optimal building practices). Siberia's permafrost has warmed more than 1°C (1.8°F) on average since the 1960s, and computer models tag vast stretches of Siberia at high risk for more structural problems as the century unfolds.

Fairbanks, **Alaska**, is a poster child for permafrost trouble. The city of 30,000 lies atop a fairly thin active layer – only a few metres deep in many locations. Record warmth in the past twenty years has percolated downward, heating permafrost across Alaska by 1–2°C (1.8–3.6°F) to its warmest levels since the last ice age in some spots. Around Fairbanks, some sections of permafrost have warmed by more than 3°C (5.4°F), bringing its temperature above –1°C (30°F) in some locations. The results are startling: drunken spruce trees, heaving bike trails and sinkholes pockmarking the landscape. Thus far the sinking has been patchy, focused where digging and construction have opened up layers of landscape to warming and melting. More widespread problems could occur if and when the bulk of Fairbanks' permafrost gets to the dangerous side of the freezing point.

Trees aren't just toppling across the Arctic; they're also burning. In 2004 and 2005, **fires** swept across more than 10% of the tundra and forest in Alaska's interior, ravaging nearly 49,000 square km (19,000 square miles). Across Siberia, which holds half of the world's evergreen forest, the fire losses have reportedly risen tenfold over the last few decades. In 2004 alone, fire consumed an area almost the size of Great Britain – 220,000 square km (85,000 square miles). Sharply warmer summers in recent years have helped dry out the landscape much sooner than usual, lengthening the fire season and making the fires that start that much more intense. Insects are playing a part, too. Epic infestations of pine and spruce bark beetles have swarmed across western North America over the last decade. The bugs have killed off millions of trees and increased the forests' vulnerability to flame. These infestations have been driven in part by winters too warm to kill off the insects (see p.172).

To be sure, a milder Arctic paves the way for the northward migration of many flora and fauna. The problem is that many parts of the ecosystem can't change with the fluidity of the climate itself. For instance, the weather in thirty or forty years may be conducive to boreal forests across some regions now covered with nutrient-poor tundra. However, it might take decades longer for decaying grasses to prime the soil so it can support tree growth. Still, projections for the next century suggest a massive increase in high-latitude forest, especially across western Siberia and north-central Canada. This would pull an increasing amount of carbon dioxide out of the atmosphere, though probably not enough to counteract the warming effect caused by the forests' propensity to absorb heat (see p.86).

The Arctic's signature land beasts – **caribou**, **reindeer** and their cousins – face their own challenges. Heavier spring runoff and rains across the North American Arctic could widen rivers enough to jeopardize the seasonal trek of the continent's vast herds of caribou, some of which hold more than 100,000 animals. In Lapland, autumn and early winter are already proving treacherous. Reindeer normally browse below the surface of early-season snow to reach the tender lichen that spread at ground level. But in the past few years, bouts of freezing rain have become more common as warmer-than-usual air fed by the Gulf Stream flows atop the Arctic's frigid surface air and rains into it. When ice encases the snow and the underlying lichen, the reindeer lose their access to food. No major die-offs have been reported, but the trend has raised concerns among ecologists and local herders.

The Arctic's patchwork of **freshwater lakes** and **rivers** is also being rewoven as the climate warms. Across a Texas-sized patch of Siberia, researchers used satellite photos to identify more than a hundred small lakes that disappeared between the 1970s and 1990s. These were victims of thermokarsts that allowed the waters to drain out, while new lakes opened to the north.

How will Greenland's fate affect ours?

The biggest chunk of land in the far north, **Greenland** has languished in frozen obscurity for centuries. Two areas on its chilly southwest coast were populated by Erik the Red and a few hundred other Danes and Icelanders around AD 1000. The Western Settlement disappeared around 1350 during a marked cool-down, while the Eastern Settlement hung on for some 200 years more before giving way. Since then, the island's brutal cold has kept human occupation limited to a scattering of Inuit towns and, in the last few decades, a sprinkling of seasonal research camps. Several of these are parked on top of the colossal **ice sheet** that covers about 85% of the island.

Greenland's vast coating of ice could end up shaping where millions of people live a few generations from now. Actually, "sheet" may not be the

Methane lurking in the muck

Among all the by-products of the melting Arctic, one stands out in its sheer horror-movie potential. Trapped within permafrost are billions of tonnes of **methane hydrates** (also known as **methane clathrates**). These are molecules in which water and the potent greenhouse gas methane are bonded under high pressure and/or low temperature. Besides their presence in permafrost, methane hydrates are even more extensive in seafloor sediments around the margins of continents across the globe. In their supercompacted form, methane hydrates are more than 150

times more concentrated than gaseous methane. Though it's not clear what will happen to methane hydrates as the permafrost continues melting, scientists point to the possibility that vast quantities of methane could be released – perhaps enough to dwarf the greenhouse gases emitted by human activity. Should that happen, global warming could go well beyond current projections. It's believed that methane hydrates may have been involved in some of the most intense warming episodes in Earth's history (see p.228). Nevertheless, governments from the US, China and Japan have reportedly investigated the potential of methane hydrates as a fuel source: there may be twice as much of it as there is coal, oil and natural gas elsewhere on Earth.

Gaseous methane burns off a lump of seabed hydrate, releasing liquid water below.

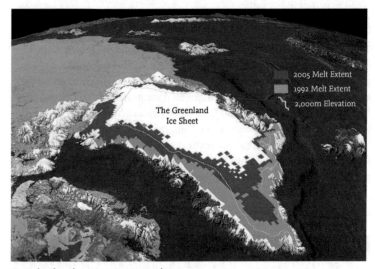

The Greenland
Ice Sheet

2005 Melt Extent
1992 Melt Extent
2,000m Elevation

Greenland melt extents in 1992 and in 2005.

best word for this icy monolith. It extends over 1.3 million square kilometres (500,000 square miles) – the size of France and Spain combined – and the bulk of it extends more than 2km (1.2 miles) high. That makes the ice itself a topographic feature on a par with the US Appalachian Mountains. Although some of the sheet probably melted in between ice ages, scientists believe it's been at least three million years since Greenland has been ice-free. Under the sheer weight of all its ice, Greenland's centre has sunk roughly 300m (1000ft) below sea level.

Some models indicate that a 2°C (3.6°F) rise in average global temperature, if sustained, would ensure the ice sheet's eventual demise. Right now, with rapid cuts in global emissions looking unlikely, we're on a trajectory that could bring global temperature well above that 2°C point this century. For Greenland's ice, the point of no return wouldn't be obvious at first. Once under way, though, the melting will become virtually unstoppable through a set of positive feedbacks. For example, as meltwater runs off from the highest elevations, the sheet's newly lowered summit would encounter warmer air temperatures. This would allow for still more melting, and so on, putting the ice in a cascade towards oblivion.

Until recently, scientists were debating whether the amount of water locked up in Greenland ice had increased or decreased over the last several decades, but new data and analyses show that the continent is indeed losing ice mass (see below). If and when the entire Greenland ice sheet were to melt into the sea, it would trigger a truly catastrophic sea-level rise of close to 7m (23ft). Such an inundation would swamp coastal cities around the world and, in some low-lying areas, close the book on millennia of human history.

Thankfully, most glaciologists believe it would take centuries for the entire sheet to melt. However, there's still cause for concern in the shorter term, because Greenland's peripheral glaciers appear to be losing ice at a faster clip than either scientists or their computer models had reckoned. Some of the ice processes now at work aren't represented well – or at all – in computer models because of their complicated nature. Yet even a small fraction of Greenland's ice entering the ocean over the next few decades could push sea-level rise well beyond the current consensus estimates.

What does seem clear is that there's an increasing amount of water cycling through the system. Since 1992, European Space Agency satellites have used laser-based altimeters to measure the ice sheet's height from above. The ESA data indicate that the sheet is gaining about 5cm (2 in) of snow a year. If so, this wouldn't be a surprise. As overall warming brings the frigid air flowing atop the sheet closer to 0°C (32°F), the air holds more moisture and snowfall is enhanced. That doesn't rule out an occasional dramatic melt at high altitudes. In 2005, a week-long thaw spread across all of southern Greenland. Even atop South Dome (elevation 2900m, or 9600ft), scientists reported three days of melting. And in the record-setting summer of 2007, melting occurred across fully half of the ice sheet's surface.

Even if snow and ice are still accumulating well inland, Greenland's ice sheet as a whole is far from healthy, according to **GRACE**, an ice-monitoring satellite launched by the US and Germany. Analyses of GRACE data by Isabella Velicogna (University of California, Irvine) show not only that the island is losing ice mass each year, but that the loss is accelerating – the rate more than doubling since the year 2002. During the period 2006–09, Velicogna estimates that Greenland lost an eye-popping one trillion metric tonnes of frozen water.

The main routes for this loss appear to be outlet glaciers, the frozen superhighways along which ice flows to the sea. Several of Greenland's largest glaciers are now moving at a pace far greater than expected. Despite the acceleration, so much melt is occurring at the forward edge of these glaciers that the net effect can be a retreat of that edge. Think of dipping a chunk of ice very slowly into a glass of hot water: even as the whole chunk goes downward, the bottom edge of the ice is melting its way upward. In fact, it appears that warmer waters are helping to erode the ice tongues that often extend from outlet glaciers, thus assisting the speeded-up ice flow. The tongues' bottom surface is especially vulnerable to melting because the added pressure of water lowers the melting-point temperature. A given amount of added heat can produce several times more bottom melting than surface melting, according to Velicogna.

The glaciers' rush to the sea is also bolstered through an important process called **dynamic thinning**. As meltwater seeps through the ice sheet into a flowing glacier, it can lubricate the base and hasten the glacier's seaward flow. Not all of Greenland's glaciers are ideally positioned for a speed-up, though. It appears that dynamic thinning is most effective on glaciers that lie atop a layer of bedrock that slopes smoothly towards the sea. Scientists from NASA and elsewhere are now working to map the topography beneath key glaciers to help predict which ones are flowing on favourable slopes.

One of the world's fastest-moving glaciers is **Kangerdlugssuaq**, along Greenland's southeast coast. Some 4% of the island's ice flows through Kangerdlugssuaq towards the Atlantic. The glacier advanced seaward at an annual pace of 5km (3 miles) in 2001, but almost three times more quickly – 14km (9 miles) – in 2005. At the same time, the glacier's forward edge pulled back a full 5km (3 miles) in 2005, by far its largest retreat on record. On the west coast, **Jakobshavn Isbrae**, the largest of Greenland's ice-draining glaciers and another fast mover, slowed slightly from 1985 to 1992 but then more than doubled its seaward velocity in the following decade. Yet in the summer of 2006, the thinning rates of Kangerdlugssuaq and another large east-coast glacier, Helheim, dropped to near zero. It's now appearing that these highly dynamic systems can move seaward in pulses that strengthen and weaken from year to year, but atop longer-term depletion apparently induced by climate change. Taken as a whole, the recent accelerations in Greenland glaciers hint at a faster-than-expected response to climate change. One study finds that the meltwater from Jakobshavn Isbrae, all by itself, has pushed up the rate of global sea-level rise by 4%. As a whole, Greenland is now contributing as much as 20% to the yearly rise.

The action on Greenland's coast appears to be shifting northward as well, consistent with long-term warming. A team led by Denmark's National Space Institute found that the rate of ice loss (as measured by mass) in far northwest Greenland has accelerated markedly since 2005. In the summer of 2010, Greenland's north coast came into the media spotlight as a gigantic "ice island" four times the size of Manhattan broke off from the **Petermann Glacier**. It was the Arctic's largest calving event since

The cleaving of the Petermann "ice island" – satellite image of Manhattan to scale inset.

1962, which also made it the largest to be portrayed through vivid satellite imagery (see photo above). Though calving is a natural process, the sheer scope of this one served as visual shorthand for the larger issue of long-term ice loss. It's fair to say that the future of Greenland's ice sheet is one of the biggest unknowns looming over our planet's coastlines.

What about Antarctica?

Strange as it may seem, Antarctica has stayed mainly on the sidelines in the saga of global warming – at least until recently. This vast, lonely land took centre stage in the mid-1980s when the infamous ozone hole was discovered a few kilometres up. The Montreal Protocol and resulting actions by world governments are expected to eliminate the hole by later in this century (see p.32).

Almost two thirds of the planet's fresh water is locked in Antarctica's ice sheet, ready to raise sea levels by spectacular amounts if the ice were to melt. Fortunately, nobody expects the entire sheet to start melting any time soon. Temperatures in the interior of Antarctica are shockingly cold: Vostok Station, for example, has never recorded a temperature above freezing.

It's been hard to determine exactly what's happening to temperatures across the great bulk of Antarctica, which has only a few reporting stations, but there are good reasons why parts of this frozen realm might run counter to Earth's long-term warming trend. For one thing, Antarctica is surrounded by the vast, cold Southern Ocean, which absorbs some of the heat that would otherwise boost air temperatures across the region. In addition, overall global warming has led to a tightening of the Antarctic vortex, the ring of upper-level winds that encircles the continent. This helps keep the coldest air focused over the continent and inhibits warmer air at mid-latitudes from working its way southward. Moreover, ozone depletion in the Antarctic stratosphere over the last several decades has helped to strengthen the Antarctic vortex. In East Antarctica – which spans the great bulk of the continent – satellite measurements of the ice sheet's temperature show cooling over some areas in recent decades, mainly during the summer and autumn. However, at the South Pole itself, surface readings that had been cooling have recently begun to warm, leading to little net change there since the 1950s.

As opposed to its eastern counterpart, West Antarctica is solidly in the warming camp – so strongly, in fact, that it more than compensates for any cooling to the east. The evidence mounted with a 2009 study in *Nature* that combined surface data with satellite-based temperature estimates

Antarctica's sea ice holds its own

If you're interested in watching sea ice grow, head south. Each winter a huge ring of ice develops around the rim of Antarctica, spanning about 14 to 16 million sq km of ocean by its September peak. That's about 10% more area than the Arctic's sea ice encompasses at its maximum each March. In contrast to the high-profile decline in northern sea ice, the south's ice extent has been quietly growing by about 1% per decade since the 1970s. Some climate models didn't see the increase coming, and many climate sceptics cite the growth as a natural compensation for the Arctic's ice loss. However, as noted by NASA's Jay Zwolly and others, Antarctic coastal temperatures that are less frigid but still below freezing could actually boost the amount of moisture in the air above the sea-ice formation zone. This, in turn, would help more snow to fall atop the newly formed ice, making it thicker and facilitating the ice's horizontal growth. Whatever its cause, the Antarctic's modest growth in sea ice pales next to the far more dramatic losses in the Arctic and the profound ecological impacts those could trigger. As Weather Underground blogger Jeff Masters points out, "Calling attention to Antarctic sea ice gain is like telling someone to ignore the fire smouldering in their attic, and instead go appreciate the coolness of the basement, because there is no fire there."

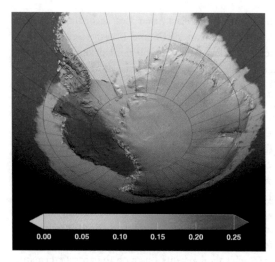

Red represents areas where temperatures have increased the most during the last 50 years, particularly in West Antarctica, where some points have warmed even more than shown here. Temperature changes are measured in degrees Celsius per decade from 1957 to 2006.

gathered since the 1980s. The combined analysis shows a continent-wide average warming of more than 0.6°C (1.1°F) since 1950 (see graphic above). NASA's ongoing analysis of global temperature trends shows similar values of warming for the Antarctic since the late 1950s. The biggest changes are taking place on the **Antarctic Peninsula**, which juts northward towards South America. Average temperatures at weather stations along the peninsula's west coast – one of the few parts of the continent that sees major thawing and refreezing each year – have soared by as much as 2.5°C (4.5°F) since the 1950s, the most rapid warming anywhere on Earth in that span.

It's also getting warmer above Antarctica, as announced in a 2006 paper in *Science* by John Turner of the British Antarctic Survey. Looking at South Pole Station plus eight coastal stations, Turner and colleagues found that the air at a height of about 5.5km (3.4 miles) above sea level had warmed by 1.5–2.1°C (2.7–3.8°F) since the 1970s. Frigid surface air often decouples from the air immediately above it, which may explain how surface cooling and upper-air warming could coexist.

Once the ozone hole begins healing in earnest later this century, it's expected that warming will accelerate across most of Antarctica, hastened by the globe-spanning effects of greenhouse gases. Across the interior, that warm-up may do little at first other than slowly build the ice sheet, in much the same way as Greenland's appears to be growing on top. Slightly warmer air (that is, air that's less bitterly cold) should bring more moisture with it, resulting in a boost to the continent's tiny snowfall amounts.

That's already been the case across East Antarctica, according to a 2005 report in the journal *Science* by a team of US and British scientists. They found the ice sheet across the East Antarctic interior thickening by about 2.2cm (0.9in) per year since the mid-1990s, somewhat more than most global computer models had predicted. However, that gain may be more than offset by a surprising amount of ice leaving East Antarctica through outlet glaciers, according to a 2009 report by scientists at the University of Texas at Austin. Using GRACE data, they found that East Antarctica appears to have entered a period of net ice loss beginning in 2006. "We had considered East Antarctica to be inviolate", said lead author Jianli Chen. "But if it is losing mass, as our data indicate, it may be an indication the state of East Antarctica has changed."

For now, the ice appears to be shifting most rapidly along the fast-warming Antarctic Peninsula. The Larsen B shelf made headlines in early 2002 when a Delaware-sized chunk – perhaps 12,000 years old – broke off. Compelling as they are, such calving events – as with the 2010 Petermann calving in Greenland noted above – have only minor overall effects on sea level, since most of the ice was already in the water. More concerning is whether such break-ups might open the door to a faster seaward flow of ice from glaciers upstream. The hydrostatic force of the ocean pushes back on large ice shelves that are grounded on offshore bedrock; this process helps stanch the flow of ice from upstream. When the shelves break up, the adjacent glacier flows more easily into the sea. A NASA-led team found that glaciers flowing into the Larsen B shelf increased their speed eightfold after the shelf's 2002 break-up.

Another group led by NASA's Robert Thomas found that glaciers in the Amundsen Sea area of West Antarctica were thinning much faster in 2002–03 than in the 1990s, as far as 300km (190 miles) inland. If the floating shelves in this area were to break up, *à la* Larsen B, they could allow for the eventual drainage of glaciers that together hold enough ice to raise sea levels by 130cm (51in). "Although these glaciers are the fastest in Antarctica, they are likely to flow considerably faster once the ice shelves are removed", wrote Thomas and his colleagues in *Science*.

Overall, according to Velicogna, GRACE data suggest that the pace of Antarctic ice loss may have doubled for the period 2006–09 compared to 2002–05. It's a good thing that climate models project the vast bulk of Antarctica's ice to stay onshore for at least the next century and probably much longer than that. Were it all to melt, it's estimated that global seas would rise by a cataclysmic 53m (174ft) – a scenario ripe for Hollywood, if nothing else.

Tropics and mid-latitudes: goodbye glaciers?

Far above the lush forests and warm seas of the tropics, a few icy sentries keep watch on our planet's climate from a unique vantage point. These are Earth's tropical glaciers, perched atop the Andes and other peaks that extend from 4600m to more than 6900m (15,100–22,600ft) above sea level. Further north, in the mid-latitudes, thousands of small glaciers and a few larger ice caps are sprinkled across parts of the Rockies and Sierras, the Himalayas and other Central Asia ranges, and the Alps.

These scattered outposts of ice aren't very good at hiding the strain they're under. Because low-latitude glaciers tend to be much smaller than their polar cousins, they often respond more quickly to climate change. A melting trend that kicked in during the mid-1800s has accelerated in recent decades across many of the tropical and mid-latitude glaciers monitored by scientists. By later in this century, some of these remnants of the last ice age could be little more than history themselves.

Lonnie Thompson: putting tropical ice on the map

He may not be able to save them, but **Lonnie Thompson** has done as much as anyone to document and publicize the plight of tropical glaciers. Based at Ohio State University, Thompson paid his first field visit to Peru's Quelccaya in 1974 as a graduate student. He has since spent more than 840 days above 5486m (18,000ft) as well as 53 consecutive days at around 6096m (20,000ft). Early on in his career, Thompson toiled for years – often working on a shoestring with few collaborators – to collect data from the Andes and other regions of vulnerable ice. Few colleagues thought the tropics had much to offer glaciologists compared to the much vaster ice fields at the poles. Thompson proved them wrong in 1983 by drilling 164m (538ft) into the Quelccaya ice cap and retrieving a unique core that documents 1500 years of climate history near the Equator. Thompson's subsequent expeditions have dug even deeper into climate's past. An ice core he retrieved in 1987 at Sajama, Bolivia, spans 25,000 years, while the Guliya cores from China's Western Kunlun range extend back more than 700,000 years.

After more than three decades of work, Thompson feels a growing urgency to capture the climate clues offered in tropical ice before it's gone. When he re-cored Quelccaya in 2003, meltwater seeping downwards had already compromised some of the ice layers from earlier years. His return to Kilimanjaro in 2006 after field work there in 2000 showed the mountain's ice continuing to dwindle. "The accelerating retreat of most tropical glaciers is very telling", says Thompson, "for it is occurring in a part of the world known for its temperature stability." Because glaciers are "great integrators of climate", as Thompson puts it, "the fact that most glaciers, from the poles to the tropics, are speaking in one voice and telling us the planet is warming should be a concern to us all."

Like rivers in slow motion, glaciers are fluid entities, with fresh water flowing through them. Snow accumulates on top and gets pushed downward and outward by the snowfall of succeeding years. For a healthy glacier, the change in its total ice mass per year – its **mass balance** – is either zero or positive. Today, however, most of the world's tropical and mid-latitude glaciers have a mass balance seriously in the red. In short, they're shrinking.

It isn't just melting that spells doom for ice. Many glaciers erode largely through a process called **sublimation**. When dry air flows over ice, the moisture in the ice can pass directly from its frozen state to its vapour state without any intervening liquid form. Sublimation can occur even when the air is below freezing, and it's the dominant process of glacier loss in some areas. Overall, however, melting is the main culprit. It takes eight times less energy for a molecule of ice to melt than it does for it to sublimate. Under the right weather conditions, direct sunshine can melt glacial ice at temperatures as cold as -10°C (14°F). And, of course, a reduction in snowfall can take its toll on a glacier even if average temperatures stay constant.

It's the unique mix of conditions at each location that shapes the behaviour of tropical and mid-latitude glaciers. Those who study glaciers fear that this complex picture sometimes gets lost in the scramble to interpret glacial loss as a simple product of warmer temperatures. Still, they agree on the main point: that climate change is eroding many glaciers at a dramatic pace.

Trouble towards the Equator

Thanks to a century of visits by adventurers, photographers and scientists, the ice cap atop Africa's **Mount Kilimanjaro** is the most widely publicized of the tropics' disappearing glaciers (see box opposite). But others tell a similar tale of woe. Not far away from Kilimanjaro, **Mount Kenya** lost seven of its eighteen glaciers over the last century; those that remain cover a paltry area of less than a square kilometre (0.4 square miles). On the Indonesian island of Irian Jaya, the glacial area atop Mount Jaya shrank by 7.5% from 2000 to 2002. Its ice now covers only about 20% of the area it did in 1950.

All of these are mere slivers next to the ice atop the **Andes** of South America. With their string of peaks and high valleys, many extending above 5000m (16,000ft), the Andes hold more than 99% of the world's tropical ice. In Peru, north of Lima, more than 700 glaciers dot Peru's **Cordillera Blanca**, or White Mountains (a name deriving mainly from the range's granodiorite rock, rather than the ice it sports). Further south

The fast-disappearing snows of Kilimanjaro

... and then they were out and Compie turned his head and grinned and
pointed and there, ahead, all he could see, as wide as all the world, great,
high, and unbelievably white in the sun, was the square top of Kilimanjaro.

The Snows of Kilimanjaro, 1938

It's a safe bet that Ernest Hemingway wasn't thinking of the greenhouse effect
when he set his tale of a man dying of gangrene near a mountain whose ice cap
is itself dying. Yet the title of Hemingway's classic short story, "The Snows of
Kilimanjaro" is now a standard piece of climate-change iconography. Africa's tallest
mountain is indeed losing its ice, and quickly, in a process that started decades
before Hemingway penned his story. Of the three peaks that make up this giant,
extinct volcano, only one (Kibo, the tallest, at 5893m/19,334ft) has anything left of
the vast ice cap that apparently covered all three peaks long ago. Kibo's ice, which is
concentrated in three fields, shrank by 80% in the twentieth century. Some reports
have its thickness going down by roughly 50cm (20in) per year since the 1960s. A
survey led by glaciologist Lonnie Thompson found that 27% of the ice cover present
on Kilimanjaro in 2000 was gone by 2009. The ice can hold its own in a good snow
year, but if these rates of melt continue, permanent ice fields would disappear from
Kilimanjaro by the 2020s.

Kilimanjaro is anything but a textbook case of "warmth equals melting", though.
In fact, it's become something of a flash point, a topic of spirited debate among
glaciologists as well as a favourite of climate-change doubters who extrapolate
from a single mountain in an attempt to disprove a world full of receding glaciers.

What makes the story of this ice so slippery? For one thing, the average air temperature over Kibo's ice cap runs well below freezing, hovering close to -7°C (21°F)
year round. That in itself doesn't rule out climate change as a factor, but it means
something more than sheer warming is probably at work. The ice cap's shape offers
more clues. Instead of being rounded, it features steep, sharp sides about 20m
(66ft) high. Georg Kaser of the University of Innsbruck argues that if air temperature alone were shifting the balance towards melting, the ice cap would quickly
revert to a smoother-edged contour. This suggests the Sun rather than the air is
the prime melting agent.

More evidence for the Sun's key role is the east-west orientation of Kilimanjaro's ice
cap. Since the mountain sits just 370km (230 miles) south of the Equator, the north-
and south-facing walls each face the midday Sun at different times of year. These
happen to coincide with the start of the peak's two dry seasons (roughly December
to February and June to September), which means few clouds sit between the Sun
and the ice. Moreover, the local winds generally stay light enough so that the Sun-
melted surface doesn't dry out and refreeze in the chilly air.

While all of this helps to explain the shape of Kilimanjaro's ice, it doesn't tell us what
caused it to start disappearing. In a 2003 paper, Kaser and colleagues offer one possible solution to the mystery. Their chief suspect is a regional drying (confirmed by

Mount Kilimanjaro on 17 February, 1993 and 21 February, 2000. Short-term conditions can affect such comparisons, but the melting trend is clear.

a drop in the level of nearby Lake Victoria) that began around 1880 and may have triggered the melting that continues today. What's unknown is how the ice cap behaved in other prolonged dry spells before 1880. Ice cores analysed by Thompson indicate that some parts of the cap have survived for more than 11,000 years. However, Kaser believes the bulk of the cap may have built up as recently as a few hundred years ago. He doesn't rule out global warming as a trigger for changes in air circulation – perhaps driven by the nearby Indian Ocean – that could be helping to dry out the peak. Three new weather stations installed on and near the ice since 2000 should help clarify how much warming and drying is going on. Meanwhile, new computer simulations are beginning to link the small-scale weather around Kilimanjaro to the regional and global atmosphere.

The interaction of people with Kilimanjaro adds more twists to the tale, one that's continuing to be investigated by scientists. At lower elevations, many trees have been illegally cleared that would otherwise be catching rain and re-evaporating the moisture. The region's drier climate – perhaps boosted by local effects – may also have played a role in the gradual loss of trees further up the mountainside.

Apart from its symbolism, the loss of Kilimanjaro's snow may not harm the region all that much. Tourism could suffer, though the mountain itself should certainly remain a draw. And it seems unlikely that locals will experience major water short-ages if the glacier disappears. Measurements show that 90% of the ice that escapes Kilimanjaro goes directly into the atmosphere. Much of the remaining runoff appears to evaporate before it reaches springs and wells downstream, which get moisture from wet-season rains.

Although the fate of the main ice fields atop Kilimanjaro appears to be sealed, Kaser believes a few of the smaller fringe glaciers – which lack Sun-facing cliffs – could hang on to life. In any event, some scientists would prefer that the spotlight be turned away from this celebrity victim and towards the thousands of other glaciers at risk from climate change. As Raymond Pierrehumbert (University of Chicago) noted on the RealClimate blog, "If Hemingway had written "The Snows of Chacaltaya", life would be much simpler."

along the Andes, not far from Bolivia, sits the vast, high **Quelccaya** ice cap, sprawling across a city-sized area of 54 square km (21 square miles).

Peru's tropical glaciers are far more than playgrounds for climbers and backdrops for photos: they carry both spiritual and practical significance. The ice melt provides water and hydroelectric power that's vital during the dry season for much of the country, including up to 80% of the water supply for the perennially parched Pacific coast. **Lima** spends months shrouded in clouds and drizzle, but it averages less than 20mm (0.8in) of rain a year, which makes it the world's second-largest desert city. Its eight million residents now depend on glacial ice that is measurably receding. So grave is Lima's risk of losing power that the city has begun expanding its portfolio of "alternative" energy sources – that is, power plants that burn natural gas. As for the potential water shortage, there's no solution in sight, though engineers have suggested a tunnel through the Andes.

What might be called "The Case of the Disappearing Tropical Glaciers" is the latest chapter in an ancient tale. Tropical ice has been on the wane ever since the end of the last ice age. The meltdown has unfolded in massive fits and starts, sometimes shaping the destiny of whole civilizations. The glaciers' most recent decline began in the mid-1800s, towards the end of the Little ice age but before human-produced greenhouse gases started to rise substantially. It's likely that a blend of forces, possibly including a slight rise in solar output, got the trend started. However, in recent decades greenhouse gases appear to have taken the reins, with assistance in the Andes from El Niño, the periodic warming of the eastern tropical Pacific. When an El Niño is in progress, temperatures generally rise across the Andean glaciers, and summer rainfall and snowfall decrease. Since the 1970s, El Niños have been unusually frequent.

Today, even the most diehard sceptics are hard-pressed to deny that tropical glaciers are vanishing before our eyes. The Andes offer especially vivid examples of low-latitude ice loss, particularly in some of the most accessible parts of the ice (which often happen to be the lowest-lying and the most tenuous to begin with):

▶ **Qori Kalis** The largest glacier flowing out of the Quelccaya ice cap, Qori Kalis was retreating at about 4.7m (15.4ft) per year in the 1960s and 1970s. Since 2000, the front of this thin ice tongue has disintegrated rapidly, producing a retreat of more than 1000m (3300ft). All told, more than a fifth of the ice extent in Qori Kalis has disappeared since 1978, and in 2007 glaciologist Lonnie Thompson (p.103) predicted that the glacier may vanish by 2012.

Qori Kalis glacier in 1978 and 2000.

▶ **Chacaltaya** For half a century, a ski lift took adventurers onto Bolivia's Chacaltaya glacier, billed as the world's highest developed ski area (the summit is at 5345m/17,530ft). But the lift wasn't used after 1998, and by 2007 Chacaltaya had lost some 80% of its surface area, with local experts predicting its demise within a year or two. As expected, the glacier vanished in early 2009.

Mid-latitude melt

The signs of glacier decline are no less dramatic as you move poleward. Patagonia is home to South America's largest icefields, as well as glaciers that cascade all the way to sea level. As with Antarctica and Greenland, the calving of glaciers that spill into the sea appears to be uncorking the flow of ice from higher up. NASA teamed up with Chile's Centro de Estudios Científicos to analyse the Patagonian glaciers in 2003. Combining space-shuttle imagery with topographic data, they found that the rate of glacial thinning had more than doubled between 1995 and 2000, making it one of the fastest thinnings on the planet. So quickly is the Southern Patagonia Icefield losing its ice that, from 2003 to 2006, the earth beneath it rose by nearly 12cm (4.8in). It's the most rapid rebound from the weight of glacial ice measured by GPS sensors anywhere on the planet. Only about half of the ice loss measured in NASA's 2003 study was attributed directly to warmer and drier conditions. The rest may be due to more-unstable glacier behaviour, such as dynamic thinning (see p.98) and enhanced calving.

The Northern Hemisphere has a much fuller palette of mid-latitude ice than the tropics, but even here things are changing quickly. Glacier decline is obvious to millions in and near the **Alps**, whose ice has lost nearly half of its surface area and overall mass since 1850. (A few individual glaciers have bucked the trend due to local variations in climate.) At

The Athabasca glacier in Jasper National Park, Canada. The glacier has receded more than 1.5km (0.93 miles) in the past 125 years, losing more than half of its volume.

heights of 1800m (5900ft), parts of the French Alps have been experiencing winters up to 3°C (5°F) warmer than those of the 1960s. Some of the Alps' smaller glaciers lost as much as 10% of their ice mass in the warm summer of 2003 alone. A linear extrapolation from recent trends indicates that some 75% of Switzerland's glaciers could be gone by the year 2050. In reality, of course, the trend could accelerate or decelerate, though there's no particular reason to expect the latter.

For ice climbers and ordinary hikers, the alpine warming carries real physical risk. Parts of the Alps saw the active layer – the zone above permafrost that melts and refreezes each year – double in depth to around 9m (30ft) in 2003. Massive ice and rock slides, such as that year's avalanche on the Matterhorn that triggered the rescue of some seventy climbers, may be increasing due to the melting of permafrost and the resulting instability of rock and ice on steep slopes.

In Montana, **Glacier National Park**, one of the jewels in the US park system, has lost nearly 80% of the 150 or so glaciers it had in 1850. Most of the remainder are likely to vanish by 2030, according to a model developed by the US Geological Survey. According to USGS ecologist Dan Fagre, recent temperature rises have outpaced that model's assumptions, which implies the glaciers might actually disappear as soon as 2020. One observer has already suggested changing the park's name to Glacier Moraine National Park, a title less likely to become obsolete.

The loss of both ice and permafrost are major issues for the Central Asian mountains – the **Himalayas**, the vast **Tibetan Plateau** and a number of nearby ranges. Together, these are home to the largest non-polar set of ice caps and glaciers on Earth: more than 100,000 in all, covering an area the size of Pennsylvania (about half of that atop the Tibetan Plateau). Across this region, sometimes dubbed the "third pole" of Earth, multiple threats to the ice are emerging. Average temperatures across parts of Nepal have climbed 1°C (1.8°F) since the 1970s. The extent of snow and ice cover across the eastern Himalayas has dropped by around 30% in that same period. And the amount of sun-absorbing black soot deposited on Tibetan glaciers has grown rapidly since the 1990s, a by-product of Asia's growing population and burgeoning industry.

Meanwhile, thanks to very moist monsoon flows, the southeast flank of the Himalayas is getting more than 20% *more* snowfall than it did in 1970, enough to keep a few glaciers steady or growing even as temperatures rise. The possibility of increased runoff from these growing glaciers poses problems for flood and water management downhill. A 1999 study by the United Nations Environment Programme (UNEP) found more than forty glacial lakes in **Bhutan** and **Nepal** primed for overflow in the next few years. (One such lake has grown sixfold since the 1950s.) While lowlands may be able to benefit from stepped-up hydroelectric power, flood risks could grow in the next few decades.

It's conceivable that the Himalayan glaciers might someday retreat enough to jeopardize streamflows and related agriculture across south and east Asia, but that prospect isn't yet firmed up in climate simulations. It's also unclear what fractions of Asian water come from glacier melt, snow melt and monsoon rains, though some recent work suggests that glaciers may contribute less than previously thought. Certainly, there's no real risk of losing these glaciers by the 2030s, an error that crept into the 2007 IPCC assessment (see p.296). However, the timing of snow and glacier melt might change as the climate warms, and that could influence water supply during the dry pre-monsoonal months. Even without such complicating factors, the sheer growth in population across this part of Asia looks set to strain water resources.

The picture is a bit more nuanced further north, in the sub-Arctic lands of North America and Europe. Certainly, **Alaska's** southern glaciers are eroding quickly, especially those that extend into the Pacific Ocean. The front edge of the vast **Columbia glacier**, at the end of Prince William

This image from NASA's Terra satellite shows the termini of the glaciers in the Bhutan-Himalaya. Glacial lakes have been rapidly forming on the surface of the debris-covered glaciers in this region during the last few decades.

Sound, has moved more than 19km (12 miles) upstream since 1980. In 2007, the glacier's end point lost its grounding on the seafloor and became a floating ice tongue, the first such transformation observed in detail by scientists. The glacier now sends huge icebergs into the sound (which, in turn, are convincing cruise ships to stay much further from the glacier's edge). In contrast, the sub-Arctic also holds some of the most publicized exceptions to global glacier retreat. Several of **Norway**'s coastal glaciers grew during the 1990s – the leading edge of **Briksdal** advanced by more than 300m (1000ft). However, these advances are easily explained by the strong westerlies and enhanced snowfall that prevailed during that decade along Norway's west coast. Further inland, virtually all of the glaciers in interior Norway have been steadily shrinking.

Many climate-change sceptics have pointed to the growth of specific glaciers as evidence against global change. However, since global warming isn't a uniform process – some regions saw little temperature rise in the twentieth century, while others saw a great deal (see p.4) – it's easy to see why a few individual glaciers scattered across the globe have grown or held their own in recent decades, especially in places where warming

A rocky future for skiers

Thanks to a warming climate, skiing and snowboarding conditions are heading downhill at some of the most popular resorts around the world. Skiing depends critically on each winter's snowfall, so there will remain both good and bad years. Overall, though, the good years are becoming less frequent and the bad ones are coming more often. One of the worst was 2006–07: many slopes in the Alps and the northeast US remained snow-free from autumn into January, which ruined countless holiday plans and scotched more than a half-dozen World Cup events.

The future looks bleakest at lower latitudes and/or lower elevations, where winter temperatures often run close to the freezing point. In the Vancouver area, the 2010 Winter Olympics came close to disaster after a record-warm January turned many slopes to slush and forced truckloads of snow to be shipped in. The games struggled through multiple days of borderline conditions, and thousands of spectators missed out when their standing-room areas were declared unsafe by heavy rains.

In Europe, the long-term outlook for skiers and boarders is ominous. A 2003 report from the United Nations Environment Programme (UNEP) estimates that the Alps' average snow line will rise 200–300m (660–990ft) by the 2040s and predicts that "many mountain villages, above all in the central and eastern parts of Austria, will lose their winter industry because of climate change." Half of Italy's ski resorts lie below 1300m (4300ft), and many German resorts also sit at relatively low altitudes. There's lots of income at stake: 8% of Switzerland's workforce, and 5% of Austria's gross national product, are ski-based. Though the highest resorts in the Alps and Colorado may hang on, a 2007 report produced for Halifax Travel Insurance foresees quota systems as demand skyrockets for the few resorts where snow remains reliable. The firm ATMOSresearch projects that ski seasons across many resorts in the western US could be 10–20 days shorter on average by 2025.

Resorts are already being forced to adapt. Switzerland's Tortin glacier, which supports the **Verbier** ski area, has receded so much that in 2005 the resort placed an insulating sheet the size of a city block atop the glacier's edge to slow summer melting. Other ski areas are thinking of moving uphill. Austrian promoters want to develop some 30km (45 miles) of slopes in the wilderness area of Piz Val Gronda, hundreds of metres above the popular **Silvretta** resort. However, the Alpine nations have signed a treaty designed to limit high-altitude development, and Austria's Tyrol government has banned all new ski areas.

hasn't yet pushed increased precipitation from snow towards rain. The bigger picture becomes clear in global analyses published annually by the World Glacier Monitoring Service, which has kept track of thirty reference glaciers across nine mountain ranges since 1980. During the period 2007–08, all but three of these reference glaciers lost mass. As the service notes, "The new data continues the global trend in strong ice loss over the past few decades."

Fans watch from the stands in the rain during the Snowboard Ladies' Parallel Giant Slalom at the Vancouver 2010 Winter Olympics.

Snowpacks across the **US West** are projected to become smaller and more variable over the coming century. Many ski resorts in the **US Northeast** – already vulnerable to winter warm spells – could become marginal. In response to all this, more than 70 of the nation's ski areas have joined forces with the Natural Resources Defense Council to promote energy efficiency and renewables and push for federal emission controls. Their motto: "Keep Winter Cool".

Skiers who flock to the modest mountains of southeast Australia savour every flake of the white stuff. But at least one part of the **Snowy Mountains** has seen a 40% drop in the depth of spring snow since the 1960s, according to Australia's Bureau of Meteorology. The future looks even gloomier for snow-lovers, according to a 2003 report prepared by the Commonwealth Scientific and Industrial Research Organization. CSIRO predicts that the area covered by snow for at least fifty days each year could shrink by 18–60% by 2020 and by 38–96% by 2050, depending on global emissions levels.

More than melting?

It's impossible to quantify just how much of the blame for the overall retreat in low- and mid-latitude glaciers can be pinned on human activity. glacier behaviour today reflects in part the conditions present years ago.

That said, tropical ice appears to be one of the most reliable of all witnesses to human-induced change. For starters, the wild temperature

Picturing glaciers as they go

It may be the most ambitious round of time-lapse photography ever attempted – and certainly the most sobering. Since 2007, 27 cameras stationed at 15 glacier-studded sites across the globe – from Greenland and Iceland to Alaska, the Rockies and the Alps – have each been taking daylight photos once per hour. The Extreme Ice Survey (www.extremeicesurvey.org) is the brainchild of US photographer James Balog, a trained geologist and regular contributor to *National Geographic*. Balog teamed with glaciologists, computer scientists and experts in still and video imagery to assemble the project's network of resilient cameras, non-polluting power sources and uplink facilities. The survey conveys ice loss to a broad audience through videos, Internet-based outreach and a 2009 book and DVD.

James Balog installs "Cliff" camera AK-3 at Columbia glacier.

swings often found at higher latitudes are muted near the Equator, especially as you go up in height. There's also little variation as you go from west to east around the globe. As a result, the average readings vary only about 1.5°C (2.7°F) across the entire equatorial belt at glacial heights of about 6km (3.7 miles). This relative evenness of temperature means that even a small trend should stand out more noticeably than it would in mid-latitudes. It also means that any warming trend should be reflected throughout the tropics, which happens to jibe with the near-universal shrinkage of low-latitude glaciers.

On the other hand, the tropics aren't the best-sampled part of the world, meteorologically speaking. A few weather stations are cropping up at the highest tropical altitudes, but observations remain sparse, both on the ground and aloft. As a result, scientists are debating exactly how much temperatures across the higher altitudes of the tropics have risen. Thus far, the jury seems to be settling on a warming somewhat less than the global surface average, but still enough to tip glaciers towards melting. Across eastern Africa, the scanty data show little if any warming trend, which means that drying and other factors may be teaming up to produce Kilimanjaro's rapid ice loss (see p.105).

Aside from temperature, there's a range of other processes that will help shape the future of tropical and mid-latitude glaciers. These can vary from region to region and even on the scale of individual mountains, which leads to a patchwork of local advances and retreats within the larger fabric of overall glacier loss.

▶ **Sunlight** not only raises the air temperature over glaciers, but it can also directly melt glacial ice even when the air remains well below freezing (as noted previously). Local changes in cloud cover could either enhance or detract from sun-induced ice loss.

▶ **Moisture** in the air can also help or hinder glacier growth, depending on location. In the tropics, where glaciers typically sit at heights well below freezing all year round, drying can allow for more snow and ice to slowly leave glaciers via sublimation (see p.104). The melting process is different in the mid-latitudes, where temperatures often warm above freezing in summer. If those balmy temperatures came with extra atmospheric moisture, they could help keep the temperature at the glacier's surface above freezing and thus hasten its melting.

As glaciers continue to shrink, their situation gets increasingly difficult, because the ratio of their surface area to their ice mass goes up. This gives the atmosphere more area to work on, causing the glaciers to change state more quickly – just as a pound of frozen peas defrosts more quickly than a pound of frozen chicken breasts.

Perhaps the only way for glaciers outside the poles to avoid death by warming would be through a boost in snowfall. That doesn't appear to be happening on a large scale, despite a general increase in atmospheric water vapour worldwide. The snowfall on glaciers atop mid-latitude mountains is fed by strong, moist, upper-level winds, such as those near the west coasts of North America or Norway. Glaciers in these areas may stand the best chance at survival, since the winds reaching them could bear more moisture in years to come. But even that might be a long shot, according to at least one study of glacier balance. It estimates that, to keep a glacier alive, a 1°C rise in average temperature would need to be counteracted by a 25% increase in average snowfall – a massive amount when it comes to long-term climatology.

Oceans

A problem on the rise

Glaciers calving like there's no tomorrow, sea ice dwindling to patches – it's easy to find drama in the warming of our planet's coldest waters. But only a small part of Earth's oceans ever produces ice. What about tropical and mid-latitude waters, the ones that together span more than half the planet? If appearances can deceive, then these oceans are master tricksters. They might not look much different to the naked eye than they did a century ago, but they've gone through quite a transformation – warming up and growing more acidic, among other changes. It appears they've hidden the true impact of our energy choices quite skilfully, absorbing some of the heat and CO_2 that would otherwise have warmed the atmosphere.

Many scientists suspected as much for years, but the first solid numbers came from a team led by Sydney Levitus of the US National Oceanic and Atmospheric Administration. After assembling a variety of deep-ocean measurements for the period 1948–98, Levitus announced that every major ocean exhibited warming down to at least 1000m (3300ft), and that the top 300m (1000ft) had warmed globally by about 0.3°C (0.54°F). You wouldn't be able to tell the difference in temperature by taking a quick dip, but the total heat added to the oceans represents as much as 90% of the energy trapped by human-produced greenhouse gases in those fifty years. Water expands as it warms, and the temperature increase translates to a rise in sea level of about 25mm (1in). Again, that may not sound like much, but – as we'll see – it's on top of the sea-level rise from all other sources.

The undersea storage of vast amounts of heat has serious implications for humanity's future. Although the topmost layer of the ocean stays in balance with the atmosphere over timescales of a month or longer, the much deeper layer below the **thermocline** is more insulated. This means the heat it slowly absorbs from above will take a long time to work its way back out. Even if greenhouse gases magically returned to their pre-industrial levels tomorrow, it would take many decades for the heat tucked away in the deep oceans to work its way out of the climate system. This idea of "climate commitment" is a key theme in research leading up to the 2007 IPCC report. Some of these studies estimate that if global greenhouse gas emissions had levelled off in 2000 (which they didn't), an additional warming of at least 0.5°C (0.8°F) beyond the year-2000 global average would be guaranteed from twentieth-century emissions alone.

Of course, as climate change unfolds, the oceans will do far more than absorb and release heat. Ocean-driven storms, primarily tropical cyclones, are showing signs of strengthening as their fuel source heats up (see pp.149–50). Sea levels are rising, too. Much of this is due to the expansion of ocean water as it warms, as noted above, but over time the rise will be increasingly enhanced by glacial melting. That melting, in turn, will shape ocean circulation in ways that push sea level even higher in some favoured locations (see p.126). The uneven pattern of ocean warming may influence how, where and when a variety of ocean-atmosphere cycles unfold. This could spell the difference between a life of plenty and a life of poverty for millions of people living close to the land in drought- or flood-prone areas.

All of this has made oceans one of the favourite trump cards of climate activists. The spectre of islands disappearing under the sea does make for a potent image. And yet the image isn't as crystal-clear as one might think. We still have limited data on the three-dimensional reality of oceans: how warm they are, where and how fast their currents are flowing, and so forth. There are big questions yet to be answered, including what "sea level" actually is.

From sticks to satellites: measuring sea level

Mean sea level (MSL), the average height of the ocean surface at a given point, sounds as if it should be an easy enough thing to calculate. Subtract the astronomical tides, factor out the storms that come and go, and *voilà* – you've got your average. In truth, however, MSL is enormously difficult to nail down. For a start, the tides depend upon the locations of the Sun and Moon within a set of cycles that takes 18.6 years to fully repeat itself. Even when you factor this out, the underlying MSL isn't the same everywhere, but varies across different sections of the open ocean by as much as 50cm (20in), depending on the surrounding topography and the warmth of the underlying waters. The effects of global warming complicate things further.

A tidal gauge in use, southeast US, 1970s.

The simplest and oldest method of measuring MSL relative to some point along the coast is through basic **tidal gauges** – essentially, sticks in the sand. A technological leap came in the 1800s with **wells** that used floating gauges to factor out the effects of waves. Similar techniques are still used at many coastal stations, though they've been joined lately by far more precise methods. Modern gauges measure the time it takes sound waves to bounce off the water's surface and back, converting that into distance and then to sea-level height.

Radar altimeters aboard satellites work in a similar way, typically sending microwave energy from space to the sea surface – the open ocean as well as coastal areas. One key advantage of satellites, when they're coupled with Global Positioning System (GPS) data, is that they provide a measure of MSL that's independent of a land-based feature. The **TOPEX/Poseidon** satellite, a collaboration between the US and French space agencies, began gauging ocean heights across much of the world in 1992, ending its long run in 2006. Its successor, **Jason-1**, went into service in 2001, and **Jason-2** followed suit in 2008. The newer Jason measures the height of virtually all of Earth's ice-free ocean surface with a vertical resolution as fine as 1cm (0.4in). Even sharper-eyed is **GRACE (Gravity Recovery and Climate Experiment)**. This European satellite system, launched in 2002, tracks MSL with an error of less than 1mm (0.04in). Because it measures subtle changes in the gravitational tug exerted by the oceans as well as Earth itself, GRACE is able to help scientists sort through the small-scale continental adjustments that may influence sea level.

Those geological tics can have a surprisingly big influence on sea level. Coastlines rise and fall with the clashing of tectonic plates. In some places, such as parts of Alaska and Japan, the motions are so pronounced that tidal gauges can't be used as a reliable index of long-term sea-level change. Other coastal locations, such as Venice and New Orleans, are plagued by gradual subsidence. Still other regions are steadily rising, on the rebound

Scandinavia's bounceback

The Gulf of Bothnia, bordering Sweden and Finland, offers a classic example of how the after effects of the last ice age can disguise global warming at work. An international team used GPS data and a computer model in 2001 to track the vertical change in both land and sea. They found that the region's mean sea level was rising at about 2mm (0.08in) per year as the ocean warmed. But that rise in MSL was obscured across much of the Gulf by a much larger rise in the coastline itself. Like a piece of dough springing back slowly from the push of a baker's thumb, the coast is still rising after the release of thousands of years of pressure from glacial ice. The rebound is as much as 10mm (0.4in) per year, which is enough to more than cancel out sea-level rise and produce a net drop in sea level from the perspective of towns on the coast.

after being squashed beneath kilometre-thick ice sheets for thousands of years. All of these ups and downs have to be factored out of sea-level data to find out how the sea itself is changing.

Balancing the sea-level budget

With all these challenges in mind, the Intergovernmental Panel on Climate Change has done its best to pool the data and estimate global sea-level rise. On average, it appears that MSL rose by somewhere close to 17cm (7in) over the twentieth century. However, this number is often specified as a range, due to regional differences, instrument variations, and other uncertainties.

The second IPCC assessment, in 1996, pegged the twentieth-century MSL rise in the range of 10–25cm (4–10in). The 2007 IPCC narrowed the range only slightly, to 12–22cm (5–9in) over the century or 1.2–2.2mm per year. That may seem like a fairly wide margin of error for measuring something that's already occurred. In part that's because, for many years, the factors known to raise sea level didn't quite add up to the gauge-measured twentieth-century rise of 1.5–2.0mm per year. The 2000 paper by Levitus, noted above, proved that the oceans had warmed deeply and dramatically, but it showed an MSL rise of a mere 0.5mm per year from ocean warming – only about 25–33% of the gauge-observed rise. Another 25–33% was attributed to melting of land-based ice. That left the remainder – between 34% and 50% of the observed century-scale rise in global MSL – unaccounted for.

In a flurry of papers starting in 2001, researchers floated several ideas on where the rest of the sea-level rise may have come from. These include:

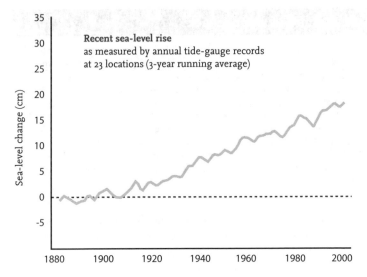

Recent sea-level rise
as measured by annual tide-gauge records
at 23 locations (3-year running average)

▶ **Incomplete data** The Levitus study lacked data for parts of the
Southern Hemisphere, as well as for water below about 3000m
(10,000ft). If warming is greater than believed in these hard-to-sample
areas, it could help solve the puzzle.

▶ **Glacier melt** Glaciers from Alaska to Patagonia appear to be releasing
more meltwater than expected. Global-scale statistics are difficult
to gather and analyse, but in a 2009 survey called the Copenhagen
Diagnosis, several experts estimated that the world's glaciers and ice
caps were contributing about 50% more to sea level than they did in
the mid-1990s.

▶ **Ice-sheet melt** A 2006 analysis led by NASA's Eric Rignot, based on
radar altimetry, shows that the Greenland ice sheet was pouring
nearly twice as much water into the sea in 2005 as it was in 1996. This
and other studies indicate that Greenland's ice sheet is now a larger
contributor to sea-level rise than previously anticipated, because
of accelerated flow in some regions as well as increased melting.
Until recently, Antarctica was expected to work against sea-level rise
this century as snowfall rates increase across its interior (see p.100).
However, recent reports indicate that the acceleration of large glaciers
that discharge ice into the sea along the Antarctic coast appears to be
making the continent a net contributor to sea-level rise.

With the more recent numbers for glacial and ice-sheet melt at hand, the 2007 IPCC report came closer to balancing the books on sea-level rise. It estimated that the Greenland and Antarctic ice sheets poured meltwater into the ocean at roughly four times the pace in 1993–2003 as compared to the longer-term average. Taking into account the increased melt from both ice sheets and glaciers, the IPCC was able to explain all but about 10% of the 1993–2003 MSL rise, though a full accounting remains elusive.

Oddly, the rate of sea-level rise appeared to slacken somewhat just as land-based ice melt kicked into overdrive. Data from TOPEX and Jason show that the MSL rise increased to roughly 3.0mm per year from 1993 to 2003, with other analyses suggesting an even higher increase – in other words, the sea rose at least 50% faster in that decade than in the twentieth century as a whole. The pace then reportedly dropped to around 2.5mm per year from 2003 to 2008, according to data from the GRACE satellite (see p.97). The finding is still being analysed, and its cause isn't yet clear. Adding to the mystery is a reduction in upper-ocean warming from 2003 to 2008 as detected by Argo, a global network of 3000 buoys that's been sampling the upper 2000m (6600ft) of the world's oceans every few days over the last few years. It may take researchers a while to sort out trends and remove possible biases from the new wealth of instruments that measure sea level, but there's no doubt that MSL is on the way up.

How high the sea?

Assuming humans continue to burn fossil fuels, it's a safe bet that sea level will continue to rise. But by how much? And what does this mean for people on or near the coast?

The 2001 IPCC report projected a rise in sea level by the year 2100 of 90–880mm (3.5–35.0in), with a midpoint of around 480mm (19in). However, in its 2007 report, the IPCC surprised many onlookers by narrowing the range of uncertainty. That meant pulling the top end of the range downward (and the bottom end upward, though that got less attention), with the central estimate dipping slightly. The panel projected a twenty-first-century MSL rise of 180–590mm (7.0–23.0in), with a midpoint of around 380mm (15in). Part of the drop was simply a matter of framing: the earlier outlook was for the year 2100, while the more recent projection is for 2090–99, in both cases compared to 1980–99 values. The newer range also omits two relatively minor factors that were included before: permafrost and very long-term melting of ice sheets unrelated to present-day climate change. As in 2001, the 2007 report excluded some aspects of rapid

Are the Maldives and Tuvalu doomed?

They sit thousands of kilometres apart, but the Maldives and Tuvalu have much in common. As two of the smallest, least populous nations on Earth, they're perched atop low-lying coral atolls, from which they watch the sea and wonder how long they can keep it from swallowing their countries whole. The world is watching too. These two tiny, once-obscure countries have been at the forefront of climate change coverage since the late 1980s. In part that's due to the eloquence and tenacity of their leaders, who have been driven onto the world stage by palpable fear of what the future may hold and anger at those deemed responsible. In 2002, Tuvalu threatened the United States and Australia with a lawsuit to be brought before the International Court of Justice, charging the two nations with reckless greenhouse-gas emissions (the lawsuit never materialized, though).

There's no doubt that water is lapping at the shores of both countries, and the IPCC projections of sea-level rise offer no reassurance. The average elevation is around 1m (3ft) in the Maldives and around 2m (6ft) in Tuvalu, while the highest terrain in each nation is around 3m (10ft) and 5m (16ft) respectively. The 2007 IPCC high-end projection for global sea-level rise of 590mm (23in) by the 2080s and 2090s – with the risk of even larger rises from dynamic thinning of glaciers – would bring major impacts to both countries. Even now, storms are capable of flooding large parts of each island chain.

The two hundred islands of the Maldives sit due south of India. They're so close to 0° latitude that they're at little risk from tropical cyclones, which need some distance from the Equator in order to develop. But even far-off storms can send swells to worrisome heights, and flooding can also occur when intense monsoon rains team up with tides. The Maldives' capital, Malé, is now fortified by a sea-wall 3m (10ft) high, built in the late 1980s and 1990s with the help of over $60 million provided by Japan. However, on another island, Kandholhudoo, over half of the residents had already committed themselves to leaving before the catastrophic Asian tsunami of 2004 struck. The waters, topping around 1.5m (6ft), ruined almost every house and sped up plans to move residents to a nearby island, which will be bolstered by landfill. Small as they are, the Maldives resemble an empire compared to Tuvalu, which is located well east of New Guinea and far to the north of New Zealand. Tuvalu's nine atolls only encompass around 26 square km (10 square miles) in land area. Stretching from 6° to 10°S, Tuvalu is far enough south to get battered by the occasional tropical cyclone. As with the Maldives, rogue waves and swells are an ever-present threat, and for both island chains the impact of such one-off events will only get worse as sea level gradually rises.

On the face of it, the residents of the two countries appear to have little choice but to either abandon their homelands or engineer their way to safety with expensive landfills and sea-walls. But not everyone is convinced the islands are doomed. That's because atolls like the Maldives and Tuvalu form on the edge of coral reefs that sweep in narrow arcs around huge lagoons. The islands themselves are made up of the pulverized remnants of corals and other reef organisms. When the climate isn't changing and the coral are healthy, the islands appear to grow

Tuvaluan men watch as waves encroach inland at high tide.

slowly over time, although there's a lot of uncertainty about how quickly this occurs. Without humans intervening, big waves occasionally overwash the islands, scouring sediment from one side of the island and depositing it on the other. In theory, the island doesn't drown so much as it shifts backward. As sea levels rise, the islands should shift position on their reefs, generally moving towards their lagoons and growing taller.

This scenario is championed by geographer Paul Kench of the University of Auckland. In studying a variety of small landforms, such as tropical atolls as well as the barrier islands common along the US Gulf and Atlantic coasts, Kench has come to believe that these lands are more resilient than they're given credit for. For a 2010 study published in *Global and Planetary Change*, Kench and colleague Arthur Webb (Pacific Islands Applied Geoscience Commission) examined changes in 27 atolls across Tuvalu, Micronesia and Kiribati. They found that only four of the atolls lost at least 3% of their land area over the periods studied, which ranged from 19 to 61 years, while nearly half of them grew by at least 3%.

However, Kench doesn't foresee an easy path ahead for the Maldives, Tuvalu and other reef-based populations. The problem is that the reefs' natural growth process, as described above, doesn't respect towns and buildings that are fixed in one spot. In addition, climate change may stunt the growth process itself. In a study partially supported by the World Bank, Kench and colleagues surveyed atoll and reef islands and adapted a model originally designed for sandy barrier islands in order to explain the life cycle of the coral-based atolls. Normally, the growing reefs provide extra sediment that helps build up the islands. As sea level rises and the ocean warms, Kench believes the heat-stressed coral (see p.138) will devote most of their energy to self-preservation, providing little or no sediment for the island. Thus, it seems, the residents will have to make do with the sediment already beneath their feet. For this reason, says Kench, "island nations should make conservation of island and nearshore sand resources a high priority".

Years of haphazard land use by indigenous residents as well as non-residents (such as the US forces that occupied Tuvalu in World War II) have triggered chronic erosion and left a difficult legacy for islanders to overcome. As for the future, if Kench's model is correct, it may spell survival for some atoll islands – at least for the next century – but constant flooding and woe for their residents. Tuvalu is exploring options for its citizens to resettle in Australia or New Zealand, where over 4000 Tuvaluans already live. In 2006 a retired Tuvaluan scientist, Don Kennedy, began lobbying to move the nation's 9000 remaining residents to the Fijian island of Kioa. There, ethnic ties with Tuvalu are strong – and the land is higher.

With the Maldives' fate just as uncertain, President Mohamed Nasheed has become a high-profile figure in the global-warming arena. He drew attention to his nation's plight in 2009 by joining eleven ministers to don scuba gear and hold a cabinet meeting underwater. Moreover, the Maldives are setting an example for larger countries by striving to become the world's first decarbonized economy by 2020. Their goal isn't just to become carbon neutral (which could be achieved simply by offsetting their emissions) but to generate virtually all of the islands' power through alternative sources: 155 wind turbines, half a square kilometre of rooftop solar panels, and a biomass plant burning coconut husks, plus a complex of batteries to bridge gaps in power supply. "Climate change is a global emergency", said Nasheed. "The world is in danger of going into cardiac arrest, yet we behave as if we've caught a common cold."

dynamic thinning of ice sheets (see p.98) that weren't considered well enough understood to incorporate in the projections. This time, the IPCC noted that these factors could add another 100–200mm to the top of the official range and that "larger values cannot be excluded". Thus, the apparently lowered range shouldn't necessarily be taken as a source of comfort.

It's also important to note that the IPCC ranges reflect many scenarios and possibilities, some of which are discussed above. For starters, we don't know exactly how much greenhouse gas will be emitted, how much global temperatures will warm and how quickly Earth's oceans and ice will respond to that warming. As higher temperatures allow more moisture to filter into Antarctica, increased snowfall could help reduce the net MSL rise by anywhere from 20mm to 120mm (0.8–4.7in), according to a mid-range emissions scenario in the IPCC's 2007 assessment. Global ocean expansion due to heating could add anywhere from 130mm to 320mm (5.1–12.6in). The other big question marks include the Greenland ice sheet, other glaciers, and melted snow running off from land.

The seemingly modest sea-level projections in the 2007 IPCC report piqued the interest of the world's press and of many scientists. Some took issue with the IPCC's assumption that dynamic thinning of Greenland and Antarctic glaciers would proceed at the same pace throughout the twenty-first century as it did in the 1990s, rather than accelerating further. During

"The future of New Orleans is that it won't be there any more."

Michael Schlesinger, University of Illinois at Urbana–Champaign

the same week it went public, the IPCC outlook was complemented by a *Science* paper from Stefan Rahmstorf of the Potsdam Institute for Climate Impact Research. He found that if sea-level rise in the twenty-first century stays proportional to the increase in global air temperature (a relation that held fairly firm in the twentieth century), then an MSL boost of 500–1400mm (20–55in) above 1990 levels could be on the cards by the century's end. Subsequent papers by Rahmstorf and other experts – all using techniques that build on the links between past temperatures and past sea-level rise – reinforce the idea that MSL could jump more than 1000mm (40in) by 2100, though some climate scientists question the validity of this approach.

What if sea-level were to rise on the order of 500mm (20in), near the high end of the 2007 IPCC range but near the low end of these recent papers? That wouldn't be enough in itself to inundate the world's major coastal cities, but it could certainly lead to an array of serious problems, especially for populous, low-lying regions. Some tropical islands may become virtually uninhabitable (see p.122) and, significantly, higher sea-levels also mean a higher base on which to pile storm surges, tsunamis and all other ocean disturbances. These are likely to be the real troublemakers as sea-level rises over the next several decades.

One region at particular risk is the Ganges Delta of eastern **India and Bangladesh**. While most of Bangladesh is expected to remain above water even if the ocean rises several metres, such a rise could still displace millions of people in this extremely low-lying country. Millions of other Bangladeshi people who remain near the coast will be vulnerable to catastrophic flooding. (See p.152 for more on surges from tropical cyclones and coastal storms.)

Another place set for serious problems from even modest sea-level rise is the glittering string of hotels, offices and condominiums along Florida's southeast coast from Miami to Palm Beach, as well as the more rustic Florida Keys. A sea-level rise of 680mm (27in) would put some 70% of the Miami area – and virtually all of the Keys – underwater. Building effective barriers against the rising ocean will be difficult if not impos-

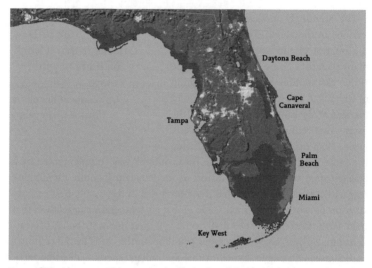

Parts of Florida susceptible to a 6m (20ft) rise in sea level are shown in red – with paler areas showing high-population zones.

sible, because seawater will tend to sneak through the porous limestone that undergirds south Florida.

The IPCC's 2007 report stressed that MSL rises won't be uniform across the world's oceans, and recent studies have driven home that point. Should the Atlantic's northward flow be impeded by meltwater streaming from Greenland, then colder, deep water wouldn't form as readily. The net warming from this process would push sea levels even higher than projected across most of the North Atlantic, according to a 2009 study in *Nature Geoscience* led by Jianjun Yin (Florida State University). The extra rise could total 10cm (4in) in London and 20cm (8in) in New York. Balancing this enhancement, sea levels across most of the North Pacific would be somewhat less than otherwise expected. Looking centuries ahead, the north–south balance of MSL rise could be influenced by the timing of the Greenland and West Antarctic ice sheet melts, because of the loss of large gravitational tugs exerted by each sheet. For example, both the Pacific and Atlantic coasts of the United States could see an additional rise of 0.8m (2.6ft) from the loss of the West Antarctic ice sheet, noted Jerry Mitrovica (University of Toronto) in a 2009 *Science* paper. By the same token, MSL rise around the rim of the North Atlantic would be reduced by the loss of the gravitational pull exerted by the Greenland ice sheet.

Swimmers and sea level

If the water level in a bathtub goes up when you get in, couldn't people in the ocean be pushing up sea level? That's the semi-serious question posed by Gregory Pasternack of the University of California, Davis. While teaching short courses for high-school teachers in Maryland, Pasternack was inspired by a state politician who claimed that an excess of boats and boaters, rather than climate change, was behind the rise in global sea level. As part of a classroom exercise he developed, Pasternack encouraged students to measure how much water they displace in a bathtub, then to extrapolate that figure to a "world of swimmers". If everyone on Earth took a dip in the sea at the same time, they might occupy a volume on the order of a third of a cubic kilometre. But spread out over the vast area occupied by oceans, that translates to a sea-level rise of a mere 0.0009mm (0.000035in) – around a hundredth of the width of a human hair.

As odd as it seems to contemplate, it's not out of the question that most of the world's coastal cities could be largely or completely underwater in a few hundred years. While the risk is seldom put in such stark terms, it's a logical outcome of the amount of warming expected to occur over the next several centuries unless major emission reductions take place. According to some estimates, sea levels ran up to 6m (20ft) above current MSL during the last interglacial period more than 100,000 years ago, enough to inundate all or most of nearly every low-lying coastal city. The warming back then was produced by changes in Earth's orbit around the Sun (see p.216), but greenhouse gases could bring as much or more warming in the next century and beyond.

James Hansen (NASA) is one of a handful of scientists cautioning that MSL rises of the order of metres are possible not only in the distant future but within this century. Hansen speculates that a 3°C (5°F) warming – near the IPCC's mid-range estimate of global temperature rise over the next hundred years – would eventually melt enough of the Greenland and Antarctic ice sheets to produce a spectacular sea-level rise of 25m (80ft). On top of that, the simple warming and expansion of ocean water could eventually add several more metres to sea level. Hansen feels the IPCC and other groups need to emphasize the risks of higher-end MSL rise, in spite of the inherent uncertainties. "In a case such as ice sheet instability and sea level rise", said Hansen, "there is a danger in excessive caution. We may rue reticence, if it serves to lock in future disasters."

Other researchers don't appear worried that truly epic sea levels will materialize over the next few decades. "The rise by the year 2100 is almost certain to be below two metres", declared researchers Jason Lowe (UK Met

The NAM, the NAO and other climate cycles

After ENSO, the next-biggest cycles of climate variability are two circulations that play out over the North and South Poles, affecting ocean as well as land. The Northern Annular Mode and Southern Annular Mode (or NAM and SAM) refer to oscillations in the strength and structure of the upper-level winds that encircle the Arctic Ocean and Antarctica, respectively. Think of bracelets of strong wind, centred a few kilometres high, that loop around the North and South Poles respectively.

The SAM is the more straightforward of the two. Unblocked by land, its westerly winds howl above the Southern Ocean, circulating around a cold vortex locked over the South Pole. The circulation alternates between a tighter, faster-flowing ring (the positive mode) and a looser, more variable ring (the negative mode) that allows cold air to spill out from the pole across the Southern Ocean more easily. Since the 1960s, the SAM has trended more towards its tighter positive mode, which is likely related to ozone depletion (see p.32). This trend is projected by climate models to continue even after the ozone hole heals later this century. Among the implications for climate are a potential drop in winter rainfall for parts of Australia (see p.78).

On the other side of the globe, the NAM (also called the Arctic Oscillation) is a more complicated beast. The NAM encounters three continents, the thick ice sheets of Greenland and the warm currents of the North Atlantic – all of which interfere with the uniformity of the vortex and help to produce a more variable circulation. Still, most climate models are consistent in projecting a tighter, more positive NAM as the century unfolds.

In and near the North Atlantic, the NAM manifests itself as the North Atlantic Oscillation (NAO), defined by pressure variations between Iceland and the Azores. It has a strong effect on European weather, especially in winter. The NAM and NAO tend to vary in tandem, shifting positive or negative for periods lasting from a week or so to a month or more. Their strongest effects on climate occur in the winter. When the NAO turns negative, the polar jet stream tends to buckle northwards across the Atlantic and then back south. This often throws northern Europe, eastern Canada and the northeast United States into extended cold spells, while milder, slow-moving storms drench the Mediterranean. A positive NAO, on the other hand, keeps Arctic air locked to the north and sends the pole-encircling winds more directly from the Atlantic across northern Europe, increasing the odds for mild and wet weather there as the Mediterranean stays dry. Thus, the projected shift of the NAM and NAO towards positive modes under global warming is one of the strongest clues to the type of climate Europeans can expect as greenhouse gases build up.

The NAM can easily confound expectations, though. Since 2000 it's trended toward its negative mode, and in the winter of 2009–10 it plummeted to unprecedented lows, with much of Europe and the eastern United States plunging into weeks of frigid temperatures and heavy snow. It was Britain's coldest winter since 1978–79, and the US corridor from Washington to Philadelphia set all-time seasonal records for snowfall. Not only did this winter onslaught afflict millions of people, it coincided with – and reinforced – a spate of scepticism about climate change (see p.297), despite the fact that the globe as a whole was experiencing near-record warmth.

Snow blankets Britain on 7 January 2010.

It's hard to assess ocean cycles that span more than a few years from peak to peak, because sea-surface temperature data before the advent of satellites in the 1970s is notoriously spotty. But many scientists are intrigued by the Pacific Decadal Oscillation (PDO), a seesaw of rising and falling temperatures across the northwest Pacific, and the Atlantic Multidecadal Oscillation (AMO), in which temperatures across much of the North Atlantic alternately warm and cool. Both cycles go for about 20–30 years before switching to their opposite phase. In both cases, the physical driver isn't yet known, though there may be a link to the global oceanic conveyor belt (see p.132). There's evidence for a link between the PDO and ENSO: when the PDO is in its positive phase, El Niños are more likely. And in the Atlantic, the AMO appears to boost the risk of Atlantic hurricanes, though it also complicates assessment of a well-publicized and much-feared risk to Europe's climate (see p.131).

Additional ocean cycles continue to pop up. The Indian Ocean Dipole (IOD) got its first widespread exposure in 1999 via the journal *Nature*. Like a smaller sibling of El Niño, the IOD involves abnormally warm and cool patches of ocean that build and decay along the Equator. A positive mode of the IOD (i.e. warm surface waters focused in the western Indian Ocean) helps boost monsoon rainfall across India. Given that some ocean cycles like the IOD are only now being analysed in depth, it'll take time to understand how climate change will affect each of them and how various cycles might interact in a warmer climate.

Office) and Jonathan Gregory (UK Met Office/University of Reading) in a 2010 paper. It's later in the century, and especially beyond 2100, that the risk of ultra-high seas will clearly rise more steeply. The catch is that the greenhouse gases we're now adding to the atmosphere at unprecedented rates will continue to warm the climate – and raise the seas – for more than a century. Thus, the greenhouse ship is becoming progressively harder to turn around, and the seas are rising around it.

Climate change and El Niño

The oceans play a vast role in shaping the vagaries of weather and climate, and much of that influence comes through a set of **ocean-atmosphere cycles**. Linked to arrangements of high and low pressure centres over various parts of the world, these cycles each alternate between two "modes", producing recognizable, repetitive weather patterns – drought, excessive rainfall, unusual warmth or cold and so forth. These can unfold half a world away from their oceanic triggers, and they're natural parts of climate. It's possible that global warming will tamper with them, but right now, scientists have more questions than answers about whether and how this will happen.

Globally, the most important ocean-atmosphere cycle is **ENSO** (El Niño/Southern Oscillation), whose two modes are known as **El Niño** and **La Niña**. ENSO is based in the tropical Pacific Ocean, which spans a third of the globe from Ecuador to Indonesia. Trade winds blow more or less continuously across this huge area from east to west, pushing warm water towards Indonesia, where the resulting balmy seas help generate persistent showers and thunderstorms. The cold, upwelled water off Ecuador and Peru, meanwhile, stabilizes the air there and produces the region's legendary aridity (Peru's capital, Lima, gets about the same amount of rain each year as Cairo). Every two to seven years, the trade winds weaken or reverse, the surface layer of warm water deepens and expands into the eastern tropical Pacific, and an El Niño sets in, typically lasting one or two years. The flip side, La Niña, occurs when trade winds are stronger than average, pushing cooler-than-usual water westward into the central tropical Pacific for a year or two. About half of the time, neither El Niño nor La Niña is in progress and the Pacific is neutral.

El Niño increases the odds of drought across Indonesia, Australia, India, southeast Africa and northern South America. It tends to produce mild, dry winters in Canada and the northern US and cool, moist winters in the US South and Southwest. It also raises the chances of hurricanes in the Atlantic, and lowers them for parts of the North Pacific. La Niña, in general, has the opposite effects.

While computer modellers are making progress in depicting El Niño and La Niña, even the best models struggle to depict the tropical Pacific's arrangement of ocean currents and rainfall patterns, so it's not yet clear how global warming will affect the mix. Some models suggest an intensified ENSO, while others point to a weaker one. Some show a sustained La Niña-like scenario. As a whole, though, the model ensembles in the 2007 IPCC report point toward a future climate that could resemble a semi-permanent El Niño, with individual El Niños and La Niñas still occurring on top of that new base state.

If there's been any trend in recent decades, it's in favour of El Niño. According to the US National Oceanic and Atmospheric Administration, the three decades from mid-1980 to mid-2010 saw six La Niñas compared to nine El Niños, including the two strongest El Niños of the last century. This trend could be part of a natural multi-decade pattern (more on this below) and/or it could reflect a warming caused by climate change. Towards Indonesia, the sea appears to be almost as warm as it can get, averaging above 29°C (84°F). Any further heating might trigger enough storminess to cool that patch of ocean back down. But, according to Gerald Meehl of the US National Center for Atmospheric Research, it's possible that the warm pool could build eastward into the central Pacific, moving the region's typical state closer to that of El Niño (as suggested by the IPCC analysis mentioned above).

Will the Atlantic turn cold on Britain?

Even as global warming appears to be proceeding full steam ahead, there are hints that the UK and parts of the European continent could be in for a countervailing force – a cooling influence that might inhibit human-induced warming in some areas. The storyline will be familiar to those who caught 2004's blockbuster *The Day After Tomorrow* (see p.293), which took the idea to cartoonish extremes. The nugget of truth in the film is that the northward flow of warm water into the far North Atlantic could slow or even shut down entirely, which would plunge much of northwest Europe into a climate more akin to that of Canada, Russia or other countries at the same latitude. In Hollywood's view, this occurs in a matter of days, flash-freezing almost everyone from Washington to Scotland. In real life, any cooling would be far less dramatic, unfolding over at least a few years, more likely decades to centuries. Even so, the possibility is worth taking seriously.

The Atlantic flow is part of the global loop of ocean circulation (see graphic above). It's a system whose mechanics are well mapped but not fully understood. One of the most important branches is often referred to as the **Atlantic thermohaline circulation**, because it's related to both temperature ("thermo") and saltiness ("haline"). Many scientists now use a different name: the Atlantic **meridional overturning circulation** (MOC). "Meridional" refers to its north-south orientation, and "overturning" to the somersault it performs in the far north. Before it gets to that point, the MOC is a vast conveyor belt of warm water that flows around South Africa and then northward. In the North Atlantic, prevailing winds sculpt the sea-surface flow into a clockwise gyre, sending warm water north along the US

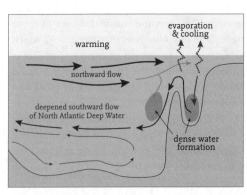

The Atlantic meridional overturning circulation.

coast in the **Gulf Stream** and bringing cool water south from Spain in the Canary Current. An extension of the Gulf Stream called the **North Atlantic Drift Current** pushes warm surface water even further northeast towards the British Isles. (Journalists often use the term "Gulf Stream" to refer to the North Atlantic Drift Current, but strictly speaking that isn't correct.)

The flip side of the conveyor occurs when the warm surface water approaches Iceland and splits around it, heading for the Labrador and Norwegian seas. Along the way, it gradually chills and sinks. Then, at several depths ranging from about 2km to 4km (6600–13,100ft), the frigid deep water begins the return trip to the South Atlantic, thus balancing the warm northward conveyor and completing the MOC.

By any measure, the MOC is a serious workhorse. As it passes between the Bahamas and the Canaries, around 25°N, the circulation brings an average of roughly a petawatt of heat energy northward, day in and day out. That's enough energy to run ten trillion 100-watt light bulbs, or more than fifty times the energy being used by all the homes, offices and factories on Earth at any moment.

Scientists have speculated for more than a century that this giant natural turbine might have more than one speed setting. Today, based on paleoclimate evidence, many ocean specialists see the MOC as having three distinct modes: essentially "faster", "slower" and "off". During the last ice age, the MOC appears to have shifted among these three modes. Since Earth began to warm up about 12,000 years ago, the MOC has stayed fairly robust, except for a couple of dramatic weakenings around 11,500 and 8200 years ago, as glacial meltwater poured into the North Atlantic and suppressed the circulation. These events were enough to cool the winter climate in and around the present-day UK by as much as 10°C (18°F) below present-day readings. In other words, London in a typical January would have felt more like St Petersburg does today.

The main worry for this century is the increasing flow of fresh water into the far North Atlantic. This comes from increased rainfall and snowfall across the Arctic and adjacent land areas, as well as from melting glaciers. Because fresh water is less dense than salty water at the same temperature, the fresh water has more trouble sinking and forming the Atlantic's cold, deep return flow. Analyses led by the UK's Centre for Environment, Fisheries and Aquaculture Science show that the upper 1000m (3300ft) of the Nordic seas – the main source of sinking water for the Atlantic – have become steadily fresher since the 1960s.

Computer models agree that the flow of fresh water into the far North Atlantic ought to continue increasing. That could put a substantial brake on the MOC of the twenty-first century. In response to this concern, the UK's Natural Environment Research Council funded a £20 million study called RAPID that ran from 2001 to 2008, with a follow-up called RAPID-WATCH slated to run through 2014. At the time of RAPID's launch, the UK Met Office Hadley Centre had produced simulations in which the MOC was halted in its tracks. The results showed a net cooldown of around 1–3°C (1.8–5.4°F) by 2050 across most of the UK, Scandinavia, Greenland and the adjacent North Atlantic, with warming elsewhere around the globe. However, as the models have become more sophisticated due to RAPID and other studies, they've produced a more modest slowdown of the MOC. A 2005 comparison of eleven different models coordinated by Jonathan Gregory of the University of Reading showed the MOC weakening by anywhere from 10% to 50%, even with a quadrupling of greenhouse gases. The 2007 IPCC report concurs with this general picture. In all of these more recent models, a net warming is shown for the British Isles and nearby countries, with greenhouse gases more than compensating for the MOC slowdown.

On top of these big changes, there are signs of a more subtle speed-up/ slowdown cycle in the MOC, with each phase peaking about every sixty or seventy years. Measurements in the deep Atlantic are too sparse to show this trend, and there's no solid physical explanation for it. Nevertheless, sea-surface temperatures across the North Atlantic show several distinct rises and falls since the late 1800s – a pattern referred to as the **Atlantic multidecadal oscillation** – and some scientists take this as evidence of the MOC waxing and waning. Interestingly, global air temperatures have risen and fallen over a roughly similar timeframe, which leads some scientists to wonder how the two processes might be intertwined.

If the sixty- to seventy-year cycle is robust, then the MOC ought to undergo a slight natural slowdown starting around the 2010s and lasting into the 2040s or thereabouts. The challenge will be to untangle this from any potentially larger human-induced slowdown. Improved measurements should help. In 2004, the RAPID project installed a set of moored instruments (maintained in collaboration with the US) that monitors the MOC continuously along 26.5°N. These data have already countered a much-publicized 2005 study published in the journal *Nature*. In that paper a team led by Harry Bryden of the UK National Oceanography Centre found that the MOC slowed by about 30% overall between 1957 and 2004. But by 2007, data from the RAPID mooring showed that the MOC's strength can vary by a factor of nine over a year's time, which suggests that natural ups and downs likely accounted for the apparent long-term drop found by the Bryden team.

As the data from RAPID continue to be analysed, we'll get an even clearer sense of the MOC's recent behaviour and how it may change with the climate. In the meantime, Britons might be best off planning for a warmer climate while learning to live with a bit of uncertainty about exactly how warm it might be.

Living in a changing sea

The creatures that make their home in the world's oceans might seem to have much more flexibility to deal with climate change than do their land-based cousins. But do they? Marine biologists are still sorting through the many interconnections among ocean species and trying to map out how things might change as the sea warms up and as its chemistry changes.

As with life on land, there will be winners and losers. Migratory fish should in principle be able to shift their ranges relatively easily. But not all components of the oceanic food chain will change in unison. A small rise in ocean temperature can affect the mix that exists among the tiniest marine plants and animals (including **phytoplankton** and the **zooplankton that consume them**). These changes, in turn, can produce mass die-offs of seabirds and fish further up the chain. A 2006 study using NASA satellite data showed a global-scale link between warmer oceans and a drop in phytoplankton abundance and growth rates over the preceding decade. But the full scope of the decline wasn't recognized until a bombshell 2010 study, led by scientists at Dalhousie University, appeared in *Nature*. Combining satellite data with nearly half a million measurements of ocean colour taken by ships, Daniel Boyce and colleagues found that phytoplankton have been on the decline in most of the world's oceans – especially across the tropics and subtropics – for more than a century. The study estimates that phytoplankton biomass has vanished since 1950 at an average rate of about 1% per year, in sync with rising ocean temperatures (as well as vastly expanded fishing). Moreover, the decline appears to have accelerated in recent years. These results are all the more stunning when you consider that phytoplankton account for half of all the biomass produced on Earth.

Scientists can zero in on shorter-term links between ocean chemistry and climate by observing the sharp regional warmings produced by El Niño. Off the coast of Peru, the warm water of El Niño

> **"This is a definite wake-up call that our oceans are becoming increasingly stressed."**
>
> Daniel Boyce, Dalhousie University, on the global decline in phytoplankton

cuts off the upwelling of colder, nutrient-rich waters that support a variety of marine life. Back in 1972, an El Niño intensified the effects of overfishing and decimated Peru's lucrative anchovy industry. Many birds that fed on the anchovies died off, and in turn the local supply of *guano* – bird droppings used as fertilizer – plummeted. Such chains of events aren't limited to the tropics. In 1997, unusually warm waters developed in the eastern Bering Sea (which lies between Alaska and Russia), killing off the type of phytoplankton that normally dominate during the summer. The disturbance worked its way up to devastate local zooplankton, seabirds and salmon.

As global warming unfolds, it's possible that some of the most dramatic impacts on ocean life will occur through intense seasonal and regional spikes in water temperature driven by El Niño and other ocean cycles. It's somewhat analogous to the way climate change makes its presence most obvious and discomforting to humans during extreme heat waves.

There's more to consider than warming alone. Some ocean life is more dependent on seasonal changes in light than on temperature, especially at higher latitudes. Other species are more purely temperature-driven. Since cycles of sunlight aren't changing, the most light-dependent creatures haven't shifted their timing, whereas many species that are more temperature-dependent are starting their yearly growth cycle at earlier points in the warm season. A 2004 study of the North Sea by Martin Edwards and Anthony Richardson of the Sir Alister Hardy Foundation for Ocean Science found that the differences between these groups is leading to a large-scale mismatch, threatening the seasonal synchrony in growth and feeding times that supports much of the sea's food chain.

Looming over this entire scene is the gargantuan threat posed by **ocean acidification**, perhaps the most underappreciated risk posed by our long-time love affair with fossil fuels. This transformation is a direct result of the enormous amounts of carbon dioxide being soaked up by Earth's oceans. Each year a net influx of roughly seven gigatonnes of CO_2 – which includes close to 25% of all the carbon dioxide produced by human activity – goes into the sea. In the well-mixed upper part of the ocean, where the absorption happens, the extra CO_2 is gradually making the water more acidic – or, to be more precise, the water is becoming less alkaline. A 2010 report by the US National Research Council notes that average pH values at the ocean surface have dropped from around 8.2 in pre-industrial times to around 8.1 today. That change may seem small, but in fact it's on the order of a hundred times faster than anything the world's oceans have seen in the last few million years.

Because pH is measured on a logarithmic scale, every change of one point is equal to a tenfold increase or decrease. The 0.1-point drop in pH already observed thus translates to a 30% increase in hydrogen ions throughout the upper ocean. Depending on the pace of global greenhouse emissions, oceanic pH could drop to the range of 7.8–7.9 by 2100. A pH value of seven is neutral, so the oceans will still be slightly alkaline, but less so than they've been for at least hundreds of thousands of years. Such values could have major impacts, according to a 2010 study led by Jason Hall-Spencer (University of Plymouth). He encountered far fewer calcifying organisms and far less diversity in foraminifera in places where pH dropped below 7.8 in the vicinity of CO_2-spewing vents in the Mediterranean.

Scientists are only beginning to probe the influence of acidification on marine life. A well-known impact is the effect on the ability of many organisms, such as corals and oysters, to secrete their calcium carbonate shells. As acidification proceeds, the carbonate ions they rely on will become more scarce. Ocean acidification has also been shown to affect such things as the availability of iron and nitrogen (both of which impede phytoplankton growth) and the ability of clownfish to sense chemicals in their environment. The full effects of such changes on marine ecosystems, however, hasn't been mapped out yet.

Even if humans can turn the tide on greenhouse-gas emissions, it will take tens of thousands of years more before the ocean returns to its preindustrial pH levels. Some engineering fixes have been proposed – for instance, adding mammoth amounts of limestone to the ocean to reduce its acidity – but these are likely to succeed only on a local level, if even then. Any such technical fix is laden with ecological question marks: for instance, could an attempt to de-acidify the oceans cause something even more harmful to occur? Similarly, it's important to keep in mind that any scheme to tamp down global warming by blocking sunlight – as proposed in several geo-engineering strategies (see p.359) – can't directly address ocean acidification, since these schemes don't reduce the amount of CO_2 entering the atmosphere and, by extension, the oceans.

Coral reefs at risk

The most famous oceanic victims of climate change may be the magnificent **coral reefs** that lace the edges of the world's subtropical and tropical waters. More than a quarter of the planet's coral reefs have already been destroyed or extensively damaged by various types of human activity and

Rich, diverse and extremely vulnerable – abundant life on the coral reef.

by spikes in water temperature that may be intensified by overall ocean warming. Nobody expects climate change to completely eliminate the world's reefs, but many individual coral species may face extinction.

It's amazing that coral reefs manage to produce the colour, texture and overall splendour that they do. After all, they grow in the oceanic equivalent of a desert: the nutrient-poor waters within 30m (100ft) of the surface. The reefs' success in this harsh environment is because of their unique status as animal, vegetable and mineral. Coral reefs consist of **polyps** – tentacled animals that resemble sea anemones – connected atop the reef by a layer of living tissue. Inside the tissue are microscopic **algae** that live in symbiosis with the polyps. The algae produce the reefs' vivid colours and feed the polyps through photosynthesis, while the polyps provide the algae with nutrients in their waste. The polyps also secrete **limestone** to build the skeletons that make up the reef itself.

Coral reefs are durable life forms – as a group, they've existed for at least forty million years – but they're also quite selective about where they choose to live. The microalgae need access to light in order to photosynthesize, and the skeleton needs a foundation, so coral reefs form in shallow, clear water atop a base that's at least moderately firm, if not bedrock.

Finally, the waters around coral reefs need to be warm, but not too warm. Generally, the average water temperature should be above 27°C (81°F), with annual lows no cooler than 18°C (62°F). If the waters spend much time at temperatures more than 1–2°C (1.8–3.6°F) above their typical summertime readings, the algae can begin to photosynthesize at a rate too fast for the coral to handle. To protect its own tissue, the coral may expel the micro-algae, allowing the white skeletons to show through the translucent polyps. When this occurs on a widespread basis, it's known as **mass bleaching** (although chlorine has nothing to do with it).

The record-setting El Niño of 1997–98 led to unusually warm water in much of the tropical Pacific, triggering a bleaching disaster unlike anything ever recorded. Some 16% of the world's coral reefs were damaged in that event alone. A weaker event in 2002 bleached up to 95% of individual reefs across parts of the western Pacific, and another strong El Niño in 2010 produced record-warm oceans and a second round of global-scale bleaching. This time the victims included some groups of coral near Indonesia that had managed to survive the 2004 tsunami. Some coral species appear to be bouncing back from the 1997–98 bleaching, although full recovery could take up to twenty years. Other species – especially those with lavishly branched forms and thin tissues – don't appear to be so lucky. Experiments hint that some coral may be able to adapt to bleaching over time by teaming up with different microalgae more suited to warmer temperatures. The upshot over the long term may be a thinning of the herd, as it were, leaving behind a hardier but less diverse and less spectacular set of coral reefs.

To complicate the picture, global warming isn't the only threat the reefs face:

▶ **Agricultural fertilizer** that washes offshore can deposit nitrogen and phosphorus that encourage the formation of seaweed or phytoplankton at the expense of polyps.

▶ **Overfishing** can increase seaweed growth atop reefs by removing many of the plant-eating fish that would otherwise keep the growth in line.

▶ **Sediment from rivers** can block needed sunlight, interfere with the polyps' feeding or even bury a reef whole. Logging and other land-based activities have produced enough sediment downstream to put more than 20% of the coral reefs off southeast Asia at risk, according to the World Resources Institute.

The ultimate impact of climate change on reefs depends in part on how extensive these other stressors become. By itself, the influence of global warming is a mixed bag. Sea-level rise is a potential positive. Many older coral reefs have built upwards to the maximum height that temperatures and other conditions allow, so a higher sea surface would give these reefs more vertical room to grow. Meanwhile, the lowest-lying coral reefs could probably grow upward quickly enough to keep up with modest sea-level rises – a typical vertical growth rate is about 1m (3ft) in a thousand years. As long as temperatures increase slowly enough to avoid coral bleaching, warmer temperatures appear to speed up the calcification process that builds reefs. And the global spread of warmer waters will add a sliver of territory around the current margins of coral-reef viability, though that extra real estate is limited by other requirements such as a firm sea surface.

Together, these pluses may not be enough to counter the negatives produced by the environmental stresses noted above, not to mention global warming itself. Along with causing bleaching, warmer waters may nourish the enemies of reefs as much or more than they help the reef itself. One study by Drew Harvell of Cornell University found that some pathogens tend to thrive at temperatures 0.6°C (1.0°F) or more higher than the optimal readings for coral themselves. Another risk is the ocean acidification discussed above, which can both limit the amount of carbonate available for reef building and perhaps erode the reef skeletons directly, in both cases slowing the rate of reef growth. And tropical cyclones may become more intense with time (see p.149), although it appears that a healthy reef can recover from hurricane or typhoon damage fairly quickly.

In short, some coral reefs should still populate the world of our grandchildren, but they may not be in the same locations or glow with the same brilliance that snorkellers and scuba divers now savour.

Hurricanes and other storms

Rough waters

Spinning their way across the warmer parts of the globe, tropical cyclones throw fear into millions of coastal dwellers each year. Depending on which ocean they're in, these huge weather systems are known as hurricanes, cyclones or typhoons. Whatever you call them, they've swept across the tropics with a vengeance over the last few years. A record ten typhoons raked Japan in 2004. Australia was hammered in early 2006 by a trio of cyclones, each packing winds above 250kph (155mph) either offshore or on the coast – the first time the region has recorded three storms of that calibre in a single season. Cyclone Nargis, which slammed into Burma in 2008, killed more than 100,000 people, making it the world's deadliest tropical cyclone in at least three decades. And the Atlantic saw a major burst of activity in 2010, with five of the year's first six hurricanes reaching Category 3 status – twice the usual number for an entire season.

For North America, 2005 produced an unforgettable tropical onslaught. Based on measurements of their lowest pressure, the year brought three of the six most powerful Atlantic hurricanes ever observed: **Wilma** (the strongest), **Rita** (fourth strongest) and **Katrina** (sixth strongest). All three made devastating strikes on coastal areas. Wilma slammed Cancún and south Florida, Rita pounded the Texas/Louisiana border, and Katrina swept through Miami before bringing mind-boggling misery and more than 1800 deaths to the New Orleans area and the Mississippi coast.

Did global warming have anything to do with all this havoc? That question couldn't be dodged in the wake of 2005's violent Atlantic season. As if on order, several major studies argued that tropical cyclones worldwide

Destroyed fishing boats lie in the port of Yangon after Cyclone Nargis, before the full scale of human casualties inland had become apparent.

are indeed getting stronger. Meanwhile, a number of hurricane experts (many of them forecasters) insisted that global warming played only a minor role at best in the recent rash of activity, and that it will be far outweighed by other factors in the future. All this has made tropical cyclones one of the liveliest arenas of climate-change debate.

To the north and south of the tropics, people tend to be more concerned with **coastal storms**. These wintry blasts of rain, snow and wind sweep through places such as New England, British Columbia and western Europe, packing less punch than hurricanes but often affecting much larger areas. Coastal storms appear to be shifting gradually polewards over time. Though the trends aren't highly dramatic, they could have major importance for the densely populated coastlines of North America and Eurasia.

Well away from the coasts, the most intense bouts of wind and rain are likely to occur with severe **thunderstorms**, which occasionally pack the tiny but powerful windstorms known as **tornadoes**. (Many hurricanes produce tornadoes as well.) Although tornadoes are being reported more often in many parts of the world, it looks as if that can be pinned on greater awareness of them. Thankfully, there's no solid evidence that the most violent tornadoes and thunderstorms have become more intense or frequent due to global warming, though some hints of future change are now showing up (see p.158).

A taste of things to come?

Two main questions loom over any discussion of future tropical cyclones: will there be more of them in a warmer world, and will they be more intense? In its 2001 report, the **IPCC** noted that no trend had yet been identified in tropical cyclone intensities, as measured by winds and pressures, adding that "past and future changes in tropical cyclone location and frequency are uncertain". Things had changed dramatically by the time of the 2007 IPCC report, which reflects a growing body of evidence that the number of intense hurricanes and typhoons has jumped in recent decades.

A hurricane rolls down towards Rio

People in Brazil aren't used to hurricanes. In fact, no hurricane had ever been reported there until 28 March, 2004, when a mysterious system packing winds up to 137kph (85mph) swept onto the coast of Santa Catarina province. Forecasters from Brazil and from the US National Hurricane Center had been tracking the storm by satellite. Although no tropical cyclones had ever been officially recorded in the South Atlantic, the swirl of clouds had the clear eye and other telltale features of a bona fide hurricane. Despite the absence of a precedent, Brazil's meteorological service issued warnings and got people out of harm's way. The storm ended up destroying many hundreds of coastal homes, but only one death was reported.

Informally christened Hurricane Catarina for its landfall location, the storm continued to stir things up even after it died. Some meteorologists questioned whether it was truly a hurricane, and a few climate-change activists pointed to Catarina as an illustration of a planet gone awry. In fact, it turns out that Catarina had some of the classic earmarks of a hurricane, but not all of them. For instance, the showers and thunderstorms circling its centre were more shallow than usual. And Catarina can't be easily pinned on climate change, because the waters over which it formed were actually slightly cooler than average for that time of year. The storm was clearly unprecedented in its landfall impacts, but analysts poring over satellite records found at least one similar (if weaker) system in 1994 that stayed well out to sea. In 2010, US and Brazilian meteorologists tracked another system with tropical characteristics that spun for two days in the same offshore region where Catarina formed. It was dubbed Tropical Cyclone Anita by Brazil's meteorological service (though the World Meteorological Organization hadn't yet added the South Atlantic to the global naming system it manages). Anita's brief life points to a growing awareness that tropical cyclones can develop in a region where few scientists were looking for them. However, in its sheer strength and impact, the much more intense Catarina remains unique in modern records – a bizarre weather event that pushed the buttons of a world increasingly jittery about climate change.

Katrina and climate change

The damage and misery wrought by 2005's Hurricane Katrina topped anything produced by a single US storm for many decades. More than 1800 people died, mostly in the New Orleans area, where public evacuation options outside of the wretched Superdome and convention centre were nearly non existent. The storm's toll in property damage soared well into the tens of billions of US dollars, and that doesn't include the cost of restoring the New Orleans levee system to its previous condition – much less the billions more needed to make the system able to withstand a direct hit from a Category 5 hurricane.

Some observers linked the Katrina debacle to climate change from the outset. In a *Boston Globe* editorial, journalist Ross Gelbspan declared, "The hurricane that struck Louisiana yesterday was nicknamed Katrina by the National Weather Service. Its real name is global warming." Others denied any connection between Katrina and climate change, a viewpoint that came to dominate US media and legislative discussion before long. It's true that no weather event can be blamed solely on climate change, and certainly a storm like Katrina doesn't require global warming

Hurricane Katrina approaches the Gulf Coast on 28 August 2005.

in order to flex its muscle. Though they're quite rare, hurricanes on a par with Katrina have developed in the Atlantic since records began. However, the gradual warming of tropical waters over the last several decades has made it easier for storms like Katrina to intensify when other conditions are right. Perhaps a more telling hint of a shift in climate wasn't Katrina itself so much as Emily, Katrina, Rita and Wilma combined. It marked the first time four, or even three, Atlantic hurricanes had reached Category 5 strength in a single year in records that date back to 1851.

As for the landfall in New Orleans, that was simply the luck (albeit the bad luck) of the meteorological draw. Several near-misses had grazed the city since 1965, when Hurricane Betsy produced flooding that killed dozens. Betsy prompted the US Corps of Engineers to build the city-girdling levee system that was in place when Katrina struck. That system was built to withstand only a Category 3 storm, a fateful move sure to be analysed by critics for years to come – especially since Katrina itself had weakened to Category 3 strength by the time it made landfall.

The notion of stronger hurricanes and typhoons in a warmer world makes physical sense. **Warm ocean waters** help give birth to tropical cyclones and provide the fuel they need to grow. Generally, the cyclones can't survive for long over sea surfaces that are cooler than about 26°C (79°F). The larger the area of ocean with waters above that threshold, and the longer those waters stay warm in a given year, the more opportunity there is for a tropical cyclone to blossom if all other conditions allow for it.

Not all areas are conducive: within about 5° of the Equator, the winds can't easily gather into a rotating pattern. And the South Atlantic doesn't seem to have the right combination of upper-level winds and warm surface water to produce hurricanes, although 2004 brought a major surprise (see box on p.143). That leaves the prime breeding waters for tropical cyclones between about 10° and 30° across the North Atlantic, North Pacific, South Pacific and Indian Oceans (see map on p.147).

The terrible 2005 trio of Katrina, Rita and Wilma shows how important warm water is for hurricanes. Katrina and Rita both surged to their peak intensity – Category 5 – just as they passed over a patch of the central Gulf of Mexico heated by an infusion of deep, warm water from the Caribbean. This channel, called the **Loop Current** for its arcing path into and back out of the Gulf, is often a hurricane booster. In 2005 it featured some of the warmest waters ever observed in the open Gulf, topping 32°C (90°F) in places. Similarly toasty waters had also overspread much of the Caribbean, where Wilma intensified in an astonishing twelve hours from Category 1 status to attain the lowest barometric pressure ever observed at sea level in the Western Hemisphere: 882 hectopascals (26.05 inches of mercury).

To keep a truly intense hurricane at top strength, the ocean's warmth needs to extend deep – ideally at least 100m (330ft). Otherwise, the powerful winds and waves can churn up enough cold water from below to dilute the fuel source and diminish a slow-moving cyclone in a matter of hours. Both Rita and Katrina began weakening after they left the Loop Current; they began to pull up cooler subsurface waters, while drier air and stronger upper-level winds from the United States started to infiltrate. By the time they struck the **Gulf Coast**, both storms were down to Category 3 strength. That didn't matter much, though, because the storms had already piled up fearsome amounts of water that slammed into the coast in the form of punishing storm surges. Katrina's surge was a US record: at least 8.2m (27ft) above sea level at its worst, with waves atop that, destroying the homes of thousands who thought they lived at safe elevations. The surge also helped lead to the failure of several levees in and around **New**

Orleans, which in turn produced the cataclysmic flood that left most of the city submerged and paralysed (see box, p.144).

The wreckage of the 2005 Atlantic hurricane season extended far beyond US shores. Ocean temperatures were markedly above average across vast stretches of the North Atlantic tropics and subtropics. All that long-lasting warmth, coupled with nearly ideal upper-level winds, led to 27 named Atlantic storms in 2005, which smashed the record of 21 set in 1933. Still another system was upgraded to tropical-storm strength post-humously, bringing the season's final total to 28. This swarm of storms was noteworthy not just for its sheer size but for its unusual geographic spread as well. Even Spain got its first-ever tropical storm, **Vince**, which made landfall on the southwest coast near Huelva.

Keeping count: will there be more cyclones in the future?

The sheer number of Atlantic hurricanes in 2005 led many to ask whether tropical cyclone frequency might already be on the rise due to climate change. One might think that a larger, more long-lasting zone of warm-enough water would be sufficient to spawn more and more tropical cyclones. And indeed, both 2005 and 2010 saw Atlantic waters that were unusually warm over vast areas for long periods, which helped seasonal forecasters accurately peg the action months ahead of time. Yet an all-star team of hurricane researchers agreed in a 2010 roundup in *Nature* that, over the last century, the average number of tropical cyclones worldwide hasn't been detectably boosted by greenhouse gases. Making such an assessment of the frequency of past tropical cyclones, and projecting their future, is a complex task for several reasons:

▶ **Other factors** Even over sufficiently warm water, only a small fraction of the weak circulations that form in a given year spin up to cyclone status. A number of factors can lead to their early demise: unfavourable upper winds, an influx of dry air or competition from nearby systems, to name a few. Warm water is critical, but it isn't enough to guarantee that a hurricane will form.

▶ **Variable records** It's hard to know exactly how many tropical cyclones roamed Earth before **satellite monitoring** became routine in the 1970s. Although ships did a good job of reporting stronger tropical cyclones in the pre-satellite era – at least across major shipping lanes – it's

World regions prone to hurricanes, typhoons and cyclones: different names for the same phenomenon.

likely that some of the weaker, smaller systems over the open ocean were missed then but are being caught now. Better observations also help explain an apparent rise in the number of short-lived Atlantic cyclones, those lasting for two days or less. If these factors are taken into account (including an estimate of the cyclones that once went uncounted), the number of Atlantic systems may not have risen significantly over the last century.

▶ **Complex global interrelations** For reasons not fully understood, the planet normally does a good job of conserving the total number of tropical cyclones. About 40–60 per year attain hurricane strength, and there's been no significant long-term trend up or down during the few decades of reliable data. The frequency stays fairly constant because the planet's ocean basins appear to trade off in their periods of peak production. Typically, when the Atlantic is producing hurricanes galore, parts of the Pacific are relatively quiet, and vice versa. Some of this seesaw effect is due to **El Niño** and **La Niña**, the cyclic warmings and coolings of the eastern tropical Pacific (see p.130). The large-scale air circulations driven by La Niña encourage Atlantic hurricanes but suppress them in some parts of the Pacific; the opposite is true of El Niño.

Until relatively recently, computers weren't of much help in assessing whether we'll see more tropical cyclones in a warmer world. That's because most hurricanes and typhoons are small enough to fall in between the grid points of the comprehensive global models that map climate out to a century or more. It's also tough for a model to pick up on the meteorological subtleties that allow one system to become a hurricane and another not to – the same unknowns that plague tropical weather forecasters on an everyday basis. But over the last decade, researchers have made headway in using models to envision the long-term future of tropical cyclones. In part, this is due to faster computers that allow new models to depict the entire globe in sufficient detail. An early sign of progress came in 2006, when a group of Japanese scientists led by Kazuyoshi Oouchi published results from a globe-spanning model with 20km (12-mile) resolution, enough to follow individual tropical cyclones in a broad-brush fashion. The results suggested that in a mid-range emissions scenario, the late-twenty-first-century Earth could see fewer tropical cyclones but a higher fraction of intense ones. Other scientists have nested fine-scale regional models inside coarser global-scale ones, or blended statistical and physical modelling. The results from these varied approaches are largely in sync with each other: we may see fewer tropical cyclones worldwide over time – anywhere from 6% to 34% fewer by late this century, according to the 2010 *Nature* roundup. It's not clear exactly why the models tend to spin up fewer storms, but one possibility is a predicted drop in relative humidity aloft, which would threaten the cocoon of moisture in which a hurricane thrives. Another is a projected weakening of the global circulation through which air rises across the tropics and sinks over the subtropics.

Even if tropical cyclones become a bit less prevalent on a global scale, they could still increase over some oceans while decreasing over others. Models haven't yet agreed on a collective portrait of region-by-region change, but a few intriguing threads have emerged. For example, a number of models suggest that a semi-permanent El Niño-like state could develop (see p.131). This could lead to a stronger warming of tropical waters in the Pacific than in the Atlantic, which in turn might trigger El Niño-like circulation patterns in the atmosphere. By examining eighteen model simulations that fed into the 2007 IPCC report, Brian Soden (University of Miami) and Gabriel Vecchi (NOAA) found that upper-level winds became stronger (i.e., more hostile for hurricanes) in the northeast Pacific and Atlantic while relaxing in the Indian and western Pacific oceans.

"While we don't know precisely how global warming will change hurricanes, that's not really the point. What matters is that today, we know enough to be worried."

Chris Mooney,
Storm World

Regardless of whether or not we see more tropical cyclones, we can count on most of them never reaching land. Even if a given ocean began to spawn more cyclones, it doesn't necessarily mean that more of them will hit the coastline. Indeed, even a slow season can bring destructive landfalls. In 1992, the Atlantic was far less busy than average, yet it still produced Florida's massively destructive **Hurricane Andrew**. Conversely, it's possible for most of a busy season's systems to churn harmlessly out to sea.

Packing more power

What about strength? Here, the tune is more ominous. Though the IPCC saw little evidence of change in its 2001 report, subsequent studies have found a jump in the power of tropical cyclones on a global basis, as noted in the 2007 report. An increase in hurricane strength is both plausible on a physical basis and apparent in some of the most recent modelling.

Two key studies pushed this topic into the headlines in 2005, where it shared space with the year's bumper crop of Atlantic super-storms. Kerry Emanuel of the Massachusetts Institute of Technology examined tropical cyclones since the 1970s, when satellite monitoring became widespread. He looked not just at the raw wind speeds but at the destructive force they

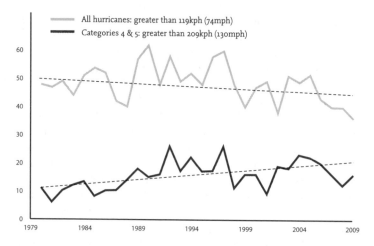

Tropical cyclone totals for the period 1980–2009: while the total number of tropical cyclones around the globe ranked Category 1 to 5 hasn't changed much in the last thirty years, the proportion of Category 4 and 5 cyclones – those that produce the worst damage – has risen markedly.

packed over their lifespan. This force, the **dissipation of power**, is proportional to the wind speed cubed. Thus, a wind of 100 kilometres per hour would be twice as damaging as a wind of 80 kilometres per hour. Small increases in wind speed, then, can produce major ramp-ups in destructive force. From this vantage point, Emanuel discovered that tropical cyclones in the Atlantic and northwest Pacific were packing nearly twice the power they did in the 1970s – the result of the strongest storms getting even stronger as well as lasting longer.

A team led by Peter Webster of the Georgia Institute of Technology found similar results to Emanuel's when they ranked tropical cyclones since the 1970s by **intensity category**. As expected, they found little change in the total number of tropical cyclones in all categories – a sign of the global balance noted above. The eye-opener was their discovery that the strongest cyclones, those at Category 4 or 5, were almost 50% more frequent globally in 1990–2004 than they were in 1975–89. Put another way, Category 4 and 5 storms went from comprising about 25% of all tropical cyclones globally to comprising about 35% of them. A follow-up analysis in 2010 by co-author Judith Curry showed the trend continuing (see graphic on previous page).

Predictably, these studies provoked a storm of controversy. They also exposed a schism between climate-change researchers, few of whom had done detailed studies of tropical cyclone behaviour, and hurricane specialists, who knew the cyclones and the data by heart but weren't always experts on the larger-scale processes related to global warming. In his 2007 book *Storm World*, science writer Chris Mooney cast the battle as only the most recent example of a tiff between observationalists and theoreticians with roots extending back to the 1800s.

One of the most vocal critics of the new findings was William Gray (of Colorado State University), who founded the field of seasonal hurricane prediction. A long-time sceptic of climate change, Gray assailed the studies on various technical grounds. Also dismissive of a climate-change link was the **US National Oceanic and Atmospheric Administration** (NOAA), whose **National Hurricane Center** (NHC) forecasts Atlantic storms. Individually, some NOAA scientists agreed that human-induced climate change has the potential for "slightly increasing the intensity of tropical cyclones", in the words of Christopher Landsea, who oversees research at NHC. However, Landsea and colleagues suggested that the apparent surge in top-strength hurricanes could be at least in part due to inconsistencies in the official record. For instance, lower-resolution satellites used in the 1970s may not have sensed the tiny areas of high, cold cloud tops associated with the strongest cyclones. Other observational tools, such as the instrument packages dropped from aircraft into hurricanes, have also evolved over the years, as have the analysis techniques that classify the storms. Landsea and colleagues are now in the midst of an effort to reanalyse several centuries of Atlantic hurricanes, which has already resulted in some hurricanes being upgraded and others downgraded. And in 2009 NOAA launched the world's first centralized, open-access database on the global history of tropical cyclones, the International Best Track Archive for Climate Stewardship (IBTrACS).

Even by restricting one's view only to years with the best data, it's apparent that change is afoot, as shown by James Elsner (Florida State University) and James Kossin (NOAA/University of Wisconsin) in studies published in 2008 and 2009. Acknowledging that observational techniques had changed, they set out to compare apples to apples by analysing only those cyclones that occurred during the period of highest-quality satellite data, which began in 1981. Through this clarifying lens, the scientists found that the peak winds produced by the globe's strongest tropical cyclones had indeed grown stronger, at a rate of about 1–2 meters per second (2–4mph) per decade. Even such a seemingly modest value adds to a storm's destructive power (as noted above). Over several decades, such a trend might be enough to push a substantial fraction of storms from one intensity category to the next.

The plot thickens even further when you consider the **Atlantic Multidecadal Oscillation**. It's long been known that, over the past century, sea-surface temperatures in the North Atlantic alternated between warming and cooling for periods of about 30–40 years each. Many have tied this waxing and waning to presumed cycles in the Atlantic thermohaline circulation, also known as the meridional overturning circulation. This circulation is only now being measured with some precision (see p.132), and its dynamics are poorly understood. Nevertheless, Gray showed that the frequency and average intensity of Atlantic hurricanes has risen and fallen every few decades in sync with the AMO – lower from the 1890s to the 1920s, higher from that point through the 1950s, then lower again from the 1960s to the 1990s. Based on this, Gray predicted an increase in Atlantic hurricane action in the mid-1990s that would continue for several decades. Thus far, his vision has proven remarkably accurate.

But even if there is a strong physical basis for the AMO, it doesn't mean climate change can be ignored. The AMO only affects hurricane formation in the Atlantic, whereas the studies mentioned above suggest a strengthening trend across several other ocean regions as well. Moreover, the recent Atlantic up-tick could be a combined product of the AMO and general ocean warming. For example, a 2006 study by Kevin Trenberth (of the US National Center for Atmospheric Research) found that global warming had more than twice the influence of the AMO in producing 2005's extra hurricane-fuelling warmth in the waters of the tropical North Atlantic. Thus, if and when the AMO cycle finally turns towards a quieter Atlantic – perhaps around 2015–25 – things may not calm down as much as promised.

Despite their limited ability to project tropical cyclones into the future, global computer models tend to agree that overall warming ought to help induce a modest increase in strength. A 2004 study led by Thomas Knutson of NOAA used nine different models and examined three regions (the Atlantic and the northeast and northwest Pacific). The consensus was for an increase of a few percent in peak tropical cyclone winds by the late twenty-first century. This translates to the typical hurricane gaining about a half-category in strength (if anything, current events might be outrunning this projection). And the newer, higher-resolution models tend to produce a more robust rise in intensity, as noted by the 2010 summary in *Nature*.

On balance, with improvements in observation taken into account, it's becoming clear that nuggets of truth lie within the much-hyped prediction of a greater number of extreme hurricanes and typhoons. This doesn't mean that more of these intensified storms will necessarily make landfall, but those that do could pack more of a wallop. One question that remains unexplored is whether the size of tropical cyclones might be changing. The biggest cyclones can bring hurricane-force winds to ten or more times more area than the smallest ones. But cyclone diameter isn't part of official global databases, and most climate-change studies to date have omitted this important dimension.

Surges and downpours

Extreme **storm surges** will continue to loom as a catastrophic possibility from tropical cyclones in the century to come, but the shape of that threat will shift as sea levels rise and the coast itself evolves. **Subsidence** and the loss of protective **wetlands** are already making many areas more vulnerable, including the Texas and Louisiana coastlines hammered in 2005. According to scientists at the UK Met Office Hadley Centre and the University of Middlesex, sea-level rise due to climate change by the year 2050 could put twenty million people around the world at risk of serious flooding who wouldn't otherwise have been affected. This assumes "busi-

Hurricane Danielle, as seen from the International Space Station on 27 August 2010 – a year in which 19 named storms raged across the Atlantic and Gulf of Mexico. The only year with more Atlantic hurricanes was 2005.

ness as usual" emissions and a mid-range estimate for sea-level rise this century of 27cm (10.6in). The stretch of coastline from India to Vietnam, plus Indonesia, would see the most people threatened.

More than any place on Earth, the **Bay of Bengal** exemplifies the risks of tropical-cyclone surge. Fully half of the world's deadliest surge tragedies have taken place in eastern India, present-day Bangladesh and Burma. Half a million Bangladeshi people died as a result of a catastrophic 1970 landfall (the event that prompted a famous relief concert by ex-Beatle George Harrison). In 2008, **Cyclone Nargis** took a track unusually far to the south, becoming the first major cyclone ever recorded in Burma's populous Irrawaddy Delta. A surge 3m (12ft) high struck at high tide and swept as far as 40km (25 miles) inland. With no experience of such a storm in living memory – and with state-controlled media downplaying the storm's threat – delta residents had no inkling of what they would face. The official death toll topped 130,000, though many thousands more may have died.

Rising sea level only adds to the Bay of Bengal's flood risk, while soaring coastal populations make even more people vulnerable to a given flood. And the intricate delta-style coastline would be extremely difficult to protect through engineering even if the nations could afford such a fix. Using a regional model for the Bay of Bengal, Jason Lowe at the Hadley

Centre found that extreme storm surges by 2050 could affect the India/ Bangladesh coast in complex ways. Some areas become more vulnerable than others in the model, as storm behaviour shifts and as the higher sea level causes changes in tidal and surge behaviour.

Landfalling tropical cyclones also produce a great deal of rain, on coastlines as well as far inland. That may get worse as moisture increases in the atmosphere and precipitation increases on a global scale. **Inland flooding** from heavy rain is already a serious threat in many hurricane-prone areas. With Katrina as the great exception, river flooding has been the leading cause of US hurricane deaths in recent decades. It's also a major problem for agriculture. Rainfall in the average tropical cyclone may have already increased by 6–8%, according to estimates by NCAR's Trenberth, and modelling studies summarized in the 2010 *Nature* roundup on tropical cyclones indicate that rainfall amounts near the core of a typical storm could rise by 20% or more by later this century.

With all this in mind, it would seem sensible to prepare for the worst despite the remaining gaps in knowledge. One major problem for planners to take on board is the explosive population growth and development taking place on many cyclone-prone coasts. Roger Pielke Jr of the University of Colorado argues that these factors could easily boost the human and financial costs of hurricane damage this century far more than any intensification related to global warming would – a reminder that demographics have everything to do with the ultimate impact of climate change. Researchers from both sides of the hurricane/climate-change debate laid down their arms in 2006 to craft a joint statement stressing the human element of the problem: "We are optimistic that continued research will eventually resolve much of the current controversy over the effect of climate change on hurricanes. But the more urgent problem of our lemming-like march to the sea requires immediate and sustained attention."

Coastal concerns beyond the tropics

Hurricanes and typhoons may cause horrific damage, but coastal storms threaten the wellbeing of millions of people who live well away from the tropics. Unlike tropical cyclones, which need a cocoon of warm, humid air in order to grow, coastal storms – part of the larger group known as **extratropical storms** – thrive on the contrasts between cold continental air and warm marine air. Moving with the polar jet stream, in which they're typically embedded, coastal storms strike many of the major coastlines at mid- and high latitudes, mostly from autumn through spring. In the

US, they're often dubbed **nor'easters**, a reference to the strong onshore winds they bring. The lack of a mountain barrier in western Europe allows coastal storms to plough well inland with heavy rain and high wind. That was the case over the last week of 1999, when an especially vicious pair of post-Christmas storms dubbed **Lothar** and **Martin** pounded much of the region with winds as strong as 219kph (136mph). The storms killed more than 140 people and ruined centuries-old treescapes from France's Versailles to Germany's Black Forest.

As with hurricanes, the challenge in projecting coastal storms is two-fold: will there be more of them, and will they become more intense? A few studies show that extratropical cyclones in general have become fewer in number but stronger on average, although the intensity change appears to be minor at best. These cyclones have also shifted polewards in the Northern Hemisphere, a sign of the jet stream shifting north in response to a warming atmosphere. Across Europe, this jibes with the observed tendency since the 1970s for milder, stormier winters occurring to the north and drier-than-average winters near the Mediterranean – the calling cards of a positive **North Atlantic Oscillation**, or NAO (see p.128).

What about the future? Climate models generally agree that jet streams and storm tracks will migrate poleward at roughly 100km (60 miles) for every 3°C rise in global temperature, according to a 2010 report by the US National Academies. Since extratropical cyclones tend to be several times larger than hurricanes – often spanning more than 1000km (600 miles) – they're more easily simulated by global computer models. A study by Jeff Yin (National Center for Atmospheric Research), carried out in support of the 2007 IPCC report, looked at the results from fifteen advanced global models. The consensus picture could be labelled "more of the same". It shows extratropical cyclones continuing to shift polewards, just as they've been doing in recent decades. The total kinetic energy within these storms increases, too, and other studies indicate that this stronger punch could play out as it has in recent years, with storms that are less frequent but stronger on average. Should all this pan out, lower-latitude coastal regions such as southern California and the Mediterranean may find themselves experiencing more prolonged and/or intense dry spells in winter, as windy, rainy, fast-moving storms track further north over time.

Any such regional projection is framed by uncertainty at this point. Smaller-scale effects that aren't well modelled or understood could overpower a global trend in particular locations. And although climate-change models tend to make the NAO increasingly positive in the decades to come, that's hardly a done deal. There's plenty of natural variation in the

NAO, both on a yearly basis (the index can slide back and forth several times in a single winter) and from year to year. In fact, the NAO has been trending away from its positive phase since the year 2000. In 2005–06 Europeans saw several spells of the bitter, snowy, negative-NAO weather that had become scarce over the preceding few years. And the winter of 2009–10 brought the most intensely negative NAO on record, along with frigid, snowy conditions to match (see p.128).

Coastal-storm flooding: a deepening problem

Even if coastal storms don't change drastically, the extreme high-water levels they generate should increase. As with tropical cyclones, this is an outgrowth of the overall rise in global sea level. One region where high population meets up with a high risk for surge from coastal storms is the **North Sea**. The Netherlands have spent centuries defending themselves from the sea with a massive series of dykes. Much of coastal England, which sits slightly higher on average, gets along with a variety of dams and other flood-control devices. Both countries found themselves unprepared for the history-making North Sea storm of 1 February 1953, which sent huge surges into the British, Dutch and Belgian coasts. Over 1800 people died in the Netherlands, where a powerful surge overpowered dykes and levees in the lower Rhine Valley. England saw more than 300 deaths on its east coast, including 58 on Canvey Island, from where more than 10,000 residents were safely evacuated.

The 1953 storm triggered a new age of massive efforts to tackle coastal flooding. The Netherlands spent over $5 billion on its **Delta Project**, completed over four decades. Its centrepiece is a pair of doors, each about 20m (66ft) high, that can close off the Rhine if a major coastal storm looms. Along the same lines, the 1953 storm inspired Britain to enhance sea-walls along its east coast and to spend over half a billion pounds on the **Thames Barrier**, which was built between 1972 and 1982. Spanning the Thames in eastern London at a width of 520m (1700ft), it includes ten massive gates, six of which rise from concrete sills in the riverbed and swing to form a single wall.

The Thames Barrier was employed only eleven times in its first decade of service. In 1999, the operating rules were changed, allowing for more frequent closures in order to keep tidal waters from blocking river runoff produced by heavy rain. This change makes it hard to assess how much climate change may have affected the closure rate. For whatever reason, the gates are swinging shut much more frequently than before – a total of 75 times in the decade 2000–09, compared to 35 closures in the 1990s and only four in the 1980s. By the 2030s, the barrier could be closing 30 times in a typical year. The region now has embarked on a hundred-year plan to protect the Thames Valley from the increased surge risk related to climate change (see box below). One of the many options put forth is a far wider barrier across the Thames estuary that would span 16km (10 miles). However, an environmental assessment released in 2009 recommended sticking with the current barrier, making repairs and improvements as needed, until at least the 2070s.

In another well-publicized initiative, Italy is building a set of 79 hinged, inflatable steel gates to protect **Venice** from rising waters. The project, dubbed **MOSE (Modulo Sperimentale Elettromeccanico)** – in a tip of the hat to Moses' parting of the Red Sea – is due to be completed by 2013 at an estimated cost of $4.5 billion. Normally lying flat on the seafloor, the gates will be pulled up as needed in a 30-minute process that would block

Trouble in the Thames

We don't know exactly how people dealt with the Thames flood of AD 9 – the earliest one for which records exist – but we do know that people living near this great British river have faced flood after flood over the succeeding two millennia. Many Brits hoped that the great Thames Barrier would be the end-all of flood control for the region. But in fact that project was only designed to keep London and other Thames cities safe until the year 2030. Thanks to a combination of subsidence and rising sea levels, the water level in the Thames estuary is reportedly increasing by as much as 3cm (1.2in) per decade. The area at risk of flooding includes 1.25 million residents (more than the entire city of Birmingham) plus many visitors, and a frenzy of new construction is luring hundreds of thousands of new residents to the Thames Gateway area. In a 2002 report, the UK Climate Impacts Programme showed that the storm surge one might expect once every fifty years across southeast England could be as much as 1.4m (4.7ft) higher by the 2080s, assuming a high-end estimate for global emissions of greenhouse gases.

With these and other factors in mind, the UK's Environment Agency extended its usual thirty-year window of concern and launched a project called TE2100, an acronym that denotes flood control for the Thames estuary up to the year 2100. The massive six-year effort took the region's climate, environment, land use and social structure into account. The project concluded in a 2009 report that climate change is less likely than previously thought to boost the strength and frequency of storm surge from the North Sea, in line with 2009 projections from the UK Climate Impacts Programme. The worst-case value for high water in the Thames estuary is now expected to be boosted by 2.7m as a result of climate change, as opposed to an earlier value of 4.2m. "We are confident that our plan can cope with a changing climate", said the report. However, it emphasized that storm surge could top the worst-case value if greenhouse gases emissions continue unabated.

Aerial view of the MOSE Project in Venice, Porto di Lido, mid-2009. The controversial project has been criticized for being expensive, experimental, monolithic and potentially obsolete even before it is completed.

three inlets from the Adriatic and help keep the city safe from increasingly threatening surges triggered by Mediterranean storms. Minor floods are now so frequent in this gradually subsiding city that visitors and locals alike often find themselves facing streets ankle-deep with water.

Venice's gates would be employed for any surge expected to exceed 110cm (3.6ft), and they're reportedly capable of handling surges up to 200cm (6.6ft). In 1996 alone, over 100 floods topped the one-metre mark, and a severe 1966 flood reached 180cm (6ft). Alas, it wouldn't take much of an increase in global sea-level rise to push a repeat of that 1966 flood over the new barriers. Some researchers are already pondering the idea of a "Great Wall of Venice" that might eventually surround the city in order to preserve it.

Tornadoes: an overblown connection?

One of the staple ingredients of many books and articles about climate change is a photo of a menacing twister, usually with a caption that goes something like this: "Tornadoes and other forms of extreme weather are expected to increase in a warming climate." Actually, there is no sign that the most violent **tornadoes** are on the increase globally. The future could always surprise us, and an important scientific maxim – "the absence of evidence does not indicate evidence of absence" – may apply here. Still, there's no compelling reason why scientists would expect tornadoes to become much more frequent or intense in the wake of human-induced greenhouse warming.

The popular confusion appears to arise because tornadoes often get lumped under the heading of "extreme weather". It's true that extremes of temperature and precipitation are already on the increase and expected to increase further in a warming climate. But tornadoes are an entirely different animal. Twisters are spawned by **thunderstorms**, so they're dependent on the same warm, moist air masses that lead to thunder and lightning. But that's not enough in itself to produce a tornado. To get the most vicious type of twister – the kind that occur with regularity in only a few parts of the globe, most commonly in east India, Bangladesh and the central and eastern United States – you need a peculiar type of thunderstorm known as a **supercell**. These long-lived storms thrive on a particular blend of conditions, including a layer of dry air about 2000–3000m high (6600–9800ft), as well as winds that increase and shift direction with height (wind shear). Many supercells have an area of rotation, called a **mesocyclone**, from which tornadoes may emerge.

The rarity of all these conditions coming together in one place helps explain why so few parts of the world are prone to violent tornadoes (those with winds topping about 320kph or 200mph). Some are reported in Europe, but generally you need a location where warm, dry air (from areas such as the US Desert Southwest or India's Thar Desert) can easily clash with warm, moist air (from such sources as the Gulf of Mexico or the Bay of Bengal). Since the geographic variables that lead to supercells won't be changing any time soon, it's unlikely we'll see much change in the preferred stomping grounds of violent tornadoes, although it's conceivable that their US range could shift northwards and perhaps affect the southern tier of Canada more often.

Another possibility is that the period considered tornado season (generally early spring in the US South and late spring to summer in the Midwest) will shift a bit earlier in the calendar as the climate warms. This may already be happening: more of the largest US outbreaks have been occurring outside the boundaries of spring from the 1990s onwards (even accounting for the tornado "inflation" discussed below).

There's a bit more uncertainty with **non-mesocyclonic tornadoes** – the far weaker, far more common variety. These tornadoes don't require the exotic blend of ingredients above: as long as the updraught in a thunderstorm is powerful enough, it may be able to produce a weak, short-lived tornado (though very few do). With the advent of video cameras and the growth of weather hobbyism, people have been reporting more and more tornadoes. The 1997 Hollywood blockbuster *Twister* appears to have played no small part in this boom, said the late Nikolai Dotzek of the German Aerospace Center (DLR). Combing through these reports, he and other scientists confirmed that, in many places – including England, Ireland, South Africa and France – tornadoes are more frequent than we once thought.

Whether the global incidence of all types of tornadoes is truly on the increase is difficult to know, but experts believe that most or all of the upward trend in reported twisters is simply because more people are reporting them. Statistics from the US and Europe bear this out. The average yearly tornado count in the US mushroomed from about 600 in the 1950s, just as the nation was implementing its watch and warning system, to 800 in the 1970s and around 1100 in the 1990s, when tornado videos became all the rage. However, US reports of violent tornadoes – the kind that were hard to miss even before storm chasing became common – haven't changed significantly in the entire century-long record, holding firm at around 10–20 per year. As for Europe, the continent as a whole, including the UK, was reporting around 160–180 tornadoes per year, according to a 2003 survey by Dotzek. That number ballooned with the advent of the European Severe Weather Database, a quality-controlled archive launched in 2006. In its first year alone, the database gathered more than 700 reports of tornadoes and waterspouts from across the continent. As was the case in the US, though, weaker tornadoes accounted for most of the jump in reports over prior years. The 2007 IPCC assessment notes the increased number of tornadoes being reported worldwide, acknowledging the issues mentioned above, but it steers clear of any projections on how the localized conditions that spawn tornadoes might change.

Scientists are now using leading-edge computer models to get a better sense of possible global changes in severe thunderstorms, the type that produce gale-force winds and hailstones (and, sometimes, tornadoes). A US–Europe network of researchers led by Harold Brooks of the US National Oceanic and Atmospheric Administration is looking at how well global models reproduce the current patterns of severe local weather. These models are too coarse to pinpoint individual thunderstorms, so Brooks and colleagues are sifting through the global projections and hunting for the ingredients noted above – such as warm, moist air and **wind shear** (winds changing with height). Generally, the models did a good job of reproducing where severe weather happens now.

As for what's to come, the models' clearest signals appear over the storm-prone US, where they point toward an increased prevalence of

Debris flies just after this violent tornado destroyed a farmhouse near Manchester, South Dakota. Thankfully, there is little reason to believe that such twisters will increase as the climate warms.

unstable air but a reduction in wind shear. According to Brooks, this would tend to tilt the balance of US storms away from those that generate violent tornadoes and large hail and toward those more likely to produce localized bouts of strong wind and heavy rain. Both types of storms could qualify as "severe", and the total number of such storms may rise, according to Jeffrey Trapp and colleagues at Purdue University. By implanting regional models inside global models to gain more detail on storm-scale weather, they've found that instability may increase enough in a future US climate to compensate for weaker wind shear. As a result, they say, cities such as Atlanta and New York could see an increase in the number of days each year conducive to severe thunderstorms – perhaps a doubling in some locations – by the late twenty-first century.

Of course, even when conditions are ripe, thunderstorms don't always form. Using a similar technique to gauge the risk for the central and eastern US, NASA's Tony Del Genio and colleagues also project a future with higher instability and weaker wind shear overall. However, their work shows the most favourable zones of instability and wind shear coinciding more often, leading to a jump in the potential frequency of the most intense thunderstorms. If further work supports these results, then some of North America's most populous areas might have to add severe weather to their list of climate-change concerns.

Ecosystems and agriculture
The future of flora, fauna and farming

If all that global warming did was to make life a bit steamier for the people who consume the most fossil fuels, then there'd be a karmic neatness to it. Alas, climate change doesn't keep its multitude of effects so nicely focused. A warming planet is liable to produce a cascade of repercussions for millions of people who have never started up a car or taken a cross-country flight. Many animal and plant species will also pay a price for our prolific burning of fossil fuels.

Food, and the lack of it, could be where a changing climate exerts some of its most troublesome impacts for society. While the changes could affect ranching and grazing as well as arable farming, much of the research to date has focused on croplands. Because of longer dry spells, hotter temperatures and more climatic uncertainty, the next century is likely to see major shifts in the crops sown and grown in various regions. Well-off countries of the North might break even or even benefit from the changes at first, if they can keep a close eye on the process and adapt their agriculture early and efficiently. Sadly, the same may not be true of the tropics, where most of the world's food crops are grown. The most problematic impacts on agriculture may wind up occurring in the poorest countries, those with the least flexibility and the most potential for catastrophic famine. On top of all this, there's global population growth and the accompanying increase in food demand.

Humans use about a third of Earth's land surface for farming and other purposes. What global warming does to the other two thirds of the land – the world's natural ecosystems – could be even more wrenching than the effects on managed lands. In his 2002 book *The Future of*

Life, famed biologist E.O. Wilson warns that, if current trends continue, half of Earth's species could be gone by the year 2100. Pollution, development and other side-effects of civilization have already put uncounted species at risk. Now climate change threatens to make the situation far worse. Here again, the tropics appear to be particularly vulnerable, because their normally muted variations in temperature imply that it would take only a small warming to push organisms into conditions they've never faced before. In a 2009 study, S. Joseph Wright (Smithsonian Tropical Research Institute) studied tropical mammals whose present range is no more than a few dozen kilometres. He found that more than 20% of those species would have to travel at least 1000km (600 miles) in order to get to a place where temperatures by the year 2100 will be comparable to readings they experienced in their habitats in the 1960s.

> "It will be difficult to anticipate future threats to biodiversity and ecosystem dynamics, even if we could know future climate change with perfect accuracy."
>
> Jonathan Overpeck,
> University of Arizona

A landmark study of the extinction risks from climate change, led by Chris Thomas of the University of Leeds and published in *Nature* in 2004, looked at regions that together span a fifth of the planet's land surface. The report found that, by the year 2050, from 15% to 37% of plant and animal species across these areas would be committed to eventual extinction should a middle-of-the-road scenario for increased emissions come to pass. If emissions are on the high side, the range jumps to 21–52%.

The IPCC's 2007 report only reinforced this grim picture. It concluded that some 20–30% of species assessed to date are at increased risk of extinction should global temperatures rise more than 1.5–2.5°C (2.7–4.5°F) above recent values. "Current conservation practices are generally poorly prepared to adapt to this level of change", the IPCC warns. Throw in the risk of climate surprises, and there is every reason to be concerned about the potential impact of greenhouse gases on the life forms around us.

The canaries of climate change

Miners once brought canaries into their dank workplaces to test whether the air was safe to breathe; if the canary died, it was time to get out. Similarly, some creatures can't help but reveal the mark of a shifting climate through their very ability – or inability – to survive.

A golden toad. The species was wiped out by a disease made more prevalent by climate change.

Amphibians and **reptiles** offer some of the most poignant examples of species at risk. As cold-blooded creatures, they must keep themselves within a fairly restricted range of air temperature, and they've got to crawl wherever they go. Those are tough constraints where the climate is warming and where highways and other human construction make it difficult to migrate. For many species of **turtle**, there's another complication: more females than males are hatched as the ambient temperature in their nesting area goes up. If a given species can't migrate quickly enough to avoid regions that are warming, then the gender imbalance could become a threat to survival. This only adds to the many other threats facing beach-nesting turtles, from commercial trade to pollution.

More than 30% of amphibian species were found to be vulnerable, endangered or critically endangered in a 2004 survey, the Global Amphibian Assessment. A 2007 study focusing on the protected rainforest of La Selva Biological Station in Costa Rica found that amphibian populations had dropped by roughly 75% since 1970 across the range of species present there, including many that had previously been considered robust. **Frogs** are among the most threatened species at many locations around the globe. A number of factors are probably involved in their decline, including the effects of pesticides and other chemicals on reproductive success. However, climate has been implicated in several cases of depletion or extinction. The atmosphere sometimes does this dirty work indirectly: climate shifts can help fungi to attack the amphibians more effectively. High atop the US Cascade Mountains, intensified dry spells have been linked to mass die-offs of the **western toad**. The droughts make for shallower ponds, which allows dangerous ultraviolet light to penetrate more fully and weakens the pond-nurtured embryos of western toads. (Ozone depletion may also be allowing slightly more UV light in, although it doesn't appear to be the main culprit in the creatures' demise.)

> "Disease is the bullet killing frogs, but climate change is pulling the trigger."
>
> Alan Pounds, Monteverde Cloud Forest Preserve

Further south, twenty out of fifty species of frog in the **Monteverde Cloud Forest Preserve** in Costa Rica appear to have vanished completely after an unusually warm, dry spring. One special species found nowhere else, the exotically coloured **golden toad,** numbered at least 1500 in the La Niña spring of 1987. Within two years, it could be found no more. Resident scientist Alan Pounds suspected that climate was involved, but the link wasn't crystal-clear: the fungus believed to have killed the embryonic toads, **chytridiomycosis,** tends to prosper in cool, moist weather, yet the Monteverde Forest has been warming overall. It took until 2005 before Pounds had solved the puzzle. As is the case in many parts of the world, the bulk of the warming at Monteverde was occurring at night, whereas days had actually cooled slightly (probably due to increased cloud cover). The reduced day-to-night range kept conditions closer to optimal for the villainous chytridiomycosis.

Butterflies are another excellent indicator of climate change on the march, partly because these widespread and aesthetically pleasing creatures have been carefully studied for centuries, especially in Britain. Both drought and flood can cause a butterfly population to crash, as happened to 5 of 21 species studied during a dry spell in the mid-1970s across California. A 2010 study found that many California butterflies are moving

The Apollo – one of the many butterfly species threatened by warming.

upward as the climate warms, with an average elevation gain per species of nearly 100m (330ft) over a 20–30 year period along the west slope of the Sierra Nevada. The **Apollo butterfly,** whose translucent wings are often seen across high-altitude areas in Europe, is no longer found at elevations below 838m (2749ft) in France's Jura mountains. Although they're skilled fliers, Apollo are finding themselves with less and less liveable terrain as they're forced to climb in elevation. A major survey of European butterflies in 2009 found that dozens of species will be unsustainable in their present locations if the warming expected by late this century comes to pass. "Most species will have to shift their distribution radically to keep pace with the changes", said chief author Josef Settele (Helmholtz Centre for Environmental Research).

Even a creature as humble as the **dormouse** can tell us something about global warming. Made famous by Lewis Carroll's *Alice in Wonderland*, this tiny, nocturnal mammal now ranges only about half as far across Britain as it did in Carroll's time. Part of this is due to the fragmentation of the trees and hedgerows where it likes to live. However, mild winters interfere with its hibernation patterns, and warm summers cause additional stress. Were it not for so much civilization in the way, the dormouse might easily be able to migrate north to escape the warm-up.

Trouble also brews when interdependent species respond in differing ways to a warming atmosphere. Those who use temperature as their cue to breed or migrate may shift their timing as warming unfolds, while those whose behaviour is driven more by the waxing and waning of sunlight may not. At the Netherlands Institute for Ecology, Marcel Visser has been studying disrupted synchrony – the mismatched timing that can occur when creatures find that a food source has departed from its usual seasonal schedule. In studying a bird called the Dutch great tit, for example, Visser and colleagues found that the birds' egg-laying behaviour hasn't kept up with the earlier springtime emergence of caterpillars, its main food source. Similarly, in California, a Stanford University team found that the progressively earlier seasonal die-off of plantains on a nearby ridge left the bay checkerspot caterpillar without its favourite food. The butterfly eventually disappeared from the ridge. Such timing mismatches could also threaten the relationship between plants and the insects that pollinate them. NASA is working to relate the amount of nectar in beehives (easily measured by weighing the hives) to the satellite-detected timing of vegetation. The agency's HoneyBeeNet reports that the peak nectar flow in Maryland hives now occurs nearly a month earlier than in the 1970s.

Do examples like the ones above add up to a **global shift**? In an extensive survey published in *Nature* in 2003, Camille Parmesan of the University of Texas and Gary Yohe of Wesleyan University found that 279 out of 677 species examined in large-scale, long-term, multi-species research showed the telltale signs of human-induced climate change. On average, these species had moved roughly 6km (3.7 miles) northward, or about 6m (20ft) up in elevation, per decade. One of the most telling clues was the "sign switching" of eight butterfly species in North America and Europe. All of them moved northwards during the warm period of the early twentieth century, shifted back south during the mid-century cool-down (see graph, inside front cover), and then headed north again most recently. None of the butterfly species contradicted this pattern. Further evidence came with an exhaustive global survey in 2009, led by NASA's Cynthia Rosenzweig and published in *Nature*, of more than 29,500 data sets that each showed a significant physical or biological change related to temperature. About 90% of the biological changes were in the direction one would expect from a warming climate.

Yohe, an economist, remained somewhat sceptical at first. Even after his own paper with Parmesan was published, he refused to claim 95% certainty that human-induced climate change was affecting ecosystems on the large scale. But in 2006, after more research had accumulated, Yohe came around: "I now feel that we have crossed the 95% confidence threshold. We can conclude with very high confidence that we have detected the fingerprint of climate change."

The big squeeze

Simply keeping up with a changing climate will be a challenge for many animals and plants. It's hard to overstate how quickly the chemical make-up of the atmosphere is changing relative to Earth's history. The unprecedented increase in greenhouse gases translates into a warming projected for the next century that may seem gradual to us but will be warp-speed by geological standards (see "The long view", p.212).

In the Arctic, where the range of life forms that can survive the harsh environment is relatively narrow, it's easy to identify the animals most at risk, chief among them **polar bears** (see p.89). Elsewhere around the planet, ecologists are keeping a close eye on **biodiversity hotspots**, where climate and geography team up to nurture a vast variety of species – far more than one might expect in such small areas. The 25 hotspots identified by researchers are scattered throughout the tropics and mid-latitudes,

Up, up and away?

For animals and plants in mountain homes that have grown too warm for comfort, there's nowhere to go but up – and that's not easy when you're already near the top. Some small mammals are at particular risk. For example, **pikas** – the endearing, hamster-sized creatures familiar to hikers across the high country of the US West – have dwindled in number across both North America and Asia. With a metabolism that can't handle air much warmer than room temperature, pikas typically hide below rocks and ice to cool off, and they can't migrate easily. The US Geological Survey reports that American pikas have vanished from more than half of 25 traditional haunts in the US Great Basin since the early twentieth century. It's unclear whether heat has killed them directly or indirectly (by cutting down on their foraging hours), or whether other factors might be involved, such as a lack of insulating winter snow. In Australia, a comparable victim is the mountain pygmy possum, whose numbers have dropped as bush rats and feral cats move into its terrain. The pika's decline prompted the US Fish and Wildlife Service to investigate whether it should be declared a threatened or endangered species – which would have been the first such declaration tied to climate change in the lower 48 US states. In 2010, though, the agency concluded that pikas still had enough high-altitude habitat to remain viable as a species.

When it comes to **plants**, the story includes both winners and losers. The losers are those attuned to the highest, most rarefied conditions: as temperatures continue to climb, many of these species seem fated to die off. In Europe, high-mountain territory covers just 3% of the landscape but accounts for some 20% of all native plants. However, as in the Arctic, a warming climate could produce a larger zone of plant-friendly territory overall. Treelines have risen some 150m (500ft) over the last century in Sweden's Scandes Mountains and up to 80m (260ft) in Russia's Ural Mountains. Swedish scientists were surprised to find large herbs at heights more than 100m (330ft) above their 1950s range, popping up in moraines only recently left behind by retreating glaciers. Some grasses may prove even more opportunistic, providing tough competition for other cold-adapted alpine plants. The potential

largely on coastlines and mountain ridges. The Monteverde Cloud Forest already mentioned is part of one such example. Most hotspots have large contrasts in temperature and moisture across small areas, which enables many different types of flora and fauna to find their niches. But the flip side of this concentrated contrast is that a seemingly small shift in climate could have major effects on plants and animals accustomed to a very specific environment. And because many hotspots are so localized – often surrounded by more uniform, human-dominated landscapes – plants and animals may not be able to easily shift their ranges outward.

What about the rest of the tropics and mid-latitudes? These areas are often classified by geographers into **biomes**, regions that share similar

American pika

"grassification" of high-altitude regions – hinted at by computer models, especially across the Alps – is a topic of keen research interest.

Beyond anecdotal reports, there's a pressing need for a more thorough survey of flora and fauna in a changing high-altitude world. That type of census may soon emerge from GLORIA (Global Observation in Alpine Environments). Launched by the Austrian government with support from the European Union, this network of observing sites now includes more than 77 alpine observing sites in Europe, the Americas, Asia, Australia and New Zealand.

As climate change pushes species upward, it may also open doors on plant and animal life of the past. Two tourists in 1991 discovered Ötzi, the Alps' now-famous 5300-year-old man, in a receding glacier. On a survey of Quelccaya in 2004, glaciologist Lonnie Thompson (see p.103) found a small, shrubby plant that he expected might have been around 5000 years old. Carbon dating soon proved it to be a moss that had been under ice for most of the last 50,000 years – and probably twice that long, making it a likely remnant of life before the last ice age.

traits of climate and landscape. Among the commonly used groupings for land areas are:

▶ **Forests** Covering roughly a third of the planet, these include the great boreal forests of Canada, Alaska and Russia; the evergreen and deciduous forests of mid-latitudes; and the tropical forests that support amazing biodiversity despite nutrient-poor soil.

▶ **Grasslands** The savannah that covers much of Africa and Australia and the temperate grasses that sprawl across the mid-sections of North America and Eurasia.

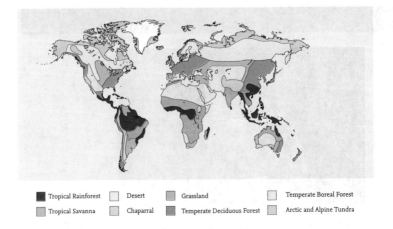

| | Tropical Rainforest | | Desert | | Grassland | | Temperate Boreal Forest |
| | Tropical Savanna | | Chaparral | | Temperate Deciduous Forest | | Arctic and Alpine Tundra |

▶ **Deserts** About a fifth of Earth's land surface. Many of the largest deserts are clustered in the zones around 30°N and 30°S, where air tends to sink and dry out. Most of Antarctica is also considered desert.

▶ **Tundra** The cold lands of high altitudes and the Arctic that support only shallow-rooted, low-rising plants.

The edges of many of these biomes are shifting already. Some studies in Europe have shown a northward push of up to 120km (75 miles) in the north edge of certain biomes over the last century. The tropics themselves – as demarcated by high-level jet streams – widened by about 225km (140 miles) between 1979 and 2005, according to satellite data analysed at the universities of Washington and Utah. The study suggests a polewards push of subtropical deserts. Likewise, one of the largest-scale shifts expected over the next hundred years is the expansion of **boreal forest** polewards into fast-thawing tundra and, in turn, the transformation of the southern edge of some boreal forests into deciduous forests or grasslands. But forests take time to establish themselves, and climate may outrun the process, leaving vast areas in transition until soils and plants can adapt to a new climate regime. To make matters worse, any climatic state we envision for the future is itself a moving target, since the greenhouse gases we're adding to the air now will be warming the climate for a century or more. Will new forests be able to take root in these ever-shifting conditions?

As for land-based creatures in the tropics and mid-latitudes, it might seem as if they can easily migrate poleward and upward as needed. But most of the planet's terrain outside hotspots and other protected areas is pockmarked with towns and cities and laced with highways, cultivated fields and other potential barriers to migration. Those creatures adapted to higher altitudes, like the pika (see p.168), may find themselves stranded on island-like patches of terrain, where climate will only constrict their range further over time.

If the exotic creatures of hotspots and the Arctic are the biggest losers in climate change, the biggest ecological winners may be the most humdrum. Animals that are common and robust could spread even further, such as the various species of deer now thriving across rural and semi-rural parts of North America and Eurasia.

A landscape of shifting climate may also offer a nearly ideal platform from which non-native plants, animals and insects can spread like wildfire. In this era of mass global travel, it's easier than ever for **invasive species** to tag along and then move pole-ward with warming temperatures. Kudzu, a vine that grows as quickly as 0.3m (1ft) per day, is a classic example. It was brought from Asia to Philadelphia in 1876 for the nation's centennial exposition, then used for decades across the US South to stabilize roadsides and feed animals before its invasive qualities were recognized. Long since out of control, kudzu has been spreading steadily northward: in 2009 Canada's first-ever kudzu colony was found on the north shore of Lake Erie.

Climate change could make particular landscapes more vulnerable to invaders than they might be otherwise. The most successful invasive species are often less picky than native species, able to tolerate a wide range of conditions, including heat and/or drought. One famous example is **tumbleweed**, or Russian thistle. Introduced to the US Great Plains in the 1870s, tumbleweed quickly spread across the US frontier, eventually becoming a symbol of its new home. Similarly, **rabbits** multiplied like, well, rabbits after 29 of them were brought to Australia in 1859 by rancher Thomas Austin. Today, the United States is dealing with the destructive and sometimes deadly **fire ant** (see p.173), a major invasive pest that's ravaged both the countryside and populated areas and even made its way to the Southern Hemisphere. When invasive flora and fauna are temperature-limited, a warming climate should only enhance their ability to spread.

For details on the effects of climate on coral reefs and other marine creatures, see p.138.

Bugged by a changing climate

It may be **insects** – adaptable creatures *par excellence* – that take the best advantage of changing climate in the coming century. Many bugs respond strongly to small shifts in temperature, rainfall and humidity. The overall warming now under way may already be affecting insect distribution, and future warming should favour the spread of a number of species. In 2006, a spider-eating French wasp showed up in Sandy, Bedfordshire, the first of its species ever observed in Britain. On a more troublesome note, a 2002 review by Drew Harvell and colleagues at Cornell University found that **parasitic diseases** are likely to become more widespread and/or severe as the planet warms and moistens.

History gives us a sneak preview of what can happen. The so-called Medieval Warm Period (see p.239) – the several centuries of mild weather across much of the Northern Hemisphere between about AD 950 and 1300 – may have helped rats and fleas to thrive across Europe, paving the way for the **Black Death** that killed a third of Europeans in 1347–52.

More than two billion people live within the range of mosquitoes that transmit the parasites that cause **malaria**. These parasites infect more than 250 million people and take one to three million lives each year. It's clear that any rise in infection rates would carry a huge toll in terms of human suffering. Yet it's been difficult for researchers to nail down exactly how climate change will affect malaria, given the variety of other factors that shape the course of the disease's spread.

For more than 30 years, malaria has been moving into previously untouched parts of Africa's central highlands, possibly due to a scarcity of cool, mosquito-killing temperatures. It was long assumed that warming

Global warming will shift the range of malaria-carrying mosquitoes.

March of the fire ants

It sounds like Hollywood horror at its most lurid: a voracious, supercharged pack of bright orange bugs chewing their way across the US. But it's happening not on the silver screen but in real life, and a warming climate isn't helping matters. The red imported fire ant, which arrived in the United States from South America in the 1930s, has conquered large swathes of the South, from North Carolina to Texas. Since the 1960s the ants have expanded their territory each year by an area the size of New Hampshire, gradually pushing north and west towards their climatological limits.

The ants have many qualities going for them. Multiple queens can team up to help launch a colony (though workers eventually kill all but one of them). The ants reproduce like mad, often pushing out native ants completely and preying on bird hatchlings. They'll also attack mammals, and have even killed a few people, most of whom have been unlucky enough to fall on or near a nest due to an accident or injury. The ants can even survive floods: one of the surreal spectacles from Hurricane Katrina was the sight of fire ants floating atop the waters in dense clots, a technique they can sustain for up to two weeks.

Temperatures much below -12°C (10°F) spell doom for fire ants, but the gradual warming of winter nights could allow their northward US spread to continue this century. They're already expected to head west into California and up that state's coast. One projection shows that if carbon dioxide increases by 1% a year over the next century, the ant could increase its US range by over 20%, populating a belt from Oklahoma to Delaware. In the meantime, the ants have been discovered near Brisbane, Australia (in 2001), and in New Zealand (in 2004). Thus far, they seem to be thriving in their new homes, according to a 2005 study led by Robert Sutherst of Australia's Commonwealth Scientific and Research Organization. He says the fire ant "has the potential to impact widely on agriculture and the native fauna of Australia."

temperatures would generally expand the range of malaria in such a way, thus widening the swath of humanity threatened. But recent models suggest that the range could actually shift rather than expand, as discussed in a 2009 study in the journal *Ecology* by Kevin Lafferty (US Geological Survey). Intensified drought, for example, may render some areas unable to sustain malaria-carrying mosquitoes even as the disease moves into other areas, such as African highlands, that were once malaria-free. The 2007 IPCC report noted that short-term variations in rainfall are a huge factor in malarial risk across many regions.

It's also possible that the influence of global warming on malaria will be outweighed by other factors, as noted by Lafferty and by Peter Gething (University of Oxford) in a 2010 *Nature* paper. The Malaria Atlas Project found that the disease's range encompassed 58% of Earth's land area in the year 1900 but only 30% in 2007. Much of the progress is likely due to insecticides, vaccines, mosquito nets and other advances in prevention. In places like Europe and the United States, where malaria was endemic for centuries, it's considered unlikely to make a comeback – even with warmer temperatures – as long as control measures and public-health systems remain strong. As reassuring as this progress might seem, mosquito populations are booming across the global South where malaria is most prevalent. And as noted above, warming temperatures may be helping the virus spread into areas where people aren't accustomed to it and may not have the resources to deal with it. Together, these factors pose vast challenges to public health efforts in chronically poor areas.

Dengue fever – caused by a group of four potentially fatal, mosquito-borne viruses – is already on the increase. Around two fifths of the world's population now live in affected areas, according to the World Health Organisation, which announced in 2010 that Asia was experiencing a sharp rise in dengue cases. The 2007 IPCC report labels it the world's most important vector-borne viral disease. As many as 100 million people are infected with dengue each year, of which around 500,000 need to be hospitalized. Children account for a large proportion of these serious cases. One study estimates that a 1°C (1.8°F) rise in average global temperature could raise the number of people at risk of dengue by up to 47%, but the complex factors that influence transmission make it hard to know if the disease would actually spread that quickly. Drought may be an important cofactor: it can increase the risk of dengue fever if it spurs people to hoard water in places where mosquitoes like to breed. Researchers are now examining the factors, including climate change, that might help dengue to spread from lower elevations surrounding Mexico City into the high-altitude urban core, where it would threaten close to 20 million people.

Unusual summertime warmth and drought helped another mosquito-borne illness make quick inroads into North America. The West Nile virus first showed up on the continent around New York City in 1999. Its mosquito host thrives in hot, dry weather, as stagnant pools of water draw infected birds and the mosquitoes that feed on them. The virus spread regionally at first and then, during the record-breaking heat of 2002, swept across much of the United States and into Canada. Up to 2005 it had killed more than one thousand North Americans. By the end of the decade, though, the numbers had declined rapidly, perhaps because birds were becoming immune to the virus or possibly because fewer humans with low-grade infections were being tested and counted.

Insects can also cause trouble for people indirectly, by attacking the food we eat, for instance. The Colorado beetle, a striped, centimetre-long US native, is a serious threat to potato crops in Europe. Each year a few of these beetles hitchhike into the UK, but to date the bug has failed to establish a foothold, thanks to vigilant inspections, public awareness efforts and eradication campaigns by the UK government. However, a study by Richard Baker of the UK Central Science Laboratory found that if Europe were to warm by an average of 2.3°C (4.1°F) by 2050, this pest could more than double its potential range in Great Britain, putting virtually all potato-farming areas at risk.

Shrinking forests

As we've seen, rainforests and their loss can have a huge impact on the global climate (see p.11), but how does global warming affect the rainforest itself? That's not so clear. In the 2009 study noted above, S. Joseph Wright found that the IPCC's mid-range emissions scenario could push 75% of the world's tropical forests into average annual temperatures higher than any now observed in closed-canopy forests. On a seasonal scale, it's precipitation more than temperature that drives the rainforest, so anything that affects when and how hard it rains can have a major impact. El Niño tends to produce drought across much of the Amazon, so a key factor will be how El Niño evolves in the future – still an open question, although the recent trend has been in favour of more and/or stronger El Niños (see p.130). A sharp warming across the Atlantic has also been identified as a likely factor in the Amazon's severe drought of 2005. But the various global climate models considered in the 2007 IPCC report aren't yet in agreement on whether a sustained drying is on the cards for the Amazon. If climate change should push the Amazon towards more drought, the trend could be exacerbated by land clearing, which causes local-scale drying due to loss of the moisture-trapping forest canopy. The fear is that these and other positive feedbacks could lead to massive rainforest loss across the Amazon, although that scenario is far from a given. Models that track climate as well as the interactions of species within a rainforest canopy are in their infancy, but they may soon help clarify how tropical forests as a complete ecosystem stand to deal with drought.

There's less uncertainty about what climate is doing to forests further towards the poles. Mountain and boreal forests are one of the largest ecosystems already being strained by a rapidly changing climate. Permafrost melt (see p.92) is one big source of high-latitude forest stress, but there are other major factors. Across western North America, the intersection of pests and a warming atmosphere has led to one of the world's largest **forest die-offs**. A series of major droughts and warm winters since the 1990s from Mexico to Alaska has transformed the landscape across huge swathes, destroying millions of hectares of forest through **beetle invasions** and **forest fires**. The extent of both is unprecedented in modern records. In 2002 alone, British Columbia lost 100,000 square km (39,000 square miles) of lodgepole pines to fire and disease. That's more than enough forested land to cover Belgium and the Netherlands. Half of British Columbia's pines are now gone, according to the Canadian government, and while the epidemic has peaked in some areas, it continues to spread in others.

The main insect behind this devastation is the **bark beetle**, which comes in various species across North America. This hardy creature burrows into trees, lays its eggs, then girdles the tree from within, blocking the nutrient flows between roots and branches. Climate has given North America's bark beetles carte blanche for explosive growth in recent decades. The spruce

Trees in the Rocky Mountains near Granby, Colorado, show the effects of an attack by the pine bark beetle.

bark beetle, one of the most common varieties in colder forests, normally requires two years to reach maturity, but an extremely warm summer can allow a generation to mature in a single year. Conversely, it takes two bitterly cold winters in a row to keep the population in check. The combination of milder winters and warmer summers since the early 1990s has allowed the beetle to run rampant.

Another variety, **the mountain pine beetle** – once limited to lower-elevation tree species such as lodgepole and ponderosa – has moved uphill as temperatures warm. It's now invading whitebark pine in Utah at elevations up to 3000m (10,000ft). "The outbreak has progressed at a truly astounding rate", says Jesse Logan of the US Forest Service. In Colorado and Wyoming, the beetles' territory had grown by 2010 to encompass 14,000 square km (5400 square miles), doubling its area in less than three years. Huge swathes of dying forest cover some of the region's most famous scenery, and hikers keep a wary eye for dead trees that could tumble into campgrounds and onto trails. In Canada, the beetles now threaten to invade **jack pine**, a critical move that could allow the insects to cross the continent and lead to massive infestations across the forests of eastern North America.

Drought weakens the ability of trees to fight off these aggressive bugs, because the dryness means a weaker sap flow that's easier for beetles to push through. Across parts of New Mexico, Arizona, Colorado and Utah, where a record-setting drought took hold in 2002, pine beetles and other drought-related stresses have killed up to 90% of the native **piñon** trees, turning the needles a sickly red before they fall off. A drought in the 1950s had already killed many century-old piñon, but the more recent drought produced "nearly complete tree mortality across many size and age classes", according to a 2005 study led by David Breshears of the University of Arizona. Even assuming the climate supports piñon regrowth, it will take decades to re-establish the landscape. Breshears sees the event as a sign that climate-driven landscape change could be far more rapid and widespread than even the experts expect.

Another destructive invader, the **spruce budworm**, has carved new territory for itself as the climate has warmed. It's now chewing up buds and needles on spruce trees as far north as Fairbanks, Alaska, where it first took hold in the early 1990s. Tree rings show no sign of any previous outbreaks in that area since the mid-1800s, and the Little ice age reigned for hundreds of years before that. In the decades to come, various budworm species – all of which benefit from warmer, drier summers – could ravage large stretches of boreal forest across Alaska and Canada.

A water-dropping helicopter attacks a wildfire in Modjeska Canyon, Orange County, California, October 2007. Such fires are expected to become more frequent and severe as the planet warms.

Certainly, both insects and fire are part of the natural ecosystems of North America. Some species of trees actually rely upon fire to propagate. Jack pine seeds are tucked into tightly sealed resinous cones that only open when a fire comes along and melts the resin, which then allows the seeds to spread and sprout in the newly cleared landscape. The question is how climate change might amplify the natural process of forest fires, especially across the high latitudes of North America and Asia. For example, some studies project an increase of up to 100% by the end of this century in the amount of land consumed by forest fire each year across Canada. These fires are a major source of pollution, and they pack a double whammy on the carbon budget: the fires themselves add significant amounts of carbon dioxide to the air (not to mention carbon monoxide, methane and ozone-producing oxides of nitrogen), and the loss of trees means less CO_2 being soaked up from the atmosphere (at least until new growth occurs).

As for the beetles and budworms, most major infestations tend to run their course after a few years, followed by new epidemics a few decades later as the next crop of tree victims matures. But nobody knows how that natural ebb and flow will intersect with the climate-driven shifting of the insects' potential range across the US and Canadian West. And it's not out of the question that a decades-long **megadrought** could set in. It's happened before, at least across the US Southwest, and a 2007 study led by Richard Seager of the Lamont-Doherty Earth Observatory warns that by 2050 that region may be in the grip of a warming-driven drought that's more or less permanent. Much of the swath from New Mexico to California has been struggling with on-and-off drought since 1999. The period from July 2006 through June 2007 was Los Angeles' driest in more than a century of record-keeping, with a mere 82mm (3.21in) of rain. That set the stage for catastrophic wildfires in and around LA that autumn. And by late 2010, Lake Mead, which slakes the thirst of the growing Las Vegas area, had reached its lowest point since the lake was created in the late 1930s, running less than 40% full.

All of these issues complicate long-running debates on how best to manage ecosystems across western North America. The overarching question is how to allow natural processes to play out while fending off the kind of large-scale, fire- or bug-produced devastation that could permanently alter the landscape.

The northward flow of maple syrup

The shift of a forest can be just as significant psychologically and economically as it is biologically. Take maples, for example. They're so closely identified with Canada that they're the centrepiece of the national flag, but for generations of Americans, it's New England that's synonymous with maple syrup. This piece of Americana appears to be in jeopardy, according to the US National Assessment of Climate Change. To produce the best syrup, maples need a series of freezing nights and milder days, together with a few prolonged cold snaps in late winter. But since the 1980s, winters across New England haven't lived up to their past performance. There have been fewer stretches of bitter cold, and nights are staying above freezing more often, even in midwinter.

Further north, maple production in Quebec was long limited by the lack of daytime thaws, as well as by the deep, sustained snow cover that kept maple harvesters from their trees. Now, improved technology allows the Quebecois to gather syrup more easily in deep snow, and the climate appears to be shifting in the Canadians' favour as well. Not only is maple production moving out of New England, but the long-term survival of maple trees themselves is in question across much of the US Northeast, according to the National Assessment. That would be a potential hit to the spectacular autumn foliage that attracts thousands of tourists each year.

Perhaps New England's best hope of keeping its stunning foliage, and its syrup, is the chance of a shift in North Atlantic currents (see p.131) that could chill the region – or at least keep it from warming as much as it otherwise might. But the chances of this happening are slim.

Crops and climate: a growing concern

In recent years, a variety of factors (no doubt including the influence of factory farming) has led to an enormous global reliance on a tiny number of staple crops. Out of some 50,000 edible plant species, a mere three – wheat, rice and maize (corn) – make up more than half of the world's current food supply. To deal with the demands of climate change, especially over the long haul, food producers around the world will need to demonstrate flexibility in the timing, location and nature of their planting. That won't be easy when there is no single, fixed "new" climate to

Will greenhouse gases boost plant growth?

Scientists have long recognized one potentially positive side effect of fossil fuel use: the stimulation of forests, crops and other vegetation by the carbon dioxide we're adding to the atmosphere. Initial work in the 1980s suggested that a doubling of CO_2 could increase crop yields by as much as 35%. However, much of the initial work involved plants in open-top chambers, which allow fresh air in through the top but don't fully replicate how plants would grow outdoors. A more true-to-life model would be to add CO_2 to the air around vegetation in a natural setting and see what happens. This is the idea behind a series of projects called Free Air CO_2 Enrichment (FACE). More than a dozen large-scale FACE sites have been established in the US, Switzerland, New Zealand, Germany, Italy and Japan since the early 1990s. At each site, a circle of vent pipes injects CO_2 into the atmosphere surrounding a plot of land 8–30m (26–100ft) in diameter. Through computer-controlled settings, the CO_2 levels can be adjusted for winds and other factors.

In a 2005 review of more than a hundred FACE studies, Elizabeth Ainsworth and Stephen Long of the University of Illinois found that the crop benefits from enhanced CO_2 fell short of expectations. For example, in studies where CO_2 was increased over present-day levels by about 50%, the yields rose by about 7% for rice crops and 8% for wheat. This is only about half the effect one would have expected by extrapolating from the doubled-CO_2 levels in the earlier chamber studies, according to Ainsworth and Long. Other FACE studies indicate that the benefits may tail off for at least some crops as CO_2 levels continue to increase. Unlike crops, trees have fared better than predicted in the FACE studies. Ainsworth and Long found that the extra CO_2 caused the total amount of dry biomass produced above ground by trees to increase by an average of 28%.

Perhaps the biggest bonus from enhanced CO_2 could be an increase in the ability of some plants to deal with drought. Ainsworth and Long found that when conditions were dry, the boosted CO_2 levels improved yields by an average of 27%; during wet conditions, the extra CO_2 had no significant effect. One possible reason for this is that the extra CO_2 constricts a plant's pores, which also helps it retain moisture.

plan for, but a climate in constant flux. The challenges will be especially great for subsistence farmers whose options are restricted even in the best of times.

The complex effects of climate change on plants are often reduced to a simplistic, and misleading, equation: plants take in carbon dioxide and use it to grow, so increasing the amount of CO_2 in the air can only be a good thing. One of the more prominent US organizations stressing the potential benefits of a world with more carbon dioxide is the Greening Earth Society. Since the 1980s, they've promoted the view that the extra CO_2 in the air will help crops and other vegetation to grow more vigor-

In order to take advantage of higher CO_2 levels, plants will need to draw on other nutrients to support their increased growth. That's possible but by no means certain. Nitrogen is a particular question mark. It's a critical part of industrialized agriculture: crops are fertilized with it so commonly, and so inefficiently, that most of the nitrogen bypasses the crop and enters the atmosphere, where it can travel thousands of kilometres before it's taken up by the soil. Nitrogen is also generated by microbes within the soil itself. It's unclear whether these two sources will provide enough nitrogen to allow natural vegetation to thrive on the extra carbon dioxide.

One of the largest and longest-running FACE studies suggests that, for grasslands at least, the extra nitrogen would be critical in order for plants to take advantage of any benefit from enhanced CO_2. Scientists from five US universities reported on the study in the journal *Nature* in 2006.. On a Minnesota prairie, the team studied six plots that contained sixteen native or naturalized types of legumes and grasses. Along with boosting CO_2 levels, they added extra nitrogen to some of the plots. The plots without enriched nitrogen fell progressively further behind their nitrogen-enriched counterparts in making use of the extra CO_2.

FACE study team leader Bruce Kimball adjusts wind sensors used to control the release of CO_2 over wheat plots near Phoenix, Arizona.

ously. Moreover, as they point out, the tendency of global warming to be most pronounced in the winter and at night should only lengthen growing seasons on average across much of the world. In 2010, the Maryland-based Smithsonian Environmental Research Center reported that a set of 55 plots of trees monitored there for 22 years grew at up to twice the expected rate.

The fertilization effect of carbon dioxide has been underscored in many climate reviews, including the IPCC's, and is supported by research in the laboratory and in the field, albeit with a few caveats (see box,, pp.180-1). The added CO_2 triggers a shrinkage of the pores on plant leaves, allowing them to retain moisture more easily. In addition, photosynthesis rates speed up for so-called C_3 plants, including wheat and rice. In experiments published in 2008, the yields of C_3 crops increased by an average of 14% when carbon dioxide was boosted to 590ppm compared to 370ppm. (Livestock that graze on C_3 grasses, which are mainly at higher latitudes, could also benefit.) In contrast, the extra CO_2 provides little benefit for the other main type of edible plant, C_4 (including maize), since it doesn't get the photosynthetic boost that C_3 plants do.

As for warming itself, scientists expect a hotter climate to enhance the overall potential for cereals and some other crops across large parts of northern Europe, Russia, China, Canada and other higher-latitude regions. This should lead to an overall expansion of the world's land areas favourable for agriculture. However, that alone doesn't guarantee tomorrow's plants will be just as healthy as today's. When crops mature more quickly – as they tend to do in a warmer, CO_2-heavy atmosphere – they generally accumulate less biomass. Thus, unless a second crop is planted (or the farmer switches to different crops), the result can be a net drop in seasonal production. Not only does biomass tend to go down with the extra CO_2, but it appears that nutrients may become more scant as well. A number of studies to date show that CO_2-enhanced crops can be significantly depleted in zinc, magnesium, or other micronutrients, perhaps because there aren't enough trace elements from the soil entering the plant to keep up with the photosynthesis boost from CO_2.

Among non-agricultural plants, woody vines such as kudzu (see p.171) appear to thrive on the extra CO_2. In part, it's because their clingy mode of growth allows the plant to devote more energy to photosynthesis as opposed to building a structure to keep it standing. A 2006 study at Duke University found that one noxious vine – poison ivy, which sends over 350,000 Americans to doctors each year – grew at three times its usual pace with doubled CO_2. To make matters worse, the ivy's itch-producing chemical became even more toxic than before.

The most recent analyses have brought down the overall agricultural benefit we can expect from carbon dioxide increases, in part because the negative influence from **near-surface ozone** will be larger than scientists once thought. One of the open-air studies carried out as part of the Free Air CO_2 Enrichment programme (see box, pp.180–1) tested the effects of enhanced carbon dioxide and ozone on soybean fields. A doubling of atmospheric CO_2 boosted soy yields by about 15%, but that benefit went away when ozone levels were increased by as little as 20%. Another open-air study, this one in Wisconsin, obtained similar results for aspen, maple and birch trees. Ozone may also cut the nutritional value of crops, as suggested in a 2007 European study. (To put a positive spin on the matter, some US studies have shown that reducing low-level ozone by less than half could preserve billions of dollars in crop value.)

Climate change itself will have an impact on crops, of course, some of it favourable. Modelling that fed into the 2007 IPCC assessment shows how **frost days** (the number of 24-hour periods when temperatures drop below freezing) might evolve over the next century. A study by Gerald Meehl and Claudia Tebaldi of the US National Center for Atmospheric Research found that the number of frost days should drop in many parts of the world, most dramatically across far western North America and northwest Europe. This is due to an expected shift in the jet stream that would pull more warm marine air onto these coastlines, helping to keep more nights above freezing. Parts of the western United States could see ten or more fewer days with frost per year by the 2080s, if the model rings true. That, in turn, should lead to a somewhat longer growing season. In fact, some of the modelling for the 2007 IPCC report supports a boost of one to three days in growing-season length across most of North America and Eurasia, although it's the year-to-year timing of frost days – a hard-to-model quantity – that matters most.

Winners and losers in farming

A battle royal is setting up for the coming decades across the world's farmlands, where the benefits of extra CO_2 and longer growing seasons will fight it out with intensified drought, spikes of extreme heat and other negatives. It looks as if the balance will have shifted toward the negative by the time mid-latitude land temperatures exceed pre-industrial values by around 3°C (5.4°F), which corresponds roughly to a global rise of around 2°C (3.6°F). One reason is that the fertilization effect of CO_2 for crops like wheat and rice tends to decrease once CO_2 is boosted beyond a certain point. Another is that most of the world's food crops are grown in the tropics. In many of these areas, dry periods are expected to grow more intense and/or more long-lasting, interspersed with periods of intensified rainfall (a paradox explored on p.65). These heightened contrasts are stressful for many crops. For example, the **groundnut** – a staple across western and southern India – produces far less fruit when temperatures consistently top 35°C (95°F) over a few days. Many other crops have similar temperature thresholds. **Wheat** that's exposed to 30°C (86°F) for more than eight hours produces less grain, and **rice pollen** become sterile after only an hour of 35°C heat.

One major study commissioned by the United Nations for 2002's World Summit on Sustainable Development compared the relative winners and losers in agriculture for a mid-range scenario of global emissions increase by the 2080s. Among the findings:

▶ **Losses in the tropics** Between 42 and 73 countries, many of them in Africa and Asia, could experience declines of at least 5% in their potential to grow cereal crops. Between one and three billion people would be living in countries that could lose 10–20% of their cereal-crop potential.

▶ **Gains in the north** In contrast, most of the world's developed countries would experience an increase in cereal productivity of 3–10%.

▶ **Agricultural GDP** In terms of agricultural gross domestic product, the biggest winners are likely to be North America (a 3–13% increase) and the former Soviet Union (up by 23%). By contrast, Africa could lose 2–9% of its agricultural GDP.

While these projections look rosy for some areas, they don't necessarily factor in the potentially destructive role of extremes, such as the agricultural damage produced by Europe's 2003 heat wave. Years such as these are more likely to contain days with higher temperature and lower water availability, with negative impacts on a variety of crops. In its 2007 report, the IPCC stressed that recent work has "highlighted the possibility for negative surprises, in addition to the impacts of mean climate change alone". Some of the benefits projected for developed areas could be offset by thirsty soils and a resulting spike in irrigation demand, as seen in Australia's recent drought. Even if the world is lucky enough to get a few decades where agricultural winners match or even outnumber losers,

Wheat harvests in the US (such as here in the Palouse, Washington) may get a boost from more CO_2, but yields are expected to drop in much of the poor world.

the resulting imbalance could put huge stresses on agricultural trade and food security, including North–South relations – and that's without considering the impact of converting food crops to fuel (see p.351), or the relentless growth in meat consumption, which increases the amount of land and water devoted to food production.

Scientists are increasingly incorporating the mix of crops and climate into computer models, but there's still much work to be done. Most studies of individual crops have been relatively small-scale, while analyses of future climate are at their strongest in depicting patterns across large sections of the globe. A natural intersection point is regional climate models, which zero in from global models to portray climate across a specific area in finer detail. For example, Andrew Challinor (Leeds University) and colleagues have developed a **General Large Area Model (GLAM)** for annual crops. GLAM is on a much larger scale than most crop models, but it's shown the ability to predict local variations (including the extremes that can harm many crops) while it's yoked to the UK Met Office Hadley Centre's regional climate model. According to recent work with GLAM, the number of wheat crop failures in China could increase by up to a factor of three even for moderate climate change (less than a 2°C increase in average local temperature). However, if agricultural practices are adapted to the new climate by using crops that better tolerate drought and/or heat, then crop failures can be kept closer to the current (and more manageable) levels.

Agricultural specialists are working on hardier strains of key crops that can deal with potential climate shifts. In the Philippines, the International Rice Research Institute (IRRI) – famed for its work in bolstering yields in the 1960s Green Revolution – is working on drought- and heat-resistant strains of rice. Half the world's population eats rice as a staple, so its fate on a warming planet is critical to world foodstocks. A 2004 study from the IRRI showed that a 1°C (1.8°F) rise in average daily temperature can reduce some rice yields by 15%.

One challenge in developing heat- and drought-suited crop strains is that they may prove less productive during good years. Another critical step in formulating crops for a warming world will be to ensure that they stay as nutritious as their predecessors, given the effects of CO_2 on micro-nutrients noted above. Otherwise, the silent threat of "hidden hunger" – nutrient deficiencies that now jeopardize the health of billions – could only get worse. And technological gains may amount to little unless social and financial arrangements are in place to help even the most humble farmer take advantage of the progress. Many subsistence farmers stick to tried-and-true crops and techniques, preferring to minimize risk given their lack of safety cushions. International alliances such as the Consultative Group on International Agricultural Research are looking into policy options such as new forms of crop insurance that could give risk-averse farmers more room to adapt to a changing climate. Farmers throughout the tropics could also benefit from locally tailored information on climatic shifts, such as those produced by El Niño and La Niña.

PART 3

THE

SCIENCE

How we know what
we know about
climate change

Night-time view of the northern Gulf coast of the US, from the International Space Station.

Keeping track

Taking the planet's temperature

It's customary to see polar bears, forest fires and hurricane wreck-age when you turn on a TV report about climate change. What you're unlikely to see are the dutiful government meteorologist launching a weather balloon, the grandmother checking the afternoon's high tem-perature in her backyard, or a satellite quietly scanning Earth's horizon from space.

These are among the thousands of bit players – human and mechani-cal – who work behind the scenes to help monitor how much the planet is warming. There's no substitute for this backstage action. Day by day, month by month, year by year, the people and instruments who measure Earth's vital signs provide the bedrock for the tough decisions that indi-viduals, states and companies face as a result of climate change.

The rise of climate science

The practice of assigning a number to the warmth or coolness of the air has royal roots. Galileo experimented with temperature measurements, but it was Ferdinand II – the Grand Duke of Tuscany – who invented the first sealed, liquid-in-glass thermometer in 1660. The Italian Renaissance also gave us the first barometers, for measuring air pressure. By the 1700s, weather observing was all the rage across the newly enlightened upper classes of Europe and the US colonies. We know it was a relatively mild 22.5°C (72.5°F) in Philadelphia at 1pm on the day that the Declaration of Independence was signed – 4 July, 1776 – because of the meticulous records kept by US co-founder and president-to-be Thomas Jefferson.

Some of the first sites to begin measuring temperature more than 300 years ago continue to host weather stations today. Most reporting sites of

Average annual temperature across central England, 1772–2010

this vintage are located in Europe. A cluster of time-tested stations across central England provides an unbroken trace of monthly temperatures starting in 1659 – the longest such record in the world – as well as daily temperatures from 1772 (see graph above for annual averages). Like the remnants of a Greek temple or a Mayan ruin, these early readings are irreplaceable traces of a climate long gone. Such stations represent only a tiny piece of the globe, however, so on their own they don't tell us much about planet-scale changes. However, they do help scientists to calibrate other methods of looking at past climate, such as tree-ring analyses. They also shed light on the dank depths of the **Little Ice Age**, when volcanoes and a relatively weak Sun teamed up to chill the climate across much of the world, especially the Northern Hemisphere. A few regional networks were set up in Europe during those cold years, starting in the 1780s with the Elector of Mannheim, who provided thermometers and barometers to any interested volunteers.

The more thorough assembling of temperature data began in earnest after the arrival of the telegraph in the mid-eighteenth century, which made possible the rapid-fire sharing of information. Overnight, it seemed, weather mapping changed from a historical exercise to a practical method of tracking the atmosphere from day to day. The embryonic art of weather forecasting, and the sheer novelty of weather maps themselves, helped feed the demand for reliable daily observations. **Weather services** were established in the United States and Britain by the 1870s and across much of the world by the start of the twentieth century. Beginning in the 1930s, the monitoring took on a new dimension, as countries began launching weather balloons that radioed back information on temperatures and winds far above ground level. Most of these data, however, weren't shared among nations until well after World War II.

Though World War II interrupted observing efforts at many weather stations, it also brought new instruments into vogue, including the radiosonde (balloon-borne instrument package) – used here in 1944 by US Army Air Force meteorologists in Iceland.

One of the products of the post-war drive for global unity was the **World Meteorological Organization**, founded in 1951 to link national weather agencies and later folded into the United Nations infrastructure. In 1961, US president John F. Kennedy urged the UN General Assembly to consider "further cooperative efforts between all nations in weather prediction and weather control". The latter goal proved to be far-fetched, but the WMO did begin work on getting the world's weather services to agree on observing practices and data protocols. Before long, a global weather observing system was in place, a model of international collaboration and data exchange that stood firm against the tensions of the Cold War and continues to this day.

During this mid-century whirlwind of progress, the focus was on improving daily weather forecasts. What about the temperature of the planet as a whole? Amazingly, hardly anyone was tracking it. The pioneering climatologist **Wladimir Köppen** took an early stab at it in 1881. There was a brief flurry of activity in the 1930s when, in separate studies, the US Weather Bureau and British engineer Guy Stewart Callendar drew on the limited data available – mostly from North America and Europe – and found that Earth had warmed since the late 1800s. Little did they know that the warming itself happened to be focused in North America and Europe, a fact that emerged only decades later. "But for this accident, it is not likely that people would have

"Climate was described as 'average weather' and climatology was looked upon merely as the dry-as-dust book-keeping branch of meteorology."

Hubert Lamb, 1959, referring to the inter-war period

So what is Earth's temperature, anyway?

If you live in Rome, Melbourne or Shanghai, you might not realize how close your climatic experience is to that of Earth as a whole, at least in the average. Each of these cities has an annual mean air temperature within two degrees Celsius of 14.4°C (57.9°F), which is roughly the planet-wide average. A hundred years ago, the global average was closer to 13.6°C (56.5°F). The current number hides a substantial difference between north and south. It's apparently taking longer for greenhouse gases to warm the ocean-dominated Southern Hemisphere; the annual average temperature there is more than 1°C (1.8°F) cooler than in the Northern Hemisphere.

paid attention to the idea of global warming for another generation", notes climate biographer Spencer Weart.

Interest in global temperature had cooled off – as had the planet itself – by 1961, when Murray Mitchell of the US Weather Bureau found a cooling trend that dated back to the 1940s. His was the most complete attempt up to that point at compiling a global temperature average. However, like his predecessors, Murray was tricked by the wealth of US and European weather stations. In the sparsely observed Southern Hemisphere, the mid-century cool-down was far weaker than it was in the north. It took until the late 1970s for temperatures to begin inching back up and concern about global warming to resurface. In the early 1980s, several groups began to carefully monitor the annual ups and downs of global air temperature across the entire planet, eventually including the oceans. These data sets are now among the most important pieces of evidence brought into the courtroom of public and scientific opinion on climate change.

Measuring the global warm-up

Among those who find human-induced warming not much to worry about, there's a fairly wide spectrum of thought. Many now accept that Earth is warming up but argue that the rise isn't catastrophic and that any further rise should be well within our ability to adapt. At the other end, a few die-hards maintain that the planet really hasn't warmed up measurably in the last century – a position dismissed by virtually all climate scientists.

Still, the die-hards' position begs an interesting question: how do the experts come up with a single global surface temperature out of the thousands of weather stations around the world? It's a far more involved process than simply adding up the numbers and averaging them out. For starters, you need to include the best-quality observations available. Modern weather stations are expected to fulfil a set of criteria established by the WMO. For example, thermometers should be located between 1.25m and 2m (49–79in) above ground level, because temperature can vary strongly with height. On a cold, calm morning, you might find frost on the ground while the air temperature at station height is 2°C (36°F) or even warmer. Also, an official thermometer should be housed in a proper instrument shelter, typically one painted white to reflect sunlight and with louvres or some other device that allows air to circulate. Anyone who's been in a car on a hot day knows how quickly a closed space can heat up relative to the outdoors.

Consistency is also critical. Ideally, a weather station should remain in the same spot for its entire lifetime. But as towns grow into cities, many of the oldest stations get moved. That means the station is sampling a different microclimate, which can inject a cool or warm bias into the long-term trend. One of the biggest agents of station change in the mid-twentieth century was the growth of air travel. Countless reporting stations were moved from downtown areas to outlying airports, where regular observations were needed for flight purposes and where trees and buildings wouldn't interfere with wind and rainfall measurement. Even when stations stay rooted, the landscape may change around them and alter their readings. Many urban locations are vulnerable to the heat island effect (see box overleaf). Large buildings help keep heat from radiating out to space, especially on summer nights, and this can make temperatures in a downtown area as much as 6°C (10°F) warmer than those in the neighbouring countryside.

On top of all this, gaps can emerge in the records of even the best-sited, longest-term stations. Natural disasters, equipment or power failures, and wars can interrupt records for days, months or even years, as was the case at some locations during World War II. Problems may also emerge after the data are sent: electronic or software glitches may corrupt the numbers in ways that aren't immediately obvious.

In a perfect world, every temperature record would be free of the blemishes noted above. That's not the case, of course, but climate scientists have found clever ways to work with what they've got. By using the oldest and best data sets available, they can cull biased stations, adjust for incomplete records, weigh each station by the area it represents and produce a homogeneous picture of long-term temperature trends.

One major task is to find out how much the global reading for a given year has departed from the recent average (which is typically calculated from data covering a thirty-year period, long enough to transcend short-term dips

Bright lights, big cities, bogus data?

One of the most calcified critiques of climate-change science is the accusation that the warming indicated by global observations is largely an artefact of urban growth. Nobody doubts the existence of the heat-island effect, by which the dense buildings and paved areas of cities absorb and trap heat in various ways. Indeed, urban heat islands are a form of climate change in themselves, and those who live in a growing urban area would probably experience a gradually warming climate even if we weren't pumping the atmosphere full of greenhouse gases.

When considered on a global scale, though, heat islands melt into the background. According to the 2007 IPCC report, they only add about 0.02°C (0.04°F) of uncertainty to the much larger warming observed over the last century. Still, mainstream climate scientists don't consider it kosher to conflate the localized effects of heat islands with the planet-wide warming produced by fossil fuel emissions. Thus, the big three data compilers described in this chapter use a variety of methods to filter out urban effects from their data sets. The first challenge is working out which stations should count as urban, and which should count as rural.

James Hansen's group at NASA started out by using population data to identify urban stations, but they soon discovered this approach couldn't always distinguish whether a reporting station was sited in the heart of a big city or just outside town. In 2001 they applied a new strategy: identifying urban areas through the lights detected by US defence satellites on clear nights (especially during new moons). Hansen found that many rural US areas generated enough night-time light to put their free-of-heat-island status into question. Eventually, the group removed about 80% of their US reporting stations from the global average. That left roughly 200 bona fide rural sites – still enough to provide a reasonable picture of US climate, says Hansen.

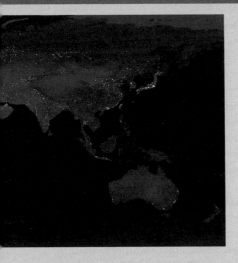

The lights of humanity in a night-time composite image of the world as seen from orbit.

In other parts of the world, where rural analogues for urban stations are scarce or where development covers entire regions, it's harder to weed out the city influence. But urban heat islands may not be affecting the measured global average as much as scientists once thought. Thomas Peterson of the US National Climatic Data Center found that, from 1880 to 1998, the rural subset of the centre's 7500-station global record actually warmed a touch faster than the full record – hardly a sign of urban contamination. Peterson also took a fine-toothed comb to 289 US stations, using the satellite-based method, and found no statistically significant urban effects. He speculated that the siting of many US weather stations in open, park-like settings (including airport grounds) helps avoid the influence of core city areas, where it might well be warmer. "We need to update our understanding of urban heat islands", says Peterson. "This phenomenon is more complex than widely believed by those not immersed in the field." Computer modelling may someday be useful in identifying and extracting heat-island effects, according to David Parker of the UK Met Office Hadley Centre, who penned a 2010 overview on the topic.

Urban heat islands are strongest on calm nights, when there's no breeze to help disperse the city-trapped air. This inspired Parker to see whether calm nights might be warming up more than windy nights globally, which would presumably be a sign of heat-island bias in the climate record. As he reported in *Nature* in 2004, that doesn't seem to be the case. For a set of 264 stations across the globe during the period 1950–2000, he found identical warming trends for nights categorized as windy, calm or lightly breezy. Parker's study added to the overall body of work showing that heat islands are a poor scapegoat for planet-wide warming. One other clue: the recent warming has manifested not only on land, but over the oceans, which aren't known for their urban sprawl.

or peaks). The three major climate centres that carry out yearly analyses of global temperature each have their own way of arriving at an average:

▶ **University of East Anglia** Phil Jones and colleagues, at the Climatic Research Unit of the University of East Anglia (UEA), pioneered a technique called the "climate anomaly method". As the name implies, this method starts with the anomaly observed at each station in a given year – how much warmer or cooler it is relative to a thirty-year base period, such as 1961–90. UEA does this calculation for about 4200 stations worldwide. It averages those station anomalies across grid boxes that span 5° each of latitude and longitude, then by hemisphere, and finally for the globe as a whole, with the UK Met Office Hadley Centre providing the ocean analyses.

▶ **The US National Aeronautics and Space Administration** A group led by James Hansen of NASA devised the "reference station method". Instead of anomalies, it uses temperatures observed at about 6000 stations, then calculates averages across a much smaller number of grid boxes than UEA (about 80 in all). The longest-term record in each grid box is dubbed the reference station and considered the gold standard for that box. Records covering a shorter timespan are adjusted so that their trends are consistent with those from the reference stations.

▶ **The US National Oceanic and Atmospheric Administration** NOAA began using a newer technique, the "first difference method", in the late 1990s. It calculates the amount of temperature change from one year to the next at each station. If data are missing for some length of time at a certain site, the difference can be taken over a longer period without much effect on the final result. Among the three groups, NOAA calls on the largest number of stations: about 7200. In 2006 NOAA introduced a new method that fills in data-sparse areas based on the long-term patterns of temperature variation observed in nearby areas, then statistically filters the results. A third major update of NOAA's global data set was being readied for release in 2010.

Each one of these techniques contains a lot of redundancy – on purpose. According to some experts, it takes no more than about a hundred strategically placed points around the world to create a decent first cut at global temperature. Everything beyond that refines and sculpts the

Amassing the figures

It was 1979 when Phil Jones and colleagues at the Climatic Research Unit of the University of East Anglia (UEA) came up with the notion of a global temperature index. The group knew that others had tried to take Earth's temperature in a systematic way but hadn't achieved truly global coverage. With support from the UK and US governments, Jones and Tom Wigley (now at the US National Center for Atmospheric Research) led the charge to sift through stacks of data, gather readings from weather services around the world, and correct as many sources of bias as possible. "It's the best thing we've ever done", Jones said in 2005.

In New York, James Hansen and colleagues at NASA's Goddard Institute for Space Studies were on the same wavelength. Hansen, a pioneer climate modeller, set out to maximize the value of climate data in poorly sampled parts of the globe. His group's technique allowed the most isolated stations to represent conditions within a radius of 1200km (750 miles), which helped expand the analysis across the station-scarce Southern Hemisphere and near the poles.

Phil Jones

As the 1980s unfolded, the NASA and UEA groups took turns making headlines with a series of landmark papers. NASA led off in 1981 with the startling announcement that the world was once again warming – and, in fact, that it had been since the 1970s. UEA concurred in a 1982 paper, and in 1986 Jones and colleagues detailed their techniques in the most thorough global analysis to date. By the time Hansen testified before the US Congress in 1988 (see p.275), the US National Oceanic and Atmospheric Administration had begun its own programme to monitor global temperature, led by Thomas Karl.

Little did the researchers know how politicized their products would become. Climate sceptics besieged Jones and others with requests for data (some of which Jones couldn't legally share) and hunted for issues and inconsistencies that might weaken the case for global warming. Then in 2009, Jones found himself at the centre of the unauthorized release of emails dubbed Climategate (see p.294). A few months later, although some concerns remained about data access, three inquiries had cleared Jones and colleagues of data-tampering accusations. And record global warmth in early 2010 underscored the vital role of the temperature archives that Jones and colleagues had launched decades earlier.

"There are many flaws in the input data for the temperature change analyses, especially in the early years, and the effect of these flaws cannot be fully removed", noted Hansen in 2005. "Despite these problems, however, the reality of global warming in the past century is no longer at issue."

estimate; important work, to be sure, but nothing that would change the result drastically.

Which of the three methods is best? That's an impossible question to answer, because there's no absolute record to compare the results against. Each centre's approach appears to have its strengths and weaknesses but, reassuringly, there doesn't seem to be all that much difference among the results. What look like important contrasts – for instance, NASA initially ranked 2005 as the warmest year on record, while UEA put it in second place – relate more to tiny differences in particular years, although some sceptics have taken the occasional reshufflings of the rankings as a sign of sloppy research and/or insignificant warming. There's strong agreement on the bigger picture: as noted in the 2007 IPCC report, the atmosphere at Earth's surface warmed more than 0.7°C (1.26°F) over the century that ended in 2005.

Estimates of full-globe temperature include **oceans**, which are a bit tricky to incorporate. Although few people live there, the oceans represent 70% of the planet's surface, so it's critical that surface temperature is measured over the sea as well as over land. There's no fleet of weather stations conveniently suspended above the sea, unfortunately, but there are many years of sea-surface data gathered by ships and, more recently, by satellites. Because the sea surface and the adjacent atmosphere are constantly mingling, their temperatures tend to stick close to each other. This means that, over periods of a few weeks or longer, a record of sea-surface temperature (SST) serves as a reasonable proxy for air temperature.

Until the mid-twentieth century, SSTs were mainly measured by dipping uninsulated buckets (originally made of wood, then canvas) into the ocean and hauling them back to deck, with the water often cooling slightly on the trip up. Around the time of World War II, it became more common for large ships – especially US ones – to measure the temperature of water at the engine intake. This tended to produce readings averaging about 0.4°C (0.7°F) warmer than the canvas-bucket method. Some ships measured air temperature directly on deck, but these readings had their own biases: the average deck height increased through the last century, and daytime readings tend to run overly high due to the Sun heating up metal surfaces on the ship. It wasn't until the 1990s that routine measurements from ships and buoys were gathered promptly, processed for quality control and coupled with newer estimates derived from satellites, which can infer SSTs by sensing the upward radiation from the sea surface where it's not cloudy. Together, all these sources now help scientists to calculate temperatures for the entire globe, oceans included. As with the land readings, there are differences in how the three groups compile and process ocean data, but these don't appear to influence the long-term trends for global temperature too much.

Even if it's clear the world is warming, it's equally evident that the current observing network is far from perfect. That point was brought home in 2009 by an energetic blogger, Anthony Watts, who rallied more than 650 volunteers through Watts Up With That?, his hugely popular, sceptic-oriented website. The investigators found hundreds of surface weather stations across the United States that were improperly located near water treatment plants, pavements, air conditioners and other sources of artificial heat. A photo-studded report, "Is the US Surface

Improving the global thermometer

As critical as they are to climate-change science, traditional weather stations and balloon-borne radiosondes were never designed to measure the subtle trends of long-term climate change. Just as a hi-fi that's fine for heavy metal might not convey all the nuance of a symphony orchestra, a weather station might capture large day-to-day temperature changes well enough but contain tiny biases that become evident only after years have passed. To address the need for better long-term data, several UN agencies teamed with the International Council for Science in 1992 to launch the Global Climate Observing System (GCOS). Through a broad web of activities, GCOS fosters the improvement of all types of data collection on climate change and its impacts, particularly on ecosystems and sea level.

With more and better data in the queue, the job of connecting these observational dots looms larger. In 2005 nearly sixty governments and the European Commission endorsed a ten-year plan to build a Global Earth Observation System of Systems (GEOSS). The name itself signals the multi-layered complexity of the task. GEOSS must find ways to incorporate upcoming satellite systems and new ground-based tools, while working to maintain the integrity of the observational network already in place. That network suffers from multiple ailments: gaps in coverage across space and over time, inadequate archiving and a lack of certainty that data from valuable yet time-limited satellite missions will continue. The GEOSS ten-year plan calls for "targeted collective action" to address these and other observational concerns.

Even the best data sets can't help much if they're inaccessible. In 2010 the World Meteorological Organization began considering a proposal from the UK Met Office to build an open-access archive of global temperature records. Right now much of the best-quality data, including hourly temperature reports, is locked up by the weather services of nations who turn a profit by selling their data to users. If the new storehouse managed to pool these observations, researchers would have more than twice the holdings than are now at their disposal. The archive would also maintain details on all adjustments and corrections in a transparent fashion, thus bolstering its usefulness while helping to ease fears about data fudging.

Temperature Record Reliable?" was published by the Heartland Institute, a nonprofit outfit that promotes free-market approaches to problem-solving. Though the report wasn't a peer-reviewed study, it made quite a splash in the US media and the blogosphere. Nearly 90% of the 860 stations examined from NOAA's US Historical Climatology Network fell short of the agency's protocol that stations should be at least 30m (100ft) from artificial objects that could produce, reflect or radiate heat. The report also claimed that the adjustments used by NOAA to homogenize the data sets simply expanded the impact of the siting problems. If the much-vaunted US system was this flawed, Watts argued, perhaps the global database was just as compromised. The findings prompted a group of US congresspeople to ask whether NOAA was adequately ensuring the quality of its climate data.

It didn't take long for researchers to demonstrate that the station-location issues, while in need of addressing, had virtually no impact on the big picture. In a 2010 paper, a NOAA team headed by Matthew Menne compared the adjusted data from stations rated by Watts to those from a gold-standard group of well-sited stations called the US Climate Reference Network. The trends were nearly identical, even for those stations whose siting was rated "poor" by Watts – an illustration of how even less-than-ideally-positioned instruments can still capture a trend if it's distributed widely. As for the unadjusted data, they did show a bias, but not in the expected direction: the poorly sited stations actually ran cooler overall. It's mainly because those sites were more likely to have newer thermometers whose installation had led to cooler daily highs, due to a slight instrument error as well as a concurrent switch in the timing of once-per-day temperature checks. Adjustments already in place accounted for these biases well, noted Menne and colleagues. They stressed the need to wade through actual numbers – "photos and site surveys do not preclude the need for data analyses", they wrote – while also praising the work done by Watts and his volunteers to document problems in station placement.

Heat at a height

When people refer to global warming, they're usually talking about the air in which we live and breathe – in other words, the air near ground or sea level. But the atmosphere is a three-dimensional entity. Conditions at the surface are affected profoundly by what's happening throughout the **troposphere** – the lowest layer of the atmosphere, often called the "weather layer". The troposphere extends from the surface to about 8–16km (5–10 miles) above sea level. It's tallest during the warmer months and towards the tropics – wherever expanding air is pushing the top of the troposphere upward. In fact, global warming appears to be literally raising the troposphere's roof. In a 2003 study, Benjamin Santer and colleagues at the US Lawrence Livermore National Laboratory showed that the average depth of the troposphere had increased by about 200m (660ft) from 1979 to 1999.

There are two ways to measure temperatures high above Earth's surface. The first – as used by Santer's team – is to use **radiosondes**, the balloon-borne instrument packages launched from hundreds of points around the globe twice each day. Sending data via radio as they rise, the devices measure high-altitude winds, moisture and temperatures. As useful as they are, radiosondes have their limits for tracking climate change: they're launched mainly over continents, with large distances separating their paths; they're produced by more than one company, and so don't behave uniformly; and they're prone to small biases that can make a big difference when assessing long-term trends. The second technique is to use **satellites**. These provide a more comprehensive global picture, but they've also been at the centre of a recently resolved issue that was, for years, the subject of one of the most heated scientific debates in climate-change science.

Shortly after climate change burst into political and media prominence in 1988, Roy Spencer and John Christy of the University of Alabama at Huntsville began using data from a series of satellites to infer global temperature in three dimensions. Throughout the 1990s and into the new millennium, their findings roiled the world of climate research and lent ammunition to global-warming sceptics. Their data suggested that the troposphere wasn't warming much if at all, despite the observed warming at ground level. Unless the fundamental understanding of lower atmosphere was off-base, then one set of measurements had to be wrong.

Spencer and Christy's data are gathered primarily by instruments called **microwave sounding units** (MSUs), stationed aboard NOAA satellites. The MSUs measure the temperature of gases at various altitudes below them by measuring the energy that they emit. Unlike a surface station or weather balloon, the MSUs provide data that's inherently averaged over large areas. The formulas are complicated, but in general, each individual temperature measurement covers anywhere from about 8000 to 38,000 square km (3100–14,700 square miles) through a depth of a few kilometres and a time period of twelve hours. All told, the satellites cover more than 90% of the planet's surface.

How the greenhouse turns

The troposphere's name stems from the Greek *tropos*, meaning to turn – which is exactly what the troposphere does. Heated largely from below, as sunlight strikes Earth, this layer of air churns and bubbles like a boiling pot of water, with calm conditions the exception more than the rule. Usually the air is blowing faster horizontally than vertically, but no matter how you look at it, the troposphere is an ever-changing domain. All this motion – up, down and sideways – helps give rise to the variety of weather we experience at ground level.

Because the troposphere is so well mixed, there should be a close connection between how much it warms near the surface and higher up. But there's not a one-to-one relationship, because greenhouse gases aren't spread through the troposphere as evenly as you might expect, and their effects vary with height.

▶ **Shorter-lived greenhouse gases,** such as low-level ozone, are chemically transformed or washed out by rain or snow within a few days. Thus, they tend to stay within the same general latitudes as their sources – often riding the jet stream from Russia to Canada or from eastern North America to Europe during their short time aloft.

▶ **Water vapour** is most prevalent at lower altitudes, especially above the oceans, but it's been increasing higher up, thanks in part to high-flying aircraft. (Planes also add fossil-fuel emissions to the thin, cold air.)

▶ **Longer-lived greenhouse gases,** such as carbon dioxide, are thoroughly mixed across the troposphere, both horizontally and vertically. That's why a single station high atop Hawaii's Mauna Loa (see p.34) can tell us how much CO_2 the global atmosphere holds.

In general, the higher in the troposphere a greenhouse gas is, the more powerful its effect on the troposphere as a whole. This is because the amount of energy a greenhouse molecule emits is directly related to its temperature. At high altitudes,

The fireworks began with the first Spencer-Christy paper, published in the journal *Science* in 1990. It showed no discernible temperature trend since 1979 in the lowest few kilometres of the atmosphere. By comparison, surface temperatures had warmed roughly 0.2°C (0.36°F) during the 1980s. Updated through the 1990s, the Spencer-Christy data continued to show virtually no warming. What's more, radiosonde analyses tended to agree with the satellites, with relatively minor warming in northern mid-latitudes and a slight cooling in the tropics.

Many people inclined to disbelieve global warming cited these reports as proof that the world wasn't heating up. But most climate scientists weren't convinced. In 1998, two scientists in California, Frank Wentz and

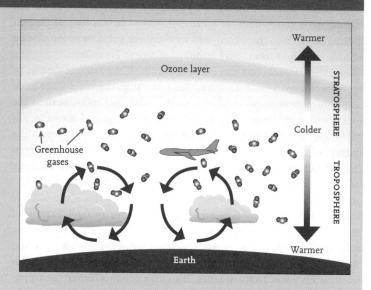

where temperatures are low, the molecules emit less of the heat energy they receive from Earth into space, resulting in more heat remaining in the troposphere.

This picture reverses in the stratosphere, where – oddly enough – human actions have led to *cooling* rather than warming. Carbon dioxide doesn't absorb much heat energy from Earth at these heights, but it continues to radiate heat to space, thus acting to cool the stratosphere. Meanwhile, the partial loss of sunlight-absorbing ozone over the last twenty or so years has exerted a cooling influence. As a result of these and other factors, temperatures in the lower stratosphere have plummeted to record-low levels.

Matthias Schabel, discovered that **atmospheric drag** might be affecting the MSU readings. They claimed that air molecules were slowing down the satellites, increasing the effect of gravity on their paths and causing them to drop slightly, hence changing their viewing angle and skewing their measurements. Spencer and Christy adjusted for the error caught by Wentz and Schabel, but also made a separate correction in the opposite direction for another newly discovered bias, so the result was little net change. Meanwhile, the global surface temperature continued to climb, so the gulf between surface and satellite trends widened even further. The persistent disagreement left policymakers and the public wondering what to believe. In 2000, the US National Academies weighed in with a special report, agreeing that the overall surface warming trend since 1979 was "undoubtedly real" but noting the possibility that high-altitude air *had* warmed more slowly than the surface, thanks to factors such as cool air filtering down from the record-cold stratosphere, sulphate pollution and volcanic emissions from 1991's eruption of Mount Pinatubo.

Since then, events have conspired to diminish the satellite data as an argument against global warming. For one thing, the years since 2000 have been notably warm compared to the 1980s and 1990s, both at the surface and aloft. Meanwhile, several other groups gained access to full MSU data, and their analyses (including new techniques for separating stratospheric cooling from tropospheric warming) tended to show even stronger satellite-based warming trends, closer to surface observations. Radiosonde trends have also crept upwards in some studies.

Overall, these and other corrections brought the surface and satellite estimates in much closer agreement with each other. In 2006, a report of the US Climate Change Science Program came to the same conclusion, as did the 2007 IPCC assessment. The IPCC noted that both surface and satellite data capture the year-to-year ups and downs associated with such events as El Niños and volcanoes. It's still unclear whether the troposphere above the tropics is heating or cooling, but for the globe as a whole, the IPCC concluded that the surface warming since 1979 (0.16–0.18°C or 0.29–0.32°F per decade) now concurs well with the MSU-derived tropospheric warming (0.12–0.19°C or 0.22–0.34°F per decade). With the dust settled, one thing is obvious: the world is warming up.

Global dimming

To get a clear idea of how our climate is changing, you've got to measure more than temperature. **Rain and snow** are an important part of the story, and they're uniquely challenging to assess (see p.65). Other variables, such as **water vapour** and the nature of **cloud cover,** have their own complexities. Satellites can measure both, but only with strict limitations. For example, some can't sense lower clouds if high clouds are in the way. A set of globe-spanning satellites called COSMIC, launched in 2006 by a US–Taiwan partnership, is now providing fuller three-dimensional portraits of temperature and water vapour. Other new satellites are keying in on cloud cover. Even so, it's likely to be years before we can assess with confidence how these features are changing.

What's in the sky has everything to do with how much sunlight reaches Earth. Sunshine and its absence weren't part of the dialogue on global

The effects of clouds, here set in a chaotic Australian sky, on climate change are complex and not fully understood.

warming until relatively recently. A 2003 paper in *Science* gave the phenomenon of **global dimming** its first widespread exposure, and by 2005 it was the star of a BBC documentary. Global dimming is real: from the 1950s through the 1980s, the amount of visible light reaching Earth appears to have tailed off by a few percent. In terms of climate, that's quite significant, though it would be difficult for the average Jane or Joe to detect it next to the on-and-off sunlight cycles that occur naturally each day. It also wasn't enough to counteract the overall global warming produced by greenhouse gases. In fact, the timing of global dimming's rise to fame is actually a bit paradoxical, since the phenomenon appears to have reversed in the 1990s, according to the 2007 IPCC report. However, as we'll see, the recent brightening brings its own set of worries.

Global dimming offers a good example of how a set of little-noticed measurements, collected diligently over a long period, can yield surprising results and help answer seemingly unrelated questions. Back in 1957, as part of the worldwide International Geophysical Year, a number of weather stations across the globe installed **pyranometers**. These devices measure the amount of short-wave energy reaching them, including direct sunshine as well as sunlight reflected downwards from clouds.

Slowly but surely, the amount of energy reaching the pyranometers dropped off. Gerald Stanhill (who later coined the phrase "global dimming") discovered a drop of 22% across Israel, and Viivi Russak noted similar reductions in Estonia. Although they and several other researchers published their findings in the 1990s, there was hardly a rush to study the problem. With the spotlight on global warming by that point, scientists might have found it counterintuitive to investigate whether less sunlight was reaching the globe as a whole. Climate models, by and large, didn't specify such dramatic losses in radiation.

Yet the data cried out for attention, especially once they were brought together on a global scale. Using a set of long-term readings from the highest-quality instruments, Stanhill found that solar radiation at the surface had dropped 2.7% per decade worldwide from the late 1950s to early 1990s. Two other studies examining much larger data sets found a smaller global drop – around 1.3% per decade from 1961 to 1990. Even these studies failed to jolt the climate research community until 2003, when two Australians, Michael Roderick and Graham Farquhar, linked the dimming to a steady multi-decade reduction in the amount of water evaporating from standardized measuring pans around the world. Warmer temperatures ought to have evaporated more water, not less, they reasoned (as suggested by global measurements of water vapour). However, Roderick and Farquhar realized that a reduction in sunlight could more than offset this effect. It's also quite possible that global evaporation actually rose in spite

The skies over northern India are filled with a thick soup of aerosol particles all along the southern edge of the Himalayan Mountains, and streaming southward over Bangladesh and the Bay of Bengal. These toxic human-produced aerosols are certainly hazardous to life, but may also act to slow global warming.

of the pan-based data, as noted in the 2007 IPCC report. Pan-based sensors can only show how much water might evaporate from a completely wet surface, but a warmer, wetter and dimmer world could still send more total moisture into the air – through the leaves of plants, for example. The lifetime of moisture in the atmosphere, which could change with the climate, is another factor that makes the evaporation issue especially hard to unravel. Still, the notion of dimming through much of the latter twentieth century is now fairly well accepted, and it's a small but significant part of the picture when scientists try to balance the global budget of incoming and outgoing radiation at Earth's surface over the last few decades. Dimming might even help explain why some Arctic tree-ring records show a divergence between temperature and tree growth since the 1950s. If these high-latitude trees are especially light-sensitive, than even a small dimming might be enough to limit their growth.

Some areas – especially urban, industrialized parts of the Northern Hemisphere – showed several times more dimming than other parts of the globe. This regional patchwork makes sense given what we know about the causes of dimming. The prime suspects are soot and other airborne particles, or **aerosols**, produced when fossil fuels are burned. For hundreds of years, people in London had first-hand experience of the sky-darkening effects of soot. Aerosol pollution does this type of darkening in several ways:

▶ **Attracting water** Many aerosols are hydrophilic, or water-attracting. When conditions are right for clouds to form, a highly polluted air mass can produce many times more cloud droplets. The moisture is spread more thinly, though, so each droplet is smaller on average. The upshot is that this larger array of droplet surfaces can reflect more sunlight upwards before it has a chance to reach the ground. From space, the clouds look brighter; from Earth, they're darker.

▶ **Extending cloud lifespans** In general, it takes longer to get rain or snow out of a cloud full of many small particles, as opposed to fewer, larger ones. For this reason, the addition of aerosols seems to lengthen the lifespan of clouds and thus increase their reflective power over time.

▶ **Reflecting light** Even on a cloudless day, aerosols themselves reflect sunlight to space. A major study of pollution across India and the adjacent ocean found that the subcontinent's thick veil of winter pollution can trim the amount of sunlight reaching the surface by as much as 10%.

The plane truth about contrails

Can those sleek, wispy lines of cloud that trail behind jets really affect the future of our climate? Researchers have already found that contrails can produce surprisingly large effects on a regional basis. The most dramatic real-world example came after the terrorist attacks on New York and Washington on 11 September 2001. With all US commercial air traffic shut down for several days, the skies were virtually contrail-free for the first time in decades. The result was slightly warmer days and cooler nights, according to David Travis of the University of Wisconsin–Whitewater. He found that the average range in daily temperature across the US (the span between each day's highest and lowest reading) increased by more than 1°C (1.8°F) during the aviation-free days. Travis suggested that the lack of contrails provided an open door for additional sunlight to reach the surface by day and for extra radiation to escape at night.

Other researchers later argued that the boosted temperature spread could have occurred by coincidence, with clear skies and dry air predominating during the period. However, the 9/11 study did provide a rare vantage point on a vexing issue. Contrails don't have a single, easy-to-summarize impact on climate, in contrast to the tone of some press and TV reports. "This is a very complicated problem, as there are multiple competing effects going on", Travis says.

Contrails are a special type of **cirrus**, which are high, thin clouds composed mainly of ice crystals and often resembling streaks or sheets. Contrails form as the water vapour and particles spewed out by an airplane's exhaust stream create water droplets that soon crystallize into ice. Pollutants in the exhaust plume serve as focal points for additional ice crystals within the contrail as well as in neighbouring clouds. (Some of these pollutants are greenhouse gases themselves, perhaps adding a few percent to overall global warming.)

Most types of clouds cool Earth's surface as they screen out sunlight. However, cirrus clouds have an overall warming effect. That's because cirrus usually allow a good deal of sunlight through but effectively trap radiation flowing up from Earth. When a contrail first forms, it behaves more like a lower-level cloud, as the water droplets reflect ample amounts of sunlight and exert a cooling effect. Larger planes tend to generate thicker, longer-lived contrails. Gradually, the contrail spreads, thins out and freezes, and its net effect soon shifts from cooling to warming. Many contrails evaporate within an hour or so, but some evolve into much larger cirrus clouds, adding to the warming influence.

The plot thickens depending on what time of day it is. At night, contrails can provide only a warming effect (since there's no sunshine to be reflected). Since night flights appear to be gaining in popularity, that could help tilt the scales towards a net warming effect. On the other hand, when sunlight hits a contrail at a low angle – in the early morning or late evening, or even at midday across high latitudes in winter – the contrail's reflectiveness appears to be boosted (which argues for flying by day rather than by night; see p.379). On top of all this, contrails aren't globally uniform; they're focused where people happen to be flying. Right now contrails are most prevalent across North America, western Europe and eastern Asia. But there's some uncertainty in their vertical distribution, because they can hide in or beneath other clouds and escape the view of cloud-detecting satellites. A 2010 study by the University of Leeds and the UK Met Office found that cloud cover negates almost half of the potential effect of contrails on climate. As for the future, it seems inevitable that increasing affluence will spawn more contrails above China and India, although it's unclear how cut-rate airline traffic in Europe and North America will evolve amid increasingly steep fuel costs.

Stir together the above ingredients, and you have a confusing mess indeed. Computer models tend to come down on the side of an overall warming effect in the regions where contrails are focused. Travis believes the models are generally on the right track, although he argues for caution (as does the 2007 IPCC report). The models may not be accounting for all of the above effects accurately, and the few observational studies thus far (including his own) don't show a significant temperature boost from contrails. "I do tend to believe that they ultimately have a warming effect. We just haven't yet found any empirical data to support this", says Travis.

Globally, the 2007 IPCC report estimates that contrails add 25–60% to the warming from aviation-related greenhouse gases. As the clouds spread out and persist, they may trap several times more additional heat, though the IPCC is less confident of this range. There's an extra complication when it comes to assessing the importance of all this: the net greenhouse impact of aviation is powerful but focused largely in the short term, whereas the effects of CO_2 from ground transport are weaker but last much longer (centuries rather than decades). Arguably, therefore, aviation is more problematic than is usually suggested because it creates most of its warmth in the near term, raising the chance of the world crossing some kind of climatic tipping point sooner rather than later.

Whatever the total impact of contrails may be, it looks set to rise as air travel continues to grow, adding further to the impact of what is already a very climate-unfriendly form of transport (see p.42).

Good evening earthshine

Measuring how much sunlight hits Earth is one thing, but how much does Earth reflect back to space? Even a tiny change in that percentage – the albedo, or reflectivity – could have huge implications for climate. Clouds are the main reflectors of the planet: as a whole, they bounce about half of the sunlight they receive back to space. The oldest and most complete satellite observations of cloud cover are part of a programme called the International Satellite Cloud Climatology Project. ISCCP can't track every cloud, but it does measure their collective impact on radiation and albedo. ISCCP data show a drop of 1–2% in Earth's albedo in the 1980s and 1990s, with a slight rise thereafter. Other cloud-based data sets differ, though, so scientists are now conducting an international intercomparison to gain more solid estimates of albedo trends.

There's another way to measure how much light is bouncing off Earth: by measuring the amount that makes it to the Moon. This is earthshine, the faint light that illuminates the dark part of a less-than-full Moon. Leonardo da Vinci recognized the source of earthshine some 500 years ago, and in the 1920s French astronomer André-Louis Danjon devised a clever way to measure it. By using a split-view telescope, Danjon dimmed the Sun-brightened part of the Moon (right) until its apparent brightness was the same as the part lit by earthshine (left). The stronger the earthshine, the less adjustment needed; thus, the adjustment itself served as a measure of earthshine strength.

Danjon and followers collected earthshine measurements in France through the 1950s. Since the mid-1990s, the Big Bear Solar Observatory has carried the torch, gathering data for global change studies as part of NASA's Project Earthshine. High in the mountains east of Los Angeles, Big Bear enjoys more than three hundred cloudless nights a year, and it experiences little of the atmospheric turbulence that can spoil sensitive measurements. Earthshine readings at Big Bear can rise or fall by a few percent over several hours as sunlight bounces off the sequence of oceans, continents, clear areas or cloudy patches presented by Earth's turning. A second earthshine sensor on the Canary Islands now allows for global coverage. Seasonal variations can run on the order of 10%. At first, the Big Bear earthshine data varied somewhat from satellite-derived readings, but the data sets are now largely in sync, with earthshine-measured albedo holding fairly constant from 2000 through at least 2007.

You couldn't pick a better single spot from which to measure albedo than the Lagrange-1 point. That's the point in space about a million miles towards the Sun from Earth, where the two bodies' gravitation fields are balanced. A satellite stationed there could keep a continuous eye on the fully lit side of Earth. Such was the idea behind Triana, one of Al Gore's most ambitious proposals as US vice president. Renamed the Deep Space Climate Observatory (DSCOVR) after Gore left office, the project languished for years and was finally terminated in 2006 after NASA had spent more than $100 million building the satellite.

Global warming itself may be responsible for some of the dimming – perhaps around a third – as it increases the amount of water vapour in the atmosphere (which makes more moisture available for cloud formation) and rearranges the distribution of clouds (lower and/or thicker clouds tend to reflect more than higher, thinner ones). As noted above, it's not entirely clear how cloud cover is evolving on a global scale, though increases of 1% or more per decade in water vapour have been recorded since the 1980s.

The big issue ahead may not be dimming but brightening. Aerosols are more visible and noxious than greenhouse gases, yet easier to control, so their reduction can be an easier political sell. Most of the world's highly industrialized nations began cleaning up their smokestacks and exhaust pipes by the 1970s, and the economic downturn of the 1990s across the former Eastern Bloc further reduced aerosol pollution. Although the breakneck pace of development across China and India has kept aerosol emissions high in those areas, the global concentration of aerosols appears to have dropped by as much as 20% since 1990, according to a 2007 NASA report. At the same time, some areas have reported an increase in solar radiation of anywhere from about 1% to 4% per decade since the early 1990s. Beate Liepert (Lamont-Doherty Earth Observatory) suspects this may reflect a recovery from the sun-shielding effects of Mount Pinatubo's 1991 eruption and from the extensive cloudiness generated by the frequent El Niño events of the early and mid-1990s. All this, in addition to greenhouse gases, may help explain why global temperatures rose as sharply as they did in the last decade of the twentieth century.

Our clean-up of aerosols comes at an ironic price. Few would complain about a world that's brighter and less fouled by aerosol pollution. Yet the overall cooling impact of aerosols will diminish as we reduce their global prevalence. Thus, in gaining a cleaner atmosphere, the world stands to lose one of its stronger buffers against greenhouse-gas warming. As the planet warms and brightens, it remains to be seen whether policymakers will be alert to this risk.

It's possible, though, that policymakers won't have much say in the matter. Commentators such as James Lovelock have raised the concern that a severe global economic downturn (caused by climate change or anything else) could lead to massively reduced aerosol emissions over a relatively short period, causing the planet to warm faster than ever just when we're least able to put money into doing anything about it.

The long view

A walk through climate history

During the four and a half billion years of our solar system, Earth has played host to an astounding array of life forms. Much of the credit goes to our world's climate, which is uniquely equable compared to conditions on the other major planets. Even if they had water and other necessities, Venus would be too hot to sustain life as we know it, and Mars and the planets beyond it too cold. But in fact, there's evidence that both planets once harboured oceans and may have been mild enough to support life before their climates turned more extreme. Likewise, during its lively history, Earth's climate has lurched from regimes warm enough to support palm-like trees near the poles to cold spells that encased the present-day locations of Chicago and Berlin under ice sheets more than half a mile thick.

The changes unfolding in our present-day climate make it all the more important to understand climates of the past. Thankfully, over the last fifty years, technology and ingenuity in this field of study have brought dramatic leaps forward in our knowledge. The clues drawn from trees, pollen, ice, silt, rocks and other sources leave many questions still to be answered. Taken as a whole, though, they paint a remarkably coherent picture. We now have a firm grasp of the changes that unfolded over the past three million years – a period ruled mainly by ice ages, with the last 10,000 years being one of the few exceptions. We also know a surprising amount about what transpired even earlier, including many intervals when Earth was a substantially warmer place than it is now.

Could those ancient warm spells serve as a sneak preview of what we might expect in a greenhouse-driven future? We know that there's already more carbon dioxide and methane in our atmosphere than at any time in at least 800,000 years, and perhaps much longer than that. Carbon dioxide has increased by more than 30% in only a century, an astonishing rate by the standards of previous eras. At the rate we're pumping greenhouse gases into the atmosphere, they could be more prevalent by the year 2050 than they've been in at least ten million years. That's one reason why **paleoclimatologists** (those who study past climates) are taking a close look at how climate and greenhouse gases interacted during previous warm intervals. The analogies are admittedly less than perfect. As we'll see, Earth's continents were positioned quite differently 200 million years ago than they are now (even ten million years ago, India was still drifting across the Indian Ocean and North and South America had yet to join up). As a result, the ocean circulations were very different as well. Also, most of the species that were then swimming, crawling or flying across our planet are now extinct. Many other aspects of the environment at that time are hard to discern today.

In spite of our less-than-ideal hindsight, we've gained enormous ground in our understanding of how climate works by piecing together the past. And advances continue to arrive at a rapid clip in the still-young field of paleoclimatology. In 2004, a new ice core from Antarctica provided a record that was almost twice as long as the previous one (see box on p.234). There's hope that a future core could yield data going back nearly 1.5 million years, including the point when ice ages switched their on-and-off tempo. If there's one message that's become increasingly clear from the ice cores, it's that natural forces have produced intense and rapid climate change in the past. That's something worth thinking about as we contemplate our future.

What controls our climate?

The great detective story that is paleoclimatology hinges on a few key plot devices. These are the forces that shape climate across intervals as brief as a few thousand years and as long as millions of years. Greenhouse gases are a critical part of the drama, of course, but other processes and events can lead to huge swings in climate. These vary greatly in terms of timespan. The sun-blocking particles spewed into the stratosphere by an unusually large volcano may cool the planet sharply, but most of those impacts are gone after a couple of years. In contrast, climate change due to slight shifts in the Sun's orbit takes tens or hundreds of thousands of years. That's still relatively fast in geological terms. The slow drift of continents, another major player in climate variation, can operate over hundreds of millions of years.

The overlap among these various processes creates a patchwork of warmings and coolings at different time intervals, some lengthy and some relatively brief. It's a bit like the mix of factors that shape sunlight in your neighbourhood. On top of the dramatic and regular shifts in sunlight you experience due to the 24-hour day and the transition from summer to winter, there are irregular shifts produced as cloudiness comes and goes.

Below is a quick summary of the biggest climate shapers. We'll come back to each of these later in the chapter.

▶ **The Sun is literally the star of the story** At its outset, the Sun shone at only about three quarters of its current strength; it took several billions of years for its output to approach current levels. One fascinating question (discussed later) is how Earth could be warm enough to sustain its earliest life forms while the Sun was still relatively weak. In another five or six billion years the Sun – classified as an "ordinary middle-aged star" – will reach the end of its life. At that point it should warm and expand dramatically before cooling to white-dwarf status.

▶ **Variations in Earth's orbit around the Sun are hugely important** These small deviations are responsible for some of the most persistent and far-reaching cycles in climate history. Each of these orbital variations affects climate by changing when and how much sunlight

James Croll

reaches Earth. As shown in the graphics overleaf, there are three main ways in which Earth's orbital parameters vary. Several visionaries in the mid-1800s, including James Croll – a self-taught scientist in Scotland – suspected that these orbital variations might control the comings and goings of ice ages. In the early 1900s, Serbian mathematician Milutin Milankovitch quantified the idea, producing the first numerical estimates of the impact of orbital variations on climate. While his exact calculations have since been refined, and there's still vigorous debate about what other factors feed into the timing of climate cycles, the basic tenets of Milankovitch's theory have held firm.

▶ **The changing locations of Earth's continents also play a huge role in determining climate** Francis Bacon observed in 1620 how easily South America and Africa would fit together if they were puzzle pieces. Many schoolchildren notice the same thing long before they're taught about **continental drift**, Alfred Wegener's bold 1915 hypothesis that Earth's continents moved slowly into their present positions. Today, satellites can monitor continental drift by tracking minuscule land movements. But scientists are also able to deduce the long-ago locations of continents thanks to various techniques – including analysing traces of magnetism generated by Earth's geomagnetic field over the millennia and preserved in volcanic magma. These and other clues tell us that much of Earth's land mass once existed as **supercontinents**. The first (and most speculative) is **Rodinia**, a clump that bears little resemblance to today's world map. Formed more than a billion years ago, Rodinia apparently broke up into several pieces. Most of these were reassembled as **Gondwana**,

The Earth, c. 250 million years ago, showing the supercontinent Pangaea

which drifted across the South Pole between about 400 and 300 million years ago. Gondwana became the southern part of **Pangaea**, an equator-straddling giant. In turn, Pangaea eventually fragmented into the continents we know now. The locations of these ancient continents are important not only because they shaped ocean currents but because they laid the literal groundwork for ice ages. Only when large land masses are positioned close to the poles (such as Gondwana in the past, or Russia, Canada, Greenland and Antarctica today) can major ice sheets develop. However, high-latitude land isn't sufficient in itself to produce an ice age. The southern poles harboured land for most of the stretch from 250 to 50 million years ago, yet no glaciation occurred, perhaps because CO_2 concentrations were far higher than today's.

Round and round we go (and so does our climate)

Everyone who lives outside the tropics knows the power of the four seasons. But a surprising number of people – including most of the Harvard graduates interviewed at random in the classic documentary *A Different Universe* – can't explain why the seasons occur. Many assume it's because Earth is closer to the Sun in summer and more distant in winter. But that doesn't wash: it's true that Earth's orbit is slightly asymmetric, but we're actually closest to the Sun each year around 4 January, when it's summer in Sydney but winter in London. The real reason for the seasons is that Earth is tilted about 23.4° relative to its orbit around the Sun (see graphic). That tilt aims each hemisphere towards the Sun in an alternating pattern that produces summers, winters, springs and autumns.

What does this have to do with climate change? Each of the factors just alluded to – Earth's tilt, the eccentricity of its orbit and the timing of its closest approach to the Sun – change slightly over time in a cyclical manner. This is due mainly to the intersecting gravitational effects of the Sun, the Moon and the other planets in our solar system. Together, the orbital cycles produce dramatic swings in climate over tens of thousands, or even hundreds of thousands, of years. By studying how these natural climate shifts come and go, scientists also find clues about how Earth might respond to change that's human-induced.

▶ **Earth's tilt** goes up and down, ranging from about 21.8° to 24.4° and back over about 41,000 years (the range widens or narrows a bit with each cycle). The tilt is now around 23.4° and on the decrease. When the tilt is most pronounced, it allows for stronger summer sun and weaker winter sun, especially at high latitudes. Ice ages often set in as the tilt decreases, because the progressively cooler summers can't melt the past winter's snow. At the other extreme, increasing tilt produces warmer summers that can help end an ice age.

Minimum tilt: 21.8°
Current tilt: 23.4°
Maximum tilt: 24.4°
0°

▶ **Earth's orbit** around the Sun is not precisely circular, but elliptical, with the Sun positioned slightly to one side of the centre point. Currently this brings Earth about 3% closer to the Sun in early January (perihelion) than in early July (aphelion), with about 7% more solar energy reaching Earth at perihelion. The **eccentricity** or "off-centredness" of the orbit varies over time in a complicated way. The net result is two main cycles, one that averages about 100,000 years long and another that runs about 400,000 years. When the eccentricity is low, there's little change through the year in the Earth–Sun distance. When eccentricity is high, the sunlight reaching Earth can be more than 20% stronger at perihelion than at aphelion. This either intensifies or counteracts seasonality, depending on the hemisphere and the time of year that perihelion and aphelion occur. The last eight ice ages have come and gone roughly in sync with the 100,000-year eccentricity cycle, but it's not yet clear how strongly the two are related (see p.244).

Perfect circle

Actual orbit (slightly elliptical)

Sun

Ellipse rotates around the Sun

▶ **Earth's axis** rotates slowly around an imaginary centre line, like a wobble in a spinning top. The main cycle of this process – known as **precession** – takes about 26,000 years, and it shifts the dates of perihelion and aphelion forward by about one day every 70 years. Thus, in about 13,000 years, Earth will be closest to the Sun in July instead of January. This will intensify the seasonal changes in solar energy across the Northern Hemisphere and weaken them in the south. The stronger summer sun is likely to bolster the African and Asian monsoons, as was the case around 10,000 years ago (see box, p.66).

Precession (rotation of the axis itself): 26,000 years

Rotation of Earth on its axis: 24 hours

Mount Pinatubo erupts on 12 June, 1991.

▶ **Volcanoes** can spew enormous amounts of ash, soot and other particles
and gases – some of them greenhouse gases – into the atmosphere.
A single large eruption can be powerful enough to loft sun-shading
particles into the stratosphere, as was the case when Mount Pinatubo
blew its top in 1991. When this happens, the high-altitude debris
can cool the planet by more than 1°C (1.8°F) for a year or more,
until gravity pulls the particles back to Earth. Tropical volcanoes are
especially effective at this cooling, since their stratospheric debris
can easily spread across both Northern and Southern hemispheres.
In the much longer term, volcanoes act as a warming influence, since
they add perhaps 0.1–0.3 gigatonnes of carbon to the atmosphere per
year on average. That's a substantial amount, though it's less than
1% of what humans now add through fossil fuel burning. At present
it appears that the volcanic addition is balanced by other natural
forces and feedbacks that remove carbon at roughly the same pace.
However, it's possible that a frenzy of volcanism early in Earth's history
pumped the atmosphere's greenhouse-gas levels far above anything
that's occurred since. Later on, undersea eruptions associated with the
spreading of continental plates may have been the main source of the
carbon dioxide that led to Earth's remarkable warmth from about 250
to 50 million years ago (see p.228).

▶ **Celestial bodies** such as **asteroids** make cameo appearances in Earth's climate saga. Most are minor. Only the very rarest of giant asteroids, which typically arrive millions of years apart, can produce climatic effects comparable to those from a major volcanic eruption. As with volcanoes, a huge asteroid might cool climate for a year or two by blasting dust and soot into the stratosphere. However, a big enough asteroid could also produce an explosive wave of heat as it enters the atmosphere, enough to set much of Earth's vegetation ablaze. Such fires would send enough carbon dioxide into the atmosphere to cause a substantial global warming that could last centuries, far longer than the brief post-collision cooling. This lurid scenario apparently played out around 65 million years ago, when an enormous asteroid struck Earth. The most likely point of impact is an area on the north end of Mexico's Yucatán Peninsula now underlaid by a giant crater of pulverized rock. Scientists combing the globe have found widespread deposits of iridium and other substances that are far more prevalent in asteroids than on Earth, all apparently deposited at the same time. Also dating from the same period are clumps of charcoal-like carbon, a hint that widespread fires occurred. More than half of all species are thought to have perished in this cataclysm and its aftermath, including most dinosaur species. (A few that survived may be the predecessors of modern-day birds and crocodiles.)

How scientists read the clues

The technology used to analyse materials and glean climate clues from them has advanced at a phenomenal rate in recent decades. The traces of past climates can show up in animal, vegetable and mineral forms, providing a wealth of evidence for paleoclimatic detective work. Below are the four main types of paleoclimate **proxy data** – records that can be used to infer atmospheric properties such as temperature or precipitation:

▶ **Biological:** tree rings, pollen, corals and fossils of plants, animals, insects and microscopic creatures such as diatoms. Their locations, concentrations and conditions point towards the atmospheric and oceanic conditions under which various life forms prospered. Many biological materials can be dated using radioactive decay rates or the presence of stable isotopes (more on these later). The chemical make-up of these samples can also yield important information.

▶ **Cryological:** cylindrical samples called ice cores, collected by drilling through dense, deep ice sheets or glaciers where air bubbles and dust are trapped within yearly layers of snowfall. The bubbles preserve the greenhouse gases and other atmospheric constituents present at a given time, and the ice cores can be analysed through radiometric dating (see p.222). The dust layers and the character of the ice itself also reveal climate processes at work.

▶ **Geological:** rock, sand dunes, ocean sediments, glacial debris, stalagmites and other materials sampled from sea and lake beds and dry land. Again, radioactive decay within these materials helps establish their age. Volcanic rock can also be dated by the traces of Earth's geomagnetic field. The location and condition of telltale geological features helps

What trees tell us about climate

It was an astronomer – who happened to be looking down instead of up – who launched the science of **dendrochronology**. In the 1920s, Andrew Ellicott Douglass pondered the rings in tree stumps he encountered while hiking through the forests of northern Arizona. Douglass discovered that the width of the tree rings in this arid area was correlated with the precipitation amounts that occurred during the trees' lifetimes. He went on to found the Laboratory of Tree-Ring Research at the University of Arizona in Tucson in 1937. Among his noteworthy accomplishments was developing a method using tree rings to determine the age of wood beams in prehistoric structures across the US Southwest.

From its roots in Arizona and elsewhere, dendrochronology has branched out across the world, with past climates now profiled at thousands of sites. To create a typical data set, the ring patterns obtained from several trees in the same area are matched, cross-dated and analysed to produce an annual, localized set of tree-ring growth indices known as a site chronology. For some types of living trees, such as bristlecone pines, these chronologies can go back thousands of years. (Tree lovers, take note: the slender cores can be extracted from live trees without harming them.) If a series of progressively older dead trees can be found whose lifespans overlap with living trees, the sequence can be extended back even further. In Germany, the University of Hohenheim has created a 12,480-year record using oak and pine trees. How much further back can dendrochronologists go? There is at least one place in the world where trees from the depths of the last ice age are preserved in relatively pristine shape. That's on the North Island of New Zealand, where bogs may hold Kauri pine dating back as far as 50,000 years.

It takes far more than counting in order to use tree rings as guides to past climates. Each site has a unique blend of characteristics – temperature, precipitation, soil establish when sea levels rose and fell or when glaciers scoured the land surface. Many ocean sediments are laden with the remnants of shells built by creatures the size of a grain of sand, including bottom-dwelling **foraminifera** as well as **plankton** that prefer near-surface waters. The prevalence and type of these shells can reveal the amount of carbon and other elements that entered the sea at various points.

▶ **Historical:** written records of atmospheric conditions and biological phenomena tied to climate. The latter might include the timing of spring blooms for a given type of tree or the poleward extent of a particular species of insect. In some areas, including England and Japan, such records go back hundreds of years with almost daily precision.

An ancient bristlecone pine from the Great Basin region of western North America.

conditions and the like – that affect how trees grow. For example, trees at arid forest borders are limited by moisture availability, so the variations in their rings will reflect ups and downs in precipitation. It's also possible to deduce catastrophic events, such as when a tree is shunted by an advancing glacier or partially knocked over by a storm. One clue is the asymmetric ring pattern produced as the tree attempts to grow upright after the injury.

Some of the most dramatic findings of dendrochronology relate to the region where the discipline was born. Connie Woodhouse of the US National Oceanic and Atmospheric Administration went to the Arizona lab in the early 1990s, and later made headlines when she and colleague Jonathan Overpeck compiled evidence, based on tree rings and other proxy records, of what Overpeck dubbed "megadroughts". These persistent dry spells, lasting several decades each, occurred across large portions of the central and western US prior to 1600. Megadroughts can affect the course of entire civilizations (see p.239), even though there may be occasional years of near-normal rain amid the dry years.

Establishing dates for climate proxies can be a challenge. Trees are among the easiest: each ring indicates a year's worth of growth, and its width hints at the state of that year's climate (see p.220). For trees that are still alive, or ice cores drawn from the surface downward, it's a simple matter to count backwards from the present and establish dates in the past, using multiple samples to ensure accuracy. For many other proxies, such as fossilized trees, you also need a benchmark, something you can use to establish the time period in which the proxy was created. A typical starting point is **radiometric dating**, which revolutionized the paleoclimatic dating game in the 1950s.

Radiometric techniques rely on the fact that elements such as **carbon** and **uranium** exist in several different forms, or **isotopes**, each with a different number of neutrons. Some of these isotopes are unstable, and they decay over time at measurable and predictable rates. For instance, neutrons hitting the upper atmosphere are continually converting nitrogen into an unstable carbon isotope known as $_{14}C$. This sifts down into Earth's atmosphere and gets absorbed by living things, decaying over time as new $_{14}C$ isotopes are created higher up. Given a certain amount of $_{14}C$, half of it will have decayed back into nitrogen after about 5700 years (its **half-life**). Knowing this, scientists can measure how much $_{14}C$ has decayed in a given substance and use that information to determine how old the substance is. Although these unstable isotopes are rare compared to stable carbon atoms, they're well distributed throughout Earth and its ecosystems. This allows geochemists to date a wide variety of substances, from carbon dioxide in an ice-core air bubble to the shells of marine creatures. One limit to using $_{14}C$ is its relatively short half-life. Since there's very little carbon left to measure after a few half-life cycles, most carbon dating only goes back about 50,000 years. For older substances, **mass spectrometry** dating is used. Uranium is the material of choice for this, because it decays into lead at a very slow rate. Uranium's half-life is about 700 million years for the isotope $_{235}U$ and an astounding 4.5 billion years – the age of Earth itself – for $_{238}U$.

As with any measurement technique, errors can creep into radiometric dating. This happens mainly when a substance is contaminated by other materials of a different age through chemical reactions, erosion or some other process. Careful cross-calibration with similar materials from other locations can help alleviate this problem.

Isotopes don't have to be unstable to be useful in dating. Stable isotopes provide another window on the climate's past. The most extensively studied is $_{18}O$, a form of oxygen. Because $_{18}O$ is heavier than standard oxygen, it condenses more readily. The difference between the two rates depends on temperature, ocean salinity and other factors. So, for example, coral reefs hold differing amounts of $_{18}O$ based on the sea-surface temperature present when they formed. In ice cores (and other places where ancient water is trapped) the relative amount of $_{18}O$ can help reveal the temperature at which the moisture condensed. This can help delineate annual layers in the ice as well as reflecting longer-term temperature trends.

Computer models are another important part of the paleoclimate toolbox. In theory, any model that simulates future climate ought to be able to reproduce the past as well, given enough information to start with. Indeed, many of the most sophisticated global models designed to project future climate are tested by seeing how well they reproduce the last century. Another useful test is to see how well the models handle modern-day seasonal transitions: in mid-latitudes, at least, going from summer to winter can serve as a microcosm of a long-term overall climate shift. For prehistoric times, there's far less data to check a model against, but even in these cases simple models can be helpful in assessing what climates might be plausible. There's often a back-and-forth between paleoclimate modelling and data: observations might point to a particular climate scenario that can be tested through models, and the model results might confirm or deny a particular interpretation of the data, sending scientists back to the drawing board (or the field).

The really big picture: from Earth's origins to the expansion of life

It takes quite a mental stretch to fully comprehend the total lifespan of Earth's atmosphere – unless, that is, you're a biblical literalist (see p.224). Most of our planet's history unfolded long before the appearance of life forms that we're now familiar with. To get a sense of this, picture Earth's 4.5 billion-year lifespan as if it were compressed into a mere 24 hours. In this imaginary day, dinosaurs wouldn't come on the scene until after 10pm. Even more impressive, the period since the last ice age ended – the time when most human civilizations emerged – would occupy only the last fifth of a second before midnight. Deducing the complete history of the atmosphere is thus a major challenge.

Earth itself offers few clues to help us figure out what climate was up to during the planet's first two billion years or so. In fact, a good deal of our knowledge about Earth's earliest climate comes from space. By studying the behaviour of stars similar to the Sun, astronomers have deduced that

Earth's age: 4.5 billion years or 6000 years?

Virtually all physical scientists accept the basic chronology of Earth's history summarized in this chapter. However, millions of Americans beg to differ. These Christian fundamentalists generally take the Bible at its word: that Earth was created in six 24-hour days about 6000 years ago. A cottage industry of experts, including some PhD scientists, promulgates this view through such enterprises as the Institute for Creation Research. They often employ the Bible's story of a giant flood to explain the formation of the Grand Canyon and other geological features. (The official gift shop at the canyon's national park sells a book along these lines called *Grand Canyon: A Different View*.) In recent years many US school boards have pondered the teaching of intelligent design (ID), a somewhat watered-down version of creation science that seeks evidence for the existence of a creator without specifying whom that might be. Most of the ID movement's energy is focused on biological rather than climate science, although its proponents seem to be philosophically inclined towards scepticism about human-induced global warming. (That's not the case for all US evangelicals – see p.288.) Public opinion polls suggest a cognitive split among many Americans that allows them to accept Earth science that contradicts the Bible even as they draw the line at the more discomforting idea of human evolution. A report by the US National Science Foundation shows that more than 80% of Americans surveyed in 2008 agreed with the statement, "The continents have been moving their location for millions of years and will continue to move." However, US surveys have shown consistently that nearly half of Americans believe God created humans in the last few thousand years.

the solar furnace (a continuous nuclear reaction that forms helium out of hydrogen) took a long time to reach its modern-day strength. It appears the Sun's luminosity started out only about 70–75% as strong as it is now, gradually increasing ever since. With the Sun relatively weak for so long, you might envision early Earth as a frozen, lifeless wasteland. Amazingly, this wasn't the case. Fossils of tiny bacteria show that life existed as far back as 3.5 billion years. Rocks from around that time show the telltale signs of water-driven erosion, another sign that Earth wasn't encased in ice. How did Earth stay warm enough for life during those early years? That's the crux of what paleoclimatologists call the **faint young sun paradox**, discovered by astronomers Carl Sagan and George Mullen in 1972. Explanations for the paradox have been floated ever since. In 2010, Danish scientist Minik Rosing and colleagues proposed that widespread oceans and relatively thin clouds allowed the early Earth to absorb much more heat than it would today. But greenhouse gases, as suggested by Sagan and Mullen, are the most commonly proposed solution to the paradox. Put enough of these gases in the air, and Earth can stay warm enough to sustain life even with a weaker Sun. The amount of carbon now stored in fossil fuels and in the oceans is more than enough to have provided such a greenhouse boost if it were in the air during the weak-Sun period. Such a boost might not have come from carbon dioxide, though, as recent geochemical analysis of ancient soil suggests that there was far too little CO_2 in the air to do the trick. Methane appears to be a strong runner-up candidate.

Volcanoes can inject huge amounts of greenhouse gas into the atmosphere in a short amount of time, so the intense volcanism of Earth's earliest days might have added enough to keep the planet from freezing. However, that begs another question: how did the greenhouse-gas levels gradually decrease over billions of years at just the right tempo to keep Earth warm, but not too warm? It's possible, but unlikely, that volcanic activity diminished at exactly the pace needed. Another possibility is that some other climate factor kicked into gear to keep greenhouse gases within a range tolerable for life on Earth.

One candidate is the steady expansion and elevation of continents during much of Earth's history. **Land masses** reduce the amount of carbon dioxide in the air through a multimillion-year process known as **chemical weathering**, which occurs when rain or snow fall on rocks that contain silicate minerals. The moisture and silicates react with carbon dioxide, pulling the CO_2 out of the air. Carbon and water then flow seaward in various forms, and most of the carbon ends up stored in ocean sediments that gradually harden into rocks. Over billions of years, this process has been responsible for some 80% of the carbon now stored underground. (Rain and snow also draw CO_2 out of the air when they leach limestone, but in that case the carbon heads seaward in a form used by marine creatures to build shells, a process that returns the carbon to the atmosphere.) Chemical weathering is such a powerful process that the growth of the Himalayas over the last 55 million years might have helped to pitch Earth into its recent series of ice ages, as we'll see later.

Apart from these vast changes, chemical weathering can also influence the atmosphere through a negative feedback process, one that tends to counterbalance climate change in either direction. In general, the warmer it gets, the more rapid chemical weathering becomes and the more carbon dioxide is pulled from the air. In part this is because the chemical reactions behind weathering operate more rapidly when it's warmer. Another reason is that warmer oceans send more water vapour into the air, and more rain and snow falls onto continents to stimulate weathering. Thus, the warmer the air gets, the more carbon dioxide is drawn out of the atmosphere through weathering, and that should help produce a cooling

Gaia and global warming

Though often associated with New Age thinking, the Gaia hypothesis (the idea that Earth and its life forms act as a self-regulating entity) isn't necessarily warm and fuzzy. The name itself, a reference to the Greek goddess of Earth, was suggested in 1969 by William Golding, author of the dystopic novel *Lord of the Flies*. Today, the father of Gaia theory, scientist **James Lovelock**, is deeply disturbed about the direction in which global warming – or as he calls it, "global heating" – is pushing the Earth system. His 2006 book *The Revenge of Gaia* warns that "we live on a live planet that can respond to the changes we make, either by cancelling the changes or by cancelling us." Even more bleak was his 2009 follow-up *The Vanishing Face of Gaia*. Billed as his "final warning" to humanity, it raises the notion that humans might render themselves extinct via climate change. "We do not seem to have the slightest understanding of the seriousness of our plight", he says.

Lovelock was inspired in the 1960s by the contrast between the lack of evidence for life on Mars and the fact that life evolved and prospered on Earth despite the slow ramp-up of solar energy (see p.244). "Together, these thoughts led me to the hypothesis that living organisms regulate the climate and the chemistry of the atmosphere in their own interest", writes Lovelock. He sees Gaia as a single, self-regulating entity that encompasses Earth's living things and the chemical and physical backdrop that sustains them. Though Lovelock doesn't view Gaia as an evolving organism in classical Darwinian terms, he does see the web of interacting feedbacks among greenhouse gases, vegetation and creatures on land and in the oceans as a system that preserves and perpetuates the conditions that foster life.

It's been a long slog for Lovelock to convince disciplinary scientists of the merit in his all-embracing theory. Partial vindication came in 2001 when a group of leading Earth scientists issued the Amsterdam Declaration on Global Change. It asserts that

trend. Conversely, if ice starts to overspread the land, weathering should decrease and carbon dioxide levels ought to increase, thus working against the glaciation.

Earth's profusion of **plant life**, which began around 400 million years ago, serves as another brake on greenhouse gases. Plants and trees take in carbon dioxide through photosynthesis and release it when they die. Generally, plant growth increases on a warmer, wetter Earth – so the hotter it gets, the more plants there are to help cool things down. Indeed, the concept of Gaia positions life itself as the main negative or balancing feedback on the entire Earth system (see box above).

Of course, there are also many **positive feedbacks** between CO_2 and climate, which tend to amplify rather than dampen a change. For instance, as carbon dioxide increases and warms the planet, more water vapour – itself "the Earth system behaves as a single, self-regulating system comprised of physical, chemical, biological and human components". Lovelock goes further by insisting this system has a goal: "to keep the Earth habitable for whatever are its inhabitants".

Like any organism, Gaia can be pushed to its limits, says Lovelock, and warming the atmosphere is asking for such trouble. If Gaia could express a preference, Lovelock believes that "it would be for the cold of an ice age, not for today's comparative warmth." One example: both drought and flood are a greater risk in warm climates than in cool ones, because water tends to evaporate more quickly the hotter it gets. Another example: oceans warmer than

James Lovelock

10°C (50°F) become stratified such that nutrients are quickly depleted in the warm surface layer. Lovelock fears that greenhouse gases could trigger an irreversible cascade of feedbacks that profoundly disrupt the self-regulating processes of Gaia. As for what to do about this, Lovelock thinks the standard renewable-energy goals of more wind farms and solar panels aren't nearly enough. He calls for an immediate ramp-up of nuclear power, as a form of energy triage, while we develop and refine alternative sources of power over the long haul. "It is much too late for sustainable development", he says; "what we need is a sustainable retreat."

a greenhouse gas – should evaporate into the atmosphere and further increase the warmth. Similarly, oceans absorb less carbon dioxide the warmer they get, which leaves more CO_2 in the air to stimulate further warming. It's the balance between negative and positive feedbacks that generally determines which way climate will turn at any given point. A big concern about the twenty-first century is that positive feedbacks, such as those now taking place in the Arctic, may overwhelm negative feedbacks, some of which (such as chemical weathering) take much longer to play out.

Whether they're looking a thousand or a billion years into the past, paleoclimatologists take it for granted that greenhouse gas concentrations are intimately linked with the rise and fall of global temperature. Ice cores and other records bear out this tight linkage over the last million years, and there's no reason to think it should be absent earlier on. There's a subtlety in this relationship, though – one often pointed out by global-warming sceptics. Because the CO_2 changes are so closely coupled in time to the growth and decay of ice sheets, it's hard to tell exactly which came first and when.

For example, there appear to be periods in which undersea volcanism produced vast amounts of CO_2, after which temperatures rose dramatically, and other periods where orbital cycles triggered ice-sheet growth and colder temperatures, with CO_2 amounts dropping as a side effect. Paleoclimatologists are still sorting out these chains of events. It appears that orbital changes play a major role in kicking off and ending ice ages, but during the growth and decay stages, there's a web of positive feedbacks at work among CO_2, ice and other climate elements. For instance, cooler oceans absorb more carbon dioxide, which acts to support further cooling. In any case, the fact that CO_2 can sometimes lag behind a major climate change doesn't mean that it can't help trigger one, which is the situation we're finding ourselves in now.

Hot in here: Earth's warm periods

What will the world look like if the more dire projections of climate change by the year 2100 come to pass? Computer models and basic physical theory tell us a lot about what to expect, but we can also gaze back into climate history for clues.

A good place to start is the **Mesozoic era**. It began after the world's greatest extinction episode, which occurred about 250 million years ago, and ended with another vast extinction 65 million years ago. In between are three periods well known to dinosaur lovers: the Triassic, Jurassic and Cretaceous. All three intervals were notably warm, with no signs of any major glaciations or even much ice at all. In part this is because Earth's land masses were located away from the North Pole, mostly clustered in the warm supercontinent of Pangaea. However, parts of Pangaea extended to the South Pole, and the warmth continued even as Pangaea broke up and the Northern Hemisphere continents shifted closer to their present-day locations.

Two of the warmest intervals were the mid- to late Cretaceous and the early Eocene (part of the Tertiary epoch), which fall on either side of the asteroid-driven extinction that occurred 65 million years ago. In both periods, average global readings soared to 5°C (9°F) or more above present values. That's near or beyond the IPCC's high-end projections of global temperature for the year 2100. Proxy data for these past periods indicate

Close calls for life on Earth

US poet Robert Frost famously observed in 1920 that some people expect the world to go out in fire, while others think the end will be an icy one. Our planet edged towards those extremes at several different points in its history. Two examples brought vastly different consequences for ecosystems.

One of the coldest periods in early Earth history ran from about 750 to 580 million years ago (pictured). There's evidence that glaciers scoured most of Earth's large land masses, including some within 10° of the Equator. The implication is that much or all of the planet was covered with ice for millions of years – a scenario dubbed Snowball Earth in 1992 by geologist Joseph Kirschvink of the California Institute of Technology. Since Earth's land was concentrated in the tropics, where precipitation is heaviest, chemical weathering (see p.225) could have pulled large amounts of carbon dioxide from the air, cooling the global atmosphere. If the planet actually did freeze over entirely, the powerful positive feedback of a "white Earth" could have reflected most

of the incoming sunlight and helped preserve the ice. How such a state would have ended isn't known, but a slow accumulation of carbon dioxide from volcanoes – on top of a cold-induced slow-down in chemical weathering– seems the most likely possibility. Whether or not ice actually covered the entire Earth, this period of widespread glaciation was followed by a sudden profusion of multi-cellular organisms. Some theorists speculate that a Snowball Earth and a subsequent warm-up could have prodded the rapid evolution of primitive life into more complex forms.

At the other end of the spectrum, an especially intense warming around 250 million years ago spelled doom for most of the life forms present at the time. The Permian/ Triassic extinction drew the curtain on more than 90% of marine species and more than two thirds of land-based creatures. The die-off unfolded quickly in geological terms, over fewer than a million years. Greenhouse gases likely soared to many times their present-day amounts, and high-latitude oceans warmed to as much as 8°C (14.4°F) above present-day readings. The warmer oceans likely enhanced the separation of surface waters from cooler deep oceans, reducing the usual mixing between the layers that distributes oxygen and nourishes many marine organisms. Researchers haven't settled on a single cause for the extinction. A modelling study led by Jeffrey Kiehl (National Center for Atmospheric Research) suggested that enhanced volcanic activity over almost a million years could produce enough atmospheric warming through added carbon dioxide to slow the ocean currents that bring oxygen to deep-water marine life. Other possibilities include a massive meteorite (although there's no sign of an impact crater dating from that era), an outpouring of the greenhouse gas methane in the form of methane hydrates released from the ocean floor into the atmosphere (see p.230), or some blend of these and/or other factors.

temperature for the year 2100. Proxy data for these past periods indicate that carbon dioxide was several times more prevalent in the atmosphere than it is today. This enhances scientists' confidence that the extra CO_2 stoked the ancient warmth and that the greenhouse gases we're now adding could do something similar in our future.

Where did the extra CO_2 come from? The most likely candidate is an active period of **undersea volcanism** (see graphic below). Earth's tectonic plates clash most visibly along the world's volcanic coastlines, such as the Pacific's "ring of fire". However, there's additional volcanic activity far beneath the oceans. Most major basins have a mid-ocean ridge, where tectonic plates separate and vast amounts of magma and carbon dioxide surge upwards to fill the void. The magma eventually solidifies to form fresh crust, slowly elevating the centre of the ridge as the older material spreads outward. It appears that, on average, the plates were separating up to 50% more quickly about 100 million years ago than they are now. This would have allowed more CO_2 to bubble upwards and pushed the plates towards continents more vigorously, stimulating more activity (and more CO_2 emissions) at coastal volcanoes.

Another source of greenhouse gas might have provided even more of an influx at key times. Hundreds of billions of tonnes of carbon are believed to be locked in cold ocean seabeds in the form of **methane hydrates**, ice crystals that hold methane in their molecular structure (also known as methane clathrates). Methane hydrates remain stable if conditions are cold enough and/or if enough pressure is exerted on them (as is the case at the bottom of the sea). But a change in ocean temperature or circulation

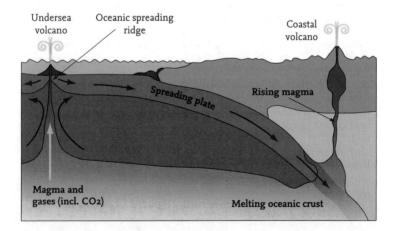

it pouring into the atmosphere. This is a long-term concern for our future climate (see p.95), and it may have played a role in at least two massive warmings during already warm times. One was the temperature spike and mass extinction that kicked off the Mesozoic era 248 million years ago. The other is a distinct blip early in the already toasty Eocene epoch called the Paleocene-Eocene Thermal Maximum (PETM), which occurred about 55 million years ago. During the PETM, high-latitude air and ocean temperatures soared by as much as 7°C (12.6°F) over a few thousand years and remained high for almost 100,000 years. Isotope studies by NASA indicate that a vast amount of greenhouse gas was pumped into the atmosphere during the PETM, though over a longer period and perhaps at a slower rate than our current emissions. Among the possible sources is a burst of volcanism in the North Atlantic, as suggested by recently analysed igneous rock formations off the coast of Norway.

One thing that's fairly clear about past warm periods is that mid- and high-latitude areas were downright balmy as compared to today. Coral reefs grew as far polewards as 40°, close to the latitudes of present-day Philadelphia and Melbourne. Crocodiles, dinosaurs and leafy trees flourished in polar regions, while other areas featured vast deserts covered with sand dunes. A 2006 study of sediments beneath the Arctic Ocean found that sea-surface temperatures (SSTs) during the PETM soared to as high as 23°C (73°F), which is far beyond earlier estimates. In the tropics, SSTs may not have been much warmer than they are now, though some intriguing evidence of readings at or above 33°C (91°F) has been collected off the coasts of present-day South America and East Africa.

The oceans were at a high-water mark during these warm times. For millions of years, **sea level** was more than 200m (660ft) higher than it is now. This put vast stretches of the present-day continents under water, including the central third of North America and all of eastern Europe from Poland southward. The oceans were so high and vast for several reasons. One is that there wasn't any water locked up in long-term ice sheets. Another is that water, like any other fluid, expands as it warms up, just as the oceans are doing now (albeit on a much more modest scale). A third reason is the shape of the oceans themselves. The faster separation of tectonic plates noted above would have produced larger mid-ocean ridges as new crustal material surged upward. This, in turn, would have raised the height of the seafloor and the resulting sea level by surprisingly large amounts – perhaps enough to account for most of the difference between the Triassic/Eocene sea levels and today's.

From greenhouse to icehouse: the planet cools down

The most recent 1% of our atmosphere's history is mainly a story of cooling. Starting in the mid-Eocene, around 50–55 million years ago, sea levels began to drop, continents shifted ever closer to their current locations, and global temperatures began to fall. The cooling appears to have temporarily reversed between about 15 and 25 million years ago, then resumed with gusto. This long process wasn't linear – there were sharp warmings and coolings in between, each lasting many thousands of years – but overall, Earth's temperature took a tumble of perhaps 4°C (7.2°F) or more. This eventually pushed the planet into the glacial/interglacial sequence that began fewer than three million years ago and continues today.

What caused the cool-down? Once again, it appears that carbon dioxide was the most direct culprit. CO_2 in the atmosphere went from several times its modern-day amounts in the mid-Eocene to barely half of its present-day concentration during the last few ice ages. The question of what drove down CO_2 levels is far more challenging. Mid-ocean ridges were separating more slowly when the cool-down began, a sign that volcanic activity was pumping less CO_2 into the atmosphere. The biggest decreases in volcanism apparently occurred prior to 40 million years ago. That would correspond to the first pulse of global cooling, but the spreading rate had increased again by the time the second pulse began, leaving that one unexplained.

The Indian subcontinent about to crash into Asia 60 million years ago.

In their search for a solution to this riddle, one group of US scientists cast their eyes towards the **Tibetan plateau**. This gigantic feature began to take shape around 55 million years ago as the Indian subcontinent joined Asia, pushing the Himalayas and the plateau upwards. The pace quickened about 30 million years ago, and today most of the Alaska-sized plateau sits at elevations greater than 3700m (12,000ft). It's bordered by mountains that soar more than twice as high.

In the late 1980s, Maureen Raymo (of Woods Hole), Flip Froelich (now at Florida State University) and William Ruddiman (of the University of Virginia) proposed that the intense chemical weathering produced by the Tibetan plateau could have pulled enough carbon dioxide from the air to trigger the global temperature drop that led to the most recent ice ages. Normally, a drop in global temperatures would be expected to act as a negative feedback, reducing precipitation and slowing down weathering. But in this case, we're faced with a change in the playing field (Earth's geography), and that could control the rate of weathering entirely on its own, overpowering any feedbacks. If ever there were a geological event conducive to weathering, it's the uplift of the Tibetan plateau – perhaps the most massive above-ground feature in Earth's history. In the absence of enough data to confirm or disprove Raymo and colleagues, the debate over the advent of Earth's ice ages goes on.

There were other tectonic developments of interest during the long cool-down. **Antarctica** separated from South America between 20 and 25 million years ago. With air and water now able to circle around Antarctica unimpeded, the resulting tight circulation might have fostered the growth of the first major Antarctic ice sheet about thirteen million years ago (though modelling studies have been inconclusive on this point).

Meanwhile, North and South America slowly approached each other, finally joining at the Isthmus of Panama nearly four million years ago. This closed the longstanding connection through which warm, salty water had flowed beneath trade winds from the tropical Atlantic into the Pacific. In a major readjustment, the Atlantic shifted into a mode much like the one we see today, with the tropical waters flowing northwards along an intensified Gulf Stream and on into the far North Atlantic. This influx of warm, saline water into the Arctic would have cut down on sea-ice formation. Ironically, it could have also furnished the moisture needed to build ice sheets over land and thus hastened glaciation.

Thanks to some combination of the factors above – plus, perhaps, others yet to be discovered – Earth inexorably slid into what some experts call a "deteriorating" climate. Antarctic ice grew in a start-and-stop fashion. Glaciers began to appear about seven million years ago in the Andes, Alaska and Greenland. Finally, **ice sheets** began spreading across North America and Eurasia about 2.7 million years ago. Since then, more than two dozen glaciations have ground their way across much of present-day Canada, northern Europe and northwest Russia. In between each glaciation was a shorter, milder period, with ice extent and temperatures closer to those at present and even warmer in some cases.

Core value: getting information out of ice

Our understanding of the last few ice ages has been boosted immeasurably by the rich vein of data obtained through ice cores from Greenland, Antarctica and several other glacier-studded areas. The task is laborious: each ice sheet is more than 3km (1.9 miles) thick, but each cylindrical core is only about 10cm (4in) in diameter, and each segment of the core must be drilled and removed a few metres at a time. Weather only permits a few short months of work each year. Even in midsummer, daytime temperatures can be colder than -20°C (-4°F) atop the Greenland ice sheet; in central Antarctica that's a mild summer day.

The first major ice cores were retrieved from both Antarctica and Greenland as part of the International Geophysical Year (IGY) of 1957–58. In Greenland, the first cores were taken at camps far from the thickest part of the ice sheet. As interest in climate change intensified and researchers clamoured for more of the ice, two parallel projects set out to gather definitive cores from the top of the sheet. By the early 1990s, these US and European teams – separated by only about 32km (20 miles) – had drilled from the summit of Greenland's ice sheet to bedrock, a depth of just over 3km (9900ft). The two cores they obtained each spanned a little more than 100,000 years of climate. In 2010 a new coring project hit bedrock in northwest Greenland, gathering data that could reach back through the last interglacial period.

Meanwhile, Russian scientists at Antarctica's Vostok Station, which was founded during the IGY, were engaged in their own decades-long drilling effort. By 1995, they'd reached a depth of around 3600m (11,800ft) before they halted to avoid contaminating a vast lake beneath the ice. The Vostok core is only about 20% longer than those from Greenland. However, since the highest reaches of the Antarctic ice sheet get so much less snowfall per year than Greenland's summit, each metre of the Vostok core details a much longer stretch of climate. The total Vostok record covers about 440,000 years.

One of the most striking findings from both poles is that carbon dioxide and methane, two key greenhouse gases, rose and fell in near-lockstep with the comings and goings of ice ages. Typically, the concentrations dropped by 20–50% during a glaciation, then quickly recovered afterwards.

The new champion of ice archives is the core completed in 2004 by the EU-sponsored European Project for Ice Coring in Antarctica (EPICA). Retrieved at a site atop Dome C, about 500km (300 miles) from Vostok, the EPICA core has already yielded climate analyses extending back 800,000 years. Researchers with EPICA hope to use the core to look back nearly a million years – around the time when ice ages mysteriously transitioned to much longer intervals than before.

A "drill-dome" on the Greenland icecap.

At first, these **interglacial** periods occurred about every 41,000 years. This corresponds remarkably well to the 41,000-year cycle in the tilt of Earth's axis (see box on p.217). When Earth's tilt is maximized, the extra solar energy received at high latitudes in summer melts ice sheets more easily (even though winters are getting less sun than usual). When the tilt is lower, summers are cooler, and ice sheets get a running start with unmelted snow from the previous year.

About a million years ago, however, the 41,000-year rhythm of ice ages ended. The ice ages got stronger and began to last much longer – about 100,000 years each, with interglacial periods on the order of 10,000 to 15,000 years long. What was behind this shift isn't yet clear. Earth's orbit also has a 100,000-year cycle in its off-centredness (see p.217), but most paleoclimatologists don't believe that this relatively weak cycle was enough in itself to wrench a change in the ice-age frequency. However, it could have interacted with some other cyclic process to provoke the timing. One of the more exotic hypotheses for the current timing of ice ages is that Earth's orbit may be passing through a disk of interplanetary dust about every 100,000 years.

Another intriguing idea, advanced by Peter Huybers (Harvard University) is that the 41,000-year cycle in Earth's tilt is still the main driver, but ice sheets are no longer responding to every peak in the cycle. By simulating ice-sheet behaviour in computer models, Huybers found that random variations – analogous to the chaotic patterns observed in day-to-day weather – could tip the system into a mode where ice sheets don't fully melt at each cyclic peak. Instead, the ice ages take hold every second or third peak by turns (about every 80,000 or 120,000 years), and this alternation leads to an average frequency of about every 100,000 years. He's now investigating whether the weight of ice released as an ice age ends could be enough to allow volcanic activity to ramp up, which in turn would help drive deglaciation by releasing large amounts of greenhouse gas. There's much yet to be learned, according to Huybers: "I think ice ages are really the outstanding mystery in Earth science", he says.

The period just before the last ice age began, about 130,000–115,000 years ago, provides a sobering illustration of what climate change could accomplish over the next few centuries. That interglacial kicked off with a sharp warming most likely driven by orbital cycles. A 2009 synthesis of paleoclimate data from around the world led by Robert Kopp (Princeton University) found that sea level was likely more than 8m (26ft) above today's global average, even though greenhouse gases were less abundant than today. Computer modelling by Jonathan Overpeck (of the University of Arizona) and Bette Otto-Bliesner (of the US National Center for Atmospheric Research) showed that major ice-sheet melting at both poles may have fed the sea-level rise. The same modelling found that Arctic summer temperatures by 2100 will be close to those achieved in the last interglacial warming. Although it could take decades or even centuries for that warming to translate into a 6m sea-level rise – which would swamp major coastal areas around the world – Overpeck worries that the process of ice-sheet melt could prove irreversible sooner than we think.

Emerging from the ice

By the time of its peak about 20,000 years ago, the last ice age had carved out a very different-looking world than the one we know now. A major effort in the 1970s to reconstruct this period's topography showed a North America and Europe half encased in ice, with global temperatures about 4°C (7.2°F) colder than today's average. South of these ice sheets, it was largely dry and windy, especially over continental areas. Tundra covered most of southern Europe, and spruce trees characteristic of high latitudes extended into the mid-Atlantic area of the present-day US. One place where rainfall exceeded today's norm is the US Great Basin: Nevada and Utah were freckled with dozens of large lakes that now exist only as salt flats. Sea levels were so low – more than 100m (330ft) below today's – that Australia extended northwards into New Guinea, and Japan and Korea were separated only by shallow water.

The ice age may not have looked exactly like this through its entire 100,000-year lifespan. Until ice cores were plucked from Greenland and the Antarctic, many scientists had pictured ice-age climates as being relatively static. Instead, the cores revealed that Earth swung through a number of distinct warmings and coolings during the start of the last ice age, about 115,000 years ago, up to and beyond its conclusion around 15,000 years ago. As the clarity and completeness of ice-core data improved in the 1990s, it was joined by analyses of ocean sediments from several dispersed basins and stalagmites from caves in China. The new evidence made it clear that warm and cold swings during the ice age were far more widespread, and could unfold far more rapidly, than scientists had previously thought (see box opposite).

There's little debate about what brought the last ice age to an end. Two of the orbital **cycles** discussed on p.217, involving Earth's tilt and its pre-cession, synchronized to produce a strong peak in the summertime input of solar energy across the Northern Hemisphere. Starting about 15,000

How fast can climate flip?

Evidence as far back as the 1970s hinted that climate during glacial periods might not have been as static as many believed. The pieces came together in the early 1990s, when scientists confirmed a sequence of major warmings and coolings that unfolded against the cold backdrop of the last ice age. Instead of furnishing a constant chill, it seems the 100,000-year ice age unfolded in a much more irregular fashion.

Gerald Bond and colleagues at the Lamont-Doherty Earth Observatory drew on data from sediments in the North Atlantic to trace the evolution of the ice age in eye-opening detail. The kind of upwards and downwards spikes they found don't seem to occur in warm regimes like our present-day, post-glacial climate. But they serve as a reminder that climate can switch from one mode to another in the geologic equivalent of a heartbeat. The great ice sheets that coated much of Europe and North America never disappeared, but conditions across many ocean and continental areas varied sharply with these warmings and coolings, each named for paleoclimatic pioneers.

▶ **Going up** On 23 different occasions during the last ice age, air temperatures quickly climbed about half of the way back to their interglacial levels, then sank back to more typical ice-age readings. The warm-ups pushed average temperatures in Greenland up by as much as 16°C (29°F) in as little as forty years, while the much slower return to glacial cold took about a thousand years. These warmings – which appear to have been concentrated in the Northern Hemisphere – are called Dansgaard-Oeschger (D-O) events, after Danish geophysicist Willi Dansgaard and Swiss geochemist Hans Oeschger.

▶ **Going down** Less frequently – at six points during the ice age – climate lurched in the other direction. Vast fields of icebergs poured from North America into the North Atlantic, disrupting the ocean circulation and cooling climate. These so-called Heinrich events are named after German scientist Hartmut Heinrich, who discovered particles of Canadian soil scraped off by ice sheets and deposited in the North Atlantic by rafts of icebergs.

Together, the Heinrich and D-O events explain much of the variability in global climate that shows up in sediments and ice cores from the last 100,000 years. Scientists are still hunting for what might lie behind the timing and occurrence of both types of climate swings. For instance, each Heinrich event occurs after three to five D-O events, implying that the extended warm spells might have progressively destabilized the growing ice sheets and led to a Heinrich-style iceberg armada.

Another interesting facet is that many of the D-O events are separated by around 1500 years, with a few spaced at about 3000 and 4500 years. "This suggests that the events are triggered by an underlying cycle ... but that sometimes a beat or two is skipped", notes Stefan Rahmstorf of the Potsdam Institute for Climate Impact Research. Other paleoclimatologists believe these sub-beats may be little more than random variations.

years ago, ice sheets began to retreat dramatically. However, the path to a warmer climate wasn't a smooth one. As the melting intensified, water occasionally poured out of glacial lakes in spectacular bursts, some of which were large enough to influence ocean circulation and trigger rapid climate shifts.

Perhaps the most dramatic was the **Younger Dryas** period, a frigid encore that brought a return to cold nearly as extreme as the ice age itself. The period is named after a tenacious alpine flower – *Dryas octopetala* – that thrives in conditions too austere for most other plant life. The flower's pollen, preserved in Scandinavian bogs, provides evidence of extremely cold periods. The most recent ("youngest") such period began nearly 13,000 years ago and persisted for some 1300 years. Ice cores from Greenland indicate that the Younger Dryas took as little as a decade both to set in and to release its grip. During the interim, average temperatures across the British Isles sank to as low as –5°C (23°F), and snowfall increased across many mountainous parts of the world. Drought and cold swept across the Fertile Crescent of the Middle East. This may even be what prodded the ancient Natufian culture to pull back from wide-ranging hunting and gathering and focus on what ranks as the world's earliest documented agriculture.

As for the cause of the Younger Dryas, most paleoclimatologists point to the massive draining of **Lake Agassiz**, a vast glacial remnant that covered parts of present-day Manitoba, Ontario, Minnesota and North Dakota. Such an influx of fresh water could disrupt the thermohaline circulation that sends warm waters far into the North Atlantic (see p.132). However, there's spirited debate about whether the Agassiz melt flowed in the right direction to affect the North Atlantic. An even more perplexing piece of the puzzle comes from the Southern Hemisphere, where records from Antarctica and other areas show that the cool-down there began a thousand years earlier – closer to 14,000 years ago.

The Younger Dryas was followed by a much shorter and less intense cooling about 8200 years ago. Once those were out of the way, the next few thousand years were, by and large, a nourishing time for civilizations. Scientists who map out the climate for this period can draw on human history as well as other biological, physical and chemical clues. (There's still a bit of a bias towards Europe, the Middle East and North America, due in part to the concentrations of researchers and documented civilizations on those continents. And, as with any climate era, the proxies aren't perfect.)

Across northern Africa and south Asia, the peak in summertime sunshine stimulated by orbital cycles during the post-glacial era produced monsoons far more intense and widespread than today's. From about 14,000 to 6000 years ago, much of the Sahara, including the drought-prone Sahel, was covered in grasses and studded with lakes that played host to crocodiles and hippopotami. Traces of ancient streams have been detected by satellite beneath the sands of Sudan, while the Nile flowed at perhaps three times its present volume. Meanwhile, from about 9000 to 5000 years ago, most of North America and Europe were ice-free and sum-mer temperatures there likely reached values similar to today, perhaps a touch warmer in some areas and at some times.

The good times couldn't last forever. Starting about 5000 years ago, global temperatures began a gradual cool-down, interspersed with warm-ings lasting a few hundred years each. This pattern is consistent with ice-core records from towards the end of previous interglacial periods, although it's far from clear when the next ice age will actually arrive (see box overleaf). One of the earliest and sharpest signs of this global cool-ing is the prolonged drought that crippled Near East civilizations about 4200 years ago. Orbital cycles had already weakened the summer sunlight that powered bountiful monsoons across Africa and Asia. At some point, the gradual drying killed off the grasses that covered much of the Sahara, which quickly pushed the region from savannah towards desert.

From AD 1000 to today (and beyond)

Because of its context-setting importance for what lies ahead, the last thousand years of climate have drawn intense scrutiny. The last millen-nium began near the peak of a 300-year stretch of widespread warmth called the **Medieval Warm Period** (or the Medieval Climatic Optimum), when temperatures in some parts of the world were close to modern-day levels. It may indeed have been an optimal climate for Europeans, or at least a liveable one, but people in parts of the Americas suffered from intense drought that ravaged entire cultures.

One example is the **Mayans**, whose empire extended from present-day Mexico across much of Central America. After nearly 2000 years of devel-opment, many Mayan cities were abandoned between about AD 750 and AD 950. Sediments in the nearby Caribbean reveal several strong multi-year droughts during this period. The Mayan collapse in eastern Mexico apparently coincided with that region's most intense and prolonged drought in over a millennium.

To the north, the region near the current US Four Corners settled by the **Anasazi** (also known as Ancestral Pueblans) experienced periods of

Sowing a savoury climate

Whether or not we're due for another ice age relatively soon is a matter of intense debate. The last three interglacials each lasted about 15,000 years, which would imply that it's nearly time for a new glacial era to begin. Some paleoclimatologists believe that orbital cycles won't trigger the next era for as long as 50,000 years. However, if in fact the next ice age is due, then human activity could be postponing its arrival. Veteran paleoclimatologist William Ruddiman argues that the development of agriculture boosted greenhouse gases and delayed a glaciation that should have already begun. Ruddiman introduced his theory with a provocative paper in 2003 entitled "The Anthropogenic Greenhouse Era Began Thousands of Years Ago". He's since elaborated it for a broader audience in the book *Plows, Plagues and Petroleum*.

Ruddiman's theory is based on ice-core records that show how greenhouse gases responded during previous interglacial periods. He noticed that in most interglacials, greenhouse gas levels tend to spike quickly and then gradually drop, setting the stage for the next icing. However, the most recent interglacial hasn't obeyed the pattern. After an initial spike and the beginnings of a drop, carbon dioxide amounts rose again by about 8% about 8000 years ago, then levelled off until the Industrial Revolution. Methane concentrations also increased, starting about 5000 years ago.

Ruddiman attributes these rises to the rapid and widespread growth of agriculture during these periods. The clearing of forests would have increased CO_2 levels, he says, and rice paddies would have emitted large quantities of methane. To bolster his case, Ruddiman points to the drop in agricultural activity and subsequent reforestation after plagues swept Europe in the 1300s and the Americas in the 1500s and 1600s. The timing, he points out, corresponds well with the reduction in both global temperature and carbon dioxide levels during the Little ice age.

Ruddiman's theory has met with enthusiastic agreement and stout resistance, both of which played out in papers submitted to a special 2011 issue of the journal *The Holocene*. Critics argue there wasn't enough farming thousands of years ago to influence CO_2 and methane, but Ruddiman cites new evidence that early farmers used far more land per capita. Others point out that an isotope associated with terrestrial carbon ($\partial^{13}CO_2$) has barely risen in the last 7000 years. This implies a cap on the CO_2 contribution from land processes, including agriculture. However, if huge amounts of carbon were taken up by boreal peatlands – as some analyses suggest – then other carbon could have poured out of newly cleared land, meaning that the $\partial^{13}CO_2$ limit isn't necessarily a deal-killer. Through it all, Ruddiman remains steadfast: "I really have not had a serious day of doubt."

erratic rain and snow during the Medieval Warm Period. Tree-ring studies indicate an especially intense drought near the end of the 1200s, when the region's emblematic cliff dwellings were abandoned. Scholars in recent years have warned against pinning the Mayan and Anasazi declines solely on drought, pointing to signs of other factors such as migration, a high demand on trees and other resources, and power struggles within each culture and with neighbouring peoples. However, climatic stresses can't have helped matters.

After the Medieval Warm Period faded, the long-term cooling resumed, kicking in around 1300 and continuing into the mid-1800s. This was the Little ice age (see box, overleaf), the coldest interval globally in thousands of years. Periodic plagues and famines ravaged Europe, and glaciers descended from the Alps to engulf a number of villages. The chill wasn't constant: as historian Brian Fagan notes, the Little ice age was marked by "an irregular seesaw of rapid climatic shifts, driven by complex and still little understood interactions between the atmosphere and the ocean."

Another influence may have been a drop in **solar energy**. Isotopes of carbon in tree rings and beryllium in ice cores show a drop-off in solar radiation during much of the Little ice age. Moreover, sunspot observations that began around 1610 show a near-absence of reported sunspots between 1645 and 1715 (the so-called Maunder Minimum). However, recent studies have brought down the relative importance of this solar effect on the Little ice age.

Also in the mix are volcanoes, which seem to have erupted more frequently after 1500 than during the Medieval Warm Period. The 1815 eruption of Indonesia's **Tambora** – the most violent ever recorded on Earth – led to a disastrously cold summer across much of the globe in 1816. That "year without a summer" brought crop failures to northern Europe as well as snows in Vermont as late as early June. (It also spawned a monster: while spending part of that dreary summer at Lake Geneva, a teenaged Mary Shelley started work on the novel that became *Frankenstein*.) Both the Medieval Warm Period and the Little ice age appear to have been strongest over the Northern Hemisphere's continents, although it's hard to completely eliminate geographic bias from these early records. Some researchers argue that both phenomena were primarily regional events, as opposed to the global-scale warming under way now.

After the mid-1800s, Earth's climate took a decided turn for the warmer. The overall climb in temperatures over the last hundred years, however, was punctuated by an apparent flattening from around the mid-1940s to the early 1980s (see graph, inside front cover). At the time, it appeared to be global cooling, because the era's worldwide temperature analyses were skewed toward the Northern Hemisphere, where temperatures actually were dropping a bit. The cause of the northern cool-down is still being debated. For years it was thought to be related to sun-reflecting pollution spewed out from North America and Europe in the industrial boom after

Painting the Little ice age

Some of northern Europe's greatest artists used oil and brush to set the mood that many associate with the Little ice age: cloudy, snowy and dank. Pieter Bruegel the Elder may have used the frigid winter of 1565 as source material for the dull, greenish sky of *Hunters in the Snow*, part of his series of seasonal depictions. This was one of the first portrayals of a snowy landscape in European art, noted William Burroughs in the British journal *Weather*. Bruegel extended the wintry theme to other topics, including *The Adoration of the Magi in the Snow*. Many Dutch artists, notably Hendrick Avercamp, took to cold-weather depictions in the mid-1600s, another period of brutal chill across the region.

The northern Renaissance also spawned a new realism in sky portraiture. Back in the early 1400s, Flemish painter Jan van Eyck was one of the first to depict cloud types that a meteorologist today might recognize and label. Hans Neuberger quantified the treatment of clouds by US and European painters in an unusual 1970 study that appeared in *Weather*. Sampling 41 museums in 9 countries, Neuberger examined more than 12,000 paintings produced between 1400 and 1967. He found that blue skies, which predominated up to 1550, gave way to low clouds in more than half of the post-1550 paintings. Neuberger didn't attempt to analyse how much of the trend was related to the Little ice age weather and how much to artistic fashion.

English landscape painters of the Little ice age held true to their island's cloudy climate. Every English sky examined by Neuberger had at least some cloudiness, and the sky was typically a pale blue at best. The English Romantic artist J.M.W. Turner specialized in foggy, misty tableaux as well as striking sunsets; the latter may have reflected the volcanic dust that added vivid hues to many sunsets in the early 1800s. Later in the century, the gigantic Krakatoa eruption of 1883 led to sunsets so striking they were noted in press reports in New York and London. According to astronomer Donald Olson of Texas State University, Krakatoa may also have inspired Edvard Munch's iconic masterpiece, *The Scream*. In describing what triggered the painting, Munch wrote of experiencing a "blood-red" sunset

World War II, though such effects are difficult to assess and quantify. Researchers are now zeroing in on the role of the oceans. According to a pair of studies in 2008 and 2010 led by David Thompson (Colorado State University), much of the mid-century hemispheric cooling can be traced back to two ocean-related factors, both of which pop into view after the ups and downs of El Niños, La Niñas and volcanic eruptions are removed. One is an apparent drop in sea-surface temperatures (SSTs) in 1945 that proved to be spurious, the result of a temporary increase in canvas-bucket measurements at the end of World War II (see p.198 for more on SST measurement techniques). The other is a major drop in Northern

Dutch painter Hendrick Avercamp (1585–1634) was a specialist in winter scenes, especially skaters on frozen rivers, canals and flood waters.

in present-day Oslo that resembled "a great unending scream piercing through nature" – though Munch didn't give a date for this experience. Although a full decade separates the eruption from *The Scream*, Olson believes that Munch may have encountered a Krakatoa sunset and waited years to depict it.

The legendary frost fairs held on the River Thames in London during occasional freeze-ups were captured in a number of paintings, including *A Frost Fair on the Thames at Temple Stairs* (1684) by Dutch painter Abraham Hondius. However, these festivals weren't as frequent as one might assume. Outside of the especially frigid mid-1600s, the Thames froze at London only about once every twenty or thirty years from the 1400s until 1814, when the last freeze-up was recorded. Moreover, it wasn't the end of the Little ice age that ended the frost fairs. When London Bridge was replaced in the 1830s, it allowed the tide to sweep further inland. This made it virtually impossible for the Thames to freeze at London, and it hasn't happened since.

Hemisphere SSTs from 1968 to 1972, focused in the North Atlantic. This may be linked to a large influx of fresh water called the Great Salinity Anomaly that was observed around the same time, though the connections aren't yet fully mapped out.

Global warming resumed with gusto in the 1980s, and by the end of the twentieth century it was becoming clear that temperatures were at least on a par with those of the Medieval Warm Period. That recognition triggered one of the highest-profile climate change debates to unfold over the last few years, one that brought unprecedented attention to the fine points of paleoclimate analysis.

In the late 1990s, Michael Mann, then at the University of Virginia, joined Raymond Bradley (University of Massachusetts) and Malcolm Hughes (Laboratory of Tree-Ring Research) in the first attempt to reconstruct global temperature from the past millennium based on varied proxy data from around the world. Initial results were published in a 1998 *Nature* paper, and in a 1999 follow-up, the three scientists (often referred to by their initials, MBH) reported that the 1990s had likely been the warmest decade of the last 1000 years. Their work gained stature through its

Four ways to think of past and future climate

The answer to whether Earth is warming or cooling depends in large part on what time frame you're considering. Each graph below includes a very rough schematic of past (and plausible future) temperatures.

▶ **Over hundreds of years** Around AD 1000, Earth was nearly as warm as it is now. After that, temperatures over much of the planet cooled by around 0.2–0.5°C (0.4–0.8°F), mostly during the period known as the Little ice age, from the 1300s to around 1850. The global average has since rebounded by nearly 1°C (1.8°F) as human-produced greenhouse gases have increased. This century is projected by the IPCC to warm by 1.4–5.8°C (2.5–10.5°F). Further warming after 2100 could be substantial if greenhouse emissions continue to increase through this century, and the delayed response of ice sheets and deep oceans to warming could produce major sea-level rise over several centuries.

▶ **Over tens of thousands of years**
The most recent ice age began about 115,000 years ago and ended about 11,500 years ago. Then came a dramatic warm-up, which lasted until about 3000 BC. Since then, Earth's temperature has changed relatively little, with a very slight cooling interrupted by warmer periods and punctuated by the last century's sharp temperature rise. More than a thousand years from now, after humans have exhausted fossil fuels and the resulting greenhouse gases have left the atmosphere naturally (mostly through slow absorption by the ocean), we may return to cooler times. If the

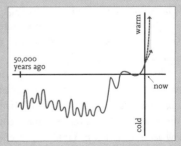

breadth and novel approach as well as its timing. It arrived just after global temperatures hit a new high in 1998 and just as the IPCC was preparing its third major assessment. The key graphic produced by Mann and colleagues, nicknamed the "hockey stick" for its distinctive shape, ended up prominently featured in the IPCC's 2001 report (p.308).

Two Canadians questioned the MBH study: a former mining engineer and government analyst, Stephen McIntyre, and an economist, Ross McKitrick of the University of Guelph. (The two are sometimes called

length of the last interglacial is any guide, we're due for another ice age in the next few thousand years, although some scientists believe it's more likely to take 20,000 years or more. It's even possible that a mix of polar melting, land-use changes and slowly waning greenhouse-gas concentrations could postpone the next ice age for an undetermined period.

▶ **Over tens of millions of years** A gradual, though sometimes erratic, cooling trend has been under way for at least 55 million years, perhaps due in part to carbon dioxide removed by weathering atop the growing Himalayas. Northern Hemisphere glaciations began about 2.7 million years ago. Warm and cool periods come and go atop this overall cooling. Simple extrapolation would keep

Earth on this extremely slow cooling trend, although there's no evidence to tell us how long that might last.

▶ **Over billions of years** On this scale, the details are blurry but the overall picture is clear. Changes in greenhouse-gas levels have so far kept our climate relatively stable (not too hot or cold for life) even though the Sun's output has risen by more than a third since the solar system was formed around 4.6 billion years ago. The Sun will continue to heat up and eventually undergo changes to its size and structure, producing a climate on Earth hot enough to evaporate the oceans and make life impossible. Eventually, the sun will shrink and start to fade, with the solar system reaching its cold "final configuration" in around twelve billion years.

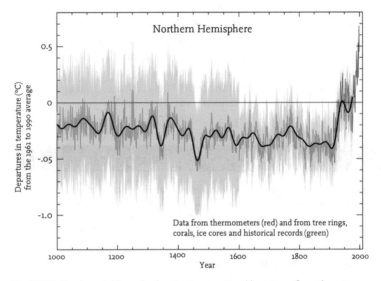

The IPCC's "hockey stick" graph, showing temperature (departures from the 1961–90 average) in the Northern Hemisphere during the past 1000 years.

M&M.) Intrigued by – and somewhat suspicious of – the drama inherent in the hockey stick, McIntyre and McKitrick attempted to replicate the MBH findings. In a 2003 paper, M&M detailed a set of errors they found in the listing of proxy data for the 1999 *Nature* article. MBH responded with a 2004 correction in *Nature*, concluding that none of the errors affected their results.

After M&M found themselves still unable to replicate those results (in part, they claimed, because they lacked access to some of the relevant software), they took a different approach. They began to criticize the use of tree-ring proxies they felt had been chosen, grouped and analysed in ways that – regardless of intent – artificially inflated the twentieth century's upward bend to the hockey stick. For example, the 1999 MBH paper incorporated tree-ring data from long-lived bristlecone pines in the western United States; some previous work had questioned the use of these trees as biological thermometers.

M&M also disputed MBH's use of collective proxy data in place of individual records, a strategy that's designed to minimize bias when a dense collection of proxies is drawn from a relatively small part of the globe (mostly North America and Europe). And M&M showed how one method of using proxy groups to assess long-term trends, a technique called principal components analysis, had the potential to generate a hockey-stick curve even when there was no such trend in the underlying data. With these and other factors taken into account, M&M asserted, the Medieval Warm Period could indeed have been warmer than recent decades. Separately, climate researcher Hans von Storch (University of Hamburg) and colleagues used computer modelling for a 2004 *Science* paper to suggest that global temperature variations over the last millennium could have been more than twice as large as MBH had found.

The debate broke open in early 2005 as McIntyre and McKitrick outlined their arguments in *Geophysical Research Letters*. Their paper galvanized attention from the media and the US Congress and set off a wave of concern, especially among those politicians and parts of the public inclined toward climate scepticism. Mann released all of the relevant data and software and a variety of other materials at the request of Congress (a request that itself raised concerns about political involvement in the scientific process). A new blog by McIntyre called Climate Audit quickly became a go-to site for climate sceptics.

A panel convened by the US National Research Council looked into the controversy. The panel's 2006 report supported the type of multiproxy research pioneered by Mann and colleagues. It agreed that some statistical techniques could produce a hockey-stick shape where none existed, but even if such were the case, additional evidence was at hand. As the panel noted, various other analyses and lines of evidence – including new global reconstructions by other scientists – supported the rapid rise at the end of the hockey stick, the idea that recent warmth has indeed been unusual against the last four centuries. The panel did stress that, before the year 1600, proxy uncertainties are greater and confidence is lower.

MBH and M&M continued to publish new work and exchange volleys. A 2008 paper by MBH and several colleagues expanded their selection of proxies and extended the reconstruction back to the year AD 200. Although it showed a somewhat warmer Medieval Warm Period than did their previous analysis, it also reaffirmed that the last several decades stand out in their warmth, even when tree-ring data are excluded. A 2007 study in *Climatic Change* led by Eugene Wahl (then at Alfred University) came to similar conclusions after taking a fresh look at the approaches of both MBH and M&M. Meanwhile, McIntyre's blog offered an exhaustive running account of his quest to change the way proxies are used and proxy-based research is conducted. Sparks flew once again in late 2009 as a trove of email exchanged among climate scientists became public (see p.294).

While this grand battle raised issues of data and software access and transparency (and showed the influence that investigators outside the climate science mainstream can wield), it also diverted attention from the unquestioned rise in both temperature and greenhouse emissions over the last several decades. Regardless of these and other dust-ups, one thing remains crystal clear: the greenhouse-gas content of our atmosphere has entered territory never before explored in human history. Given what we know about these gases, the implications for our climate remain profound.

Circuits of change

Modelling the future climate

In 1957, Roger Revelle and Hans Seuss of the Scripps Institution of Oceanography described the intersection of fossil fuels and global climate change in words that proved far more resonant than the authors might have imagined at the time. Their landmark study, published in the journal *Tellus*, included this oft-quoted passage:

> "... human beings are now carrying out a large scale geophysical experiment of a kind that could not have happened in the past nor be reproduced in the future. Within a few centuries we are returning to the atmosphere and oceans the concentrated organic carbon stored in sedimentary rocks over hundreds of millions of years."

Like any good scientists, they were curious about the outcome: "This experiment, if adequately documented, may yield a far-reaching insight into the processes determining weather and climate."

It may seem callous to describe our current situation – glaciers melting, cities baking, hurricanes churning, species disappearing – as an interesting experiment. It's also quite accurate. Each year our emissions push the chemistry of the atmosphere a little further into territory never before explored in human history. If there were a million Earths in our solar system, each with a different concentration of greenhouse gases, we'd have a better idea of just what to expect in the future. Unfortunately, as Revelle and Seuss point out, we can't replicate the experiment that we're now conducting on planet Earth. The next best thing is to simulate it, and the only place that's feasible is inside a computer.

Reproducing the atmosphere with **computer models** has been a key activity in computer science ever since the field was born, just after World War II. At first the emphasis was on weather prediction. By the 1960s, forecasters were using 3-D models of atmospheric flow, or **general circulation models**, to gaze into the future of weather patterns several days out.

Climate, weather and chaos theory

One of the most common gripes about climate-change projections is the idea that a computer model can tell us anything worthwhile about climate in the distant future. It's often phrased this way: "If they can't get the forecast right for next week, how can they predict the climate a hundred years from now?" Even the late Michael Crichton employed this tactic often (see p.284).

But a weather forecast and a climate projection are two different beasts. Weather models can't see clearly beyond a few days because of inherent limits to the predictability of small-scale weather (i.e. what will happen in your neighbourhood, as opposed to the globe as a whole). Weather observing stations are situated a few kilometres or miles apart at best, and in much of the world you can go 100km (60 miles) between stations, not to mention the largely unsampled oceans. Small disturbances missed in the data can influence larger weather events over time – the "butterfly effect", as discovered in the 1960s by Edward Lorenz, the father of chaos theory. We'll never have enough weather stations to catch every one of these tiny weather makers. That's why specific local forecasts – while steadily improving – will always be problematic for days more than about two weeks in the future.

Climate models aren't interested in the vagaries of individual weather events so much as the influence of long-term climate shapers such as greenhouse gases, solar variations and polluting aerosols. Chaotic variations in weather normally average out over years and decades, so they don't corrupt a long-term climate projection the way they might a short-term weather forecast. The downside is that global models can't provide the precision in time and space that many policy-makers and the public want to see, although progress is being made through techniques such as "downscaling" from global output to regional depictions.

Since then, breakthroughs in weather and computer science, together with Moore's Law (the maxim that computer processor speed doubles about every eighteen months), have allowed for steady improvements in weather forecasting models and thus in forecast skill. Revelle and Seuss might have been happy to have a solid three-day forecast; now, ten-day outlooks are commonplace.

Models focusing on climate long lagged behind their counterparts in the weather field, and with good reason. The most obvious weather changes emerge from the interplay among a fairly limited set of ingredients in the atmosphere: pressure, temperature, moisture, wind and clouds. Weather is also shaped by other factors in the surrounding environment, such as ocean temperature, vegetation and sea ice. But these latter factors don't change much over the course of a few days, so a weather model can focus on tracking the atmosphere while keeping the rest of the environment constant.

Climate models don't have that luxury. As the weeks roll into months, vegetation thrives and decays. Sea ice comes and goes. Greenhouse gases accumulate. The ocean gradually absorbs heat from our warming atmosphere, sometimes releasing it in giant pulses during El Niño events. In the very long term, even the topography of Earth changes. All of these variations feed into the atmosphere and influence weather – often so subtly that the effect isn't obvious until it's been playing out for years. For all of these reasons, it's extremely difficult to go beyond a weather model's focus on the atmosphere and to depict Earth's whole environment accurately in a **global climate model**.

Difficult, but not impossible. Climate modelling has undergone a rapid transformation in the last twenty years. As recently as the 1980s, the most sophisticated climate models portrayed the atmosphere and the land surface – and that was about it. Gradually, scientists have forged models that depict other parts of the Earth system. Many of these are incorporated in ever-growing global climate models. The best of the bunch now include land, ocean, sea ice, vegetation and the atmosphere, all interacting in a fairly realistic way. Global models are far from ideal; as sceptics often point out, there are still major uncertainties about climate processes that no model can resolve. However, the skill of recent global models in replicating twentieth-century climate, as we'll see below, is one sign that the models are doing many things right.

How computer modelling got started

The best analogy to a climate model may be a state-of-the-art computer game. Instead of cars tearing down a racecourse, or warlords fighting over territory, this game follows the heating, cooling, moistening and drying of the atmosphere. It's not quite as sexy but far more important to the future of the planet.

The action unfolds in a virtual space that's typically divided into a three-dimensional grid of rectangles. Imagine the steel latticework in a building under construction, but extended across the entire surface of the planet (see graphic on p.253). At the centre of each rectangle formed by the intersection of the imaginary beams is a **grid point**, where atmospheric conditions are tracked as the model's climate unfolds. In a typical global model, each grid point accounts for a horizontal area of roughly 1.25° longitude by 1.25° latitude. At mid-latitudes (40°N), that represents an area of around 140km by 100km (87 x 62 miles) – about the size of Connecticut or roughly half the size of Sicily. Thus, each grid point represents the atmos-

From great ideas to global views

The computer modellers now helping the world decide how to confront climate change owe a big debt to L.F. Richardson, a far-seeing British scientist of the 1920s. The story begins in World War I, when a team of meteorologists in Bergen, Norway, came up with the first three-dimensional theory of how weather worked. They named the boundaries that separated cold and warm air "fronts", after the battle fronts then raging in Europe. Meanwhile, Richardson, then an ambulance driver for the French army, was hatching a scheme to calculate the future of the atmosphere.

After the war, Richardson returned to England and set to work. Building on ideas that emerged from the Bergen group, he came up with seven equations based on physical principles from Isaac Newton, Robert Boyle and Jacques Charles. If these equations could be solved, Richardson believed, one could not only describe the current weather but extend it into the future. He envisioned a "forecast factory", where hundreds of clerks would carry out the adding, subtracting, multiplying and dividing needed to create a forecast by numbers. Richardson and his wife spent six weeks doing their own number-crunching in order to test his ideas on a single day's weather. The resulting outlook's skill was abysmal, but Richardson's test demonstrated to the world that one could – in principle, at least – carry out calculations and forecast the weather using more than intuition and rules of thumb that were standard in the 1920s.

Richardson's equations materialized again almost thirty years later in the world's first general-purpose electronic computer. The ENIAC (Electronic Numerical Integrator and Computer) was created in the US at the close of World War II. In order to show the machine's prowess, its developers hunted for a science problem that could benefit from raw computing power. Weather prediction fitted the bill. Richardson's equations were updated and translated into machine language, and in Princeton, New Jersey, on 5 March 1950, the ENIAC began cranking out the first computerized weather forecast. The computer took almost a week to complete its first 24-hour outlook, but ENIAC did a better-than-expected job of portraying large-scale weather a day in advance.

As weather modelling became established in the 1950s and 1960s, a small group of US climate experts began to adapt some of the weather models for climate and to create new ones from scratch. One centre of action was the Geophysical Fluid Dynamics Laboratory, based in Princeton. Another was the Courant Institute in New York and a third was the National Center for Atmospheric Research, newly established in Boulder, Colorado. In the mid-1960s, NCAR's Warren Washington and Akira Kasahara built one of the first general circulation models that spanned the globe. (Many models at the time were two-dimensional or covered only the Northern Hemisphere.) Generations later, the seeds of that early model remain in NCAR's Community Earth System Model, used by university researchers around the world.

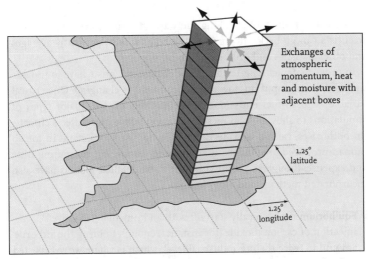

Exchanges of atmospheric momentum, heat and moisture with adjacent boxes

1.25° latitude

1.25° longitude

A climate model divides the atmosphere into 3-D chunks and measures the change in each one as weather unfolds over days, years and decades of model time.

phere over a fairly large area – much more so than for a weather model. Toward the poles, the width of each rectangle shrinks toward zero, which calls for special computational treatment. Some experimental models are now replacing the 3-D global latticework with a hexagonal structure, resembling a honeycomb, that could make for smoother and more efficient computing.

The typical global model also includes perhaps 25 or so vertical layers. Each layer spans anywhere from 100–500m (330–1650ft) close to Earth's surface, with the vertical distance between grid points increasing as you go up.

A climate model operates in a parallel universe of sorts, one in which time runs far more quickly than in real life. Each model advances in **time steps** that move the simulated climate forward by anywhere from five minutes to a half hour, depending on the model's complexity. At each time step, the model calls on formulas derived from the laws of physics to compute how all the atmospheric and environmental variables have changed since the last time step – the amount of sunlight hitting each spot on Earth, the convergence of winds, the formation and dissipation of clouds and so forth. Each step only takes a few seconds to process, but it can still take weeks of dedicated time on a supercomputer to depict a century of climate in detail. The output from all this number-crunching is saved in various formats along the way. After the processing is done, scientists can go back, sift through the results, calculate yearly averages and other statistics and produce graphics.

In order to depict long-term climate change, scientists have to provide the model with input on how the environment is changing. One critical variable is fossil-fuel emissions. Modellers rely on two major types of simulations to show how an increase in greenhouse gases affects climate. (In both cases, carbon dioxide is traditionally used as a stand-in for all human-produced greenhouse gases in order to save on computing time and expense, though in recent years the most detailed models have also incorporated methane and other key gases.)

▶ **Equilibrium** runs typically start off with an immediate, massive injection of carbon dioxide (for instance, enough to bring the airborne amount to twice the pre-industrial level, which is where we're expected to be by later in this century). The model's climate gradually responds to this extra CO_2 and eventually attains an equilibrium at some new (and presumably warmer) state. The idea isn't to see how long the climate takes to respond – since we'd never add that much CO_2 instantaneously in the real world – but where the climate ends up. To save on time and expense, these runs typically use simplified oceans.

▶ **Transient** runs more closely resemble reality. In these simulations, CO_2 is added in smaller increments that resemble the actual amounts added by human activity – typically 1% per year, compounded year over year. In this case, scientists are interested as much in the process as the outcome. For instance, does the model climate warm smoothly, or does the temperature go up in fits and starts? The various global models have tended to show closer agreement for transient runs than for equilibrium runs – an encouraging sign, since the transient scenario is closer to how greenhouse gases increase in real life.

Of course, you can't simply add CO_2 to a global model and leave everything else alone. As discussed in "Ecosystems and agriculture" (see p.162), a large part of that CO_2 gets absorbed by vegetation and the oceans. There are also other processes that kick in as carbon dioxide warms Earth. Consider the sea ice that melts as the atmosphere warms up. As we saw in The big melt (see p.82), the loss of that ice means less reflection of sunlight and an amplifying, positive-feedback loop that leads to even more warming. The positive feedback from water vapour also has to be portrayed accurately:

the warming produced by the extra CO_2 helps evaporate more moisture from the sea, and that water vapour raises global temperature further.

These environmental processes are quite different from the physics that drives day-to-day weather, and they can't be added to a climate model in an easy or simple way. The original global climate models of the 1960s and 1970s couldn't hope to account for these processes in a realistic fashion. But as computing power has improved, the models have grown steadily in sophistication and detail. They've incorporated a growing number of sub-models, dealing with each of the major climate-influencing elements of what scientists call the **Earth system**. When an atmospheric model is yoked to one or more of these sub-models, the new hybrid is called a **coupled model** (even when it has three or more components).

In a simple coupling, a sub-model runs side by side with a global model and passes it information – a one-way dialogue, as it were. For instance, a global model might be fed information on airborne nitrogen levels from a chemistry model, but the resulting changes in climate wouldn't feed back into nitrogen production, as they would in real life. More realistic are **integrated** couplings, which allow two-way exchanges of information

Hot spots for climate modelling

It takes vast amounts of computing power to run a comprehensive global model at a reasonable pace. Also needed are dozens of physical and computer scientists to ensure that a model's software is accurate and efficient. As such, the world's major climate models are produced by a fairly small number of institutions, most of them operated or funded by national governments. They include Australia's **Bureau of Meteorology Research Center** and its **Commonwealth Scientific and Industrial Research Organization**, both in the Melbourne area; the **Canadian Centre for Climate Modelling and Analysis**, based at the University of Victoria; Japan's **Frontier Research Center for Global Change** in Yokohama; the **Hadley Centre**, part of the UK Met Office, based in Exeter; Germany's **High Performance Computing Centre for Climate- and Earth-System Research** (Deutsches Klimarechenzentrum) and the **Max Planck Institute for Meteorology**, both in Hamburg; and the **Laboratory of Dynamic Meteorology** in Paris, part of France's Centre National de la Recherche Scientifique. In the US, there's NOAA's **Geophysical Fluid Dynamics Laboratory** in Princeton, New Jersey, NASA's **Goddard Institute for Space Studies** in New York City, several facilities of the **Department of Energy**, and the **National Center for Atmospheric Research** (NCAR) in Boulder, Colorado. NCAR's Community Earth System Model is one of the few open-source, high-end global models that anyone with sufficient computing power can download and use.

between model and sub-model. However, it can take years to develop such linkages and verify their accuracy.

Models also grow in the level of detail they provide. Model resolution – the fineness of the spacing between grid points – has gradually increased over the years as computing power has grown. Today, some cutting-edge global models are beginning to resemble weather models. For instance, a team from Japan's Meteorological Research Institute has simulated the frequency of tropical cyclones in a doubled-CO_2 world using a model with a resolution on the order of 20km (12 miles), with 60 vertical layers. To accommodate the sharp detail of these models, their time frame has to be restricted. Typically the models start out with a "future" atmosphere (such as one with doubled carbon dioxide) and simulate only a decade or so rather than a century or more.

Climate modellers have typically avoided incorporating current weather conditions when projecting the climate decades into the future, since there's so much day-to-day variability in weather. However, bringing current ocean conditions into the model could help nail down how global temperature is likely to unfold over the next decade. The UK Met Office Hadley Centre used this strategy to issue a novel kind of forecast unveiled in a 2007 paper in *Science*. By running a global model in two ways – either including or omitting the patchwork of present-day ocean temperatures – the group showed that incorporating current ocean conditions, and the "memory" they bring to the atmosphere, can add measurable benefit to a climate outlook. The model captures realistic-looking peaks and valleys of global warm and cool spells caused by natural variability, but with an underlying rise that projects at least half of the period from 2009 to 2014 to be warmer than the record-setting year of 1998. Decadal-scale projections similar to this one, recreating climate as far back as the 1960s and projecting it forward in time slices such as 2026–35, were being produced by modelling centres around the world in 2010 in preparation for the 2013–14 IPCC report.

Especially when the focus is on a weather or climate feature with restricted geographic extent, such as hurricanes, it makes sense to limit the model's extra-sharp precision (and the related computing burden) to the areas where it's most needed. This can be done using a **nested grid**, whereby the globe as a whole is simulated at a lower resolution and the areas of interest are covered at higher resolutions.

Similarly, when studying how climate might evolve in a fixed area, such as parts of a continent, global models are often coupled to **regional climate models** that have extra detail on topography. The output from nested global models and regional models is especially useful in creating the climate assessments that are sorely needed by policymakers. However, bumping up a model's resolution isn't a simple matter: it involves a careful look at the physics of the model to ensure that processes which don't matter so much in a low-resolution simulation, such as showers and thunderstorms, are realistically depicted at the new, sharper resolution.

Whatever its design, every global model has to be calibrated and tested against some known climate regime in order to tell if it's functioning properly. The easiest way to do this is through a **control run** with greenhouse gases increasing at their twentieth-century pace. By simulating the last century of climate, scientists can see if a model accurately depicts the average global temperature as well as regional and seasonal patterns, ocean circulations and a host of other important climate features. Once it's passed this test, a model can go on to project the climate of the future – or the distant past, the goal of paleoclimate modelling.

Another safety check is that the world's leading models are regularly checked against each other through such activities as the Program for Climate Model Diagnosis and Intercomparison, conducted at the US Lawrence Livermore National Laboratory. When the models agree, that bolsters confidence in the results; if one model deviates markedly from the others, it's a flag that more work is needed.

The limitations

In both weather and climate models, it's not easy to calculate every iota of atmospheric change. **Clouds**, for example, are still a tremendous challenge. If they change in a systematic way, clouds can have major implications for how much radiation gets reflected and absorbed from both the Sun and Earth. But clouds are far more varied and changeable in real life than any model can accurately represent, so modellers have to simplify them in their simulations. There's currently no gold standard for how to do this, and new techniques are being explored all the time. Typically, each grid point in a model will be assigned an amount or percentage of cloud cover based on how much water vapour is present and whether the air is rising or sinking (rising air is conducive to cloud formation). The type of cloud and its reflective value are estimated from the height and temperature of the grid point.

To a greater or lesser degree, climate models also include **aerosols** – pollutants, dust and other particles that can shield sunlight and serve as nuclei for cloud droplets and ice crystals. This allows scientists to see

Modelling for the masses: ClimatePrediction.net

Until recently, climate modelling was restricted to the supercomputers of a few universities and research labs. That all changed with ClimatePrediction.net, an online project that has set global models churning in more than 100,000 homes and offices worldwide. It's the brainchild of Myles Allen, a physicist and climate analyst at Oxford University. Allen was inspired in the late 1990s by the success of the SETI@ home project, which has involved more than a million people around the world dedicating standby time on their PCs to assist in the search for extraterrestrial life.

The amount of computing time needed to analyse radio-telescope data for SETI is vast, but it can be parcelled out in small doses. Each volunteer agrees to download a program that runs on her or his PC at times when the computer would otherwise be resting (the software never interferes with other tasks). These "screensaver scientists" get the satisfaction of contributing to a grand scientific endeavour with a minimal commitment of time, as well as the fun of joining a like-minded online community.

Right away, Allen saw the potential value to climate science of this distributed approach to computing. On an up-to-date home computer, a fifty-year simulation of global climate could be run in six months, he figured. If thousands of volunteers were involved, the virtual horsepower would be far more than any research centre could afford on its own. Moreover, this setup would be an ideal way to run ensembles – sets of closely related simulations, each with slightly different starting conditions (see p.264). To drum up interest, Allen pitched the idea in a commentary for *Nature* in 1999. "Anyone respectable enough to sit on a peer review committee", he wrote, "will probably find the idea of getting schoolchildren to run full-scale climate models on their parents' PCs completely daft. But there will be others for whom the idea is as natural as Amazon.com." After several years of experiments and effort, Allen and colleague David Stainforth obtained financial backing from the UK's National Environmental Research Council (NERC), settled upon the name ClimatePrediction.net and launched the experiment in 2003.

The greatest participation in CP.net is in North America and Europe: more than half of the 90,000-plus computers involved as of 2010 were in the US, UK and Germany. But there's a healthy sprinkling of users all across the globe, with participants in 138 nations. The CP.net team works to engage volunteers with its extensive website, which includes background on climate science, techniques for comparing one volunteer's results against others and instant-messaging tools for the volunteer community.

how changes in pollution might affect cloud formation and temperature, as in the case of the global dimming observed in the late twentieth century (see p.204). There's been substantial progress in modelling of the direct effects of aerosols (on sunlight, for example), but the indirect effects on clouds remain more difficult to capture. In order to test whether their cloud schemes are working well, scientists run their models for present-

The first science goal for ClimatePrediction.net was to examine sensitivity, the amount of global warming that a doubling of carbon dioxide would produce (see p.261). The results grabbed headlines. Out of 2500 simulations using a version of the UK Met Office's Unified Model, most were in the ballpark of previous studies, but a few showed a climate sensitivity as high as 11.5°C (20.7°F). Allen believes the possibility of such high values was downplayed in previous research because traditional schemes for measuring sensitivity in any process (climatic or otherwise) don't portray the high part of the range well. From its initial work, CP.net has moved on to investigate such hot topics as the likelihood of a shutdown of the Atlantic thermohaline circulation (see p.132), the climate of 6000 years ago, and the impact of sulphate aerosols and geoengineering on climate. With support from Microsoft Research, the project has completed more than half a million regional simulations for Europe, western North America and southern Africa.

day climate and then compare the overall distributions of cloud cover to the real-world situation observed by satellite. (These data sets have their own limitations, though, since most satellites can't yet delineate multiple cloud layers very well.) The results are also compared to special cloud-resolving models that are too expensive to run routinely but provide a valuable cross-check.

Oceans are also tricky. Real-life oceans feature a **mixed layer** on top, where the turbulence of waves and close contact with the atmosphere help keep the temperature fairly constant. The mixed layer may be as shallow as 10m (33ft) or as deep as 200m (660ft), depending on season, location, time of day and weather conditions. Underneath the mixed layer is a zone of deeper, colder water that interacts with the mixed layer much more slowly. The earliest climate models depicted the mixed layer as a uniform slab but lacked a deep layer, a problem for long-term simulations. Oceans were a frequent source of **climate drift**, which occurs when a model-depicted climate gradually veers into unrealistically warm or cold territory. If unaddressed, climate drift could render any depiction of global warming suspect. Modellers traditionally handled climate drift through special computational schemes that kept the global climate from being thrown off balance.

Today's global models are coupled with much more realistic ocean sub-models that allow for the two-way exchange of heat, moisture, chemistry and momentum among the surface, the mixed layer and deeper levels, with 25 or more layers now depicted in some models. Still, the complicated circulations of the real ocean remain hard to portray. For instance, the North Atlantic's flow is often characterized as a simple loop, with warm water flowing north off the US east coast towards Europe, cold water streaming back south towards Africa, and warmer waters flowing west with the trade winds towards the Caribbean. But in reality, the northward flow branches into several components as it approaches Greenland and the British Isles, and the water descends far undersea into a cold return flow that moves southward in several distinct layers. Where the contrasts in temperature and salinity are strong, small but powerful eddies can spin off. All this complexity is hard enough to measure, much less portray in a model.

The tropical Pacific remains one of the most difficult areas for climate models to depict in detail, not only because of ocean circulation but because of the gigantic areas of rainfall spawned by the warm waters, especially in and near Indonesia. This warm pool periodically expands eastward or contracts westward during El Niño and La Niña (see p.130). Most global models can now depict El Niño and La Niña, and some are able to represent the frequency of their comings and goings fairly well, but their geography and long-term behaviour are still hard to pin down. Even in neutral conditions, it's a challenge for models to realistically capture the amount of heat and moisture pumped into the upper atmosphere by tropical showers and thunderstorms.

A sensitive topic: predicting the warm-up

The question of what will happen in a greenhouse-warmed world is so essential to climate modelling that it's spawned an index that's widely used to assess model performance. This is the **sensitivity** of a global model, most often defined as the amount of temperature change produced in the model when carbon dioxide is doubled all at once in an equilibrium-style run, as discussed above. The climate eventually settles down at a new global temperature, and the difference between that temperature and the old one is the sensitivity. It provides a convenient way to compare models and how readily they respond to greenhouse gas.

This definition isn't cast in concrete: one might consider a tripling or quadrupling of CO_2 instead, and various starting or ending points can be used, such as the peak of the last ice age. However, the instantaneous doubling of CO_2 since pre-industrial times (generally taken as the mid-1800s) serves as a useful and time-tested convention familiar to all climate modellers.

At the current rate of increase, greenhouse gases will have roughly doubled in the atmosphere by the mid- to late twenty-first century (though we might be able to postpone the doubling, or perhaps even forestall it entirely, through a concerted effort to reduce greenhouse emissions). The doubling of CO_2 used in model sensitivity gives us a good idea of how much warming to expect once the atmosphere and oceans have adjusted to the extra greenhouse gas, which could take centuries longer than the doubling itself. Should there be big changes along the way – such as the partial or total loss of major ice sheets – then it could take many thousands of years more for global temperature to settle down. This potential longer-term rise is usually analysed separately from the more conventional definition of sensitivity used below.

Even though sensitivity is expressed as a single number, it's actually the outgrowth of all the processes that unfold in a model. That's why equally sophisticated climate models can have different sensitivities. That said, what's striking about sensitivity is how little the average value has changed as models have improved. Some of the earliest global climate models showed a doubled-CO_2 sensitivity of around 2.5°C (4.5°F). As more models of varying complexity came on line, the range of possibilities was estimated to be **1.5–4.5°C (2.7–8.1°F)**. This range – something of a best guess at the time – was first cited in a landmark 1979 report on climate change by the US National Academy of Sciences. It was held constant by the IPCC in its first three assessments and was narrowed only slightly in the 2007 report (to 2.0–4.5°C or 3.6–8.1°F). Most of the major coupled models now show sensitivities clustered within a degree of **3.0°C (5.4°F)**, which is the best estimate pegged in the latest IPCC report. The fact that this value hasn't changed much after nearly thirty years of modelling is, arguably, testimony to how well modellers have captured the most important processes at work.

A sceptic might grouse that the models have simply been tuned – their physics adjusted – to produce the result that people expect. Indeed, modellers have to guard against making so many adjustments to a model that it gets a "right" answer for the wrong reasons. But there's also a physical rationale behind the 3.0°C number. It's fairly easy to show that, if the atmosphere had no clouds or water vapour, the warming from doubled

What exactly does probability mean?

There's an important difference between subjective and objective probabilities. In weather forecasting, a computer model might calculate a 60% chance of rain. That's an **objective** probability – it depends only on the physics or statistics in the model and the data it's working from. However, if a human forecaster raised the value to 80% based on information that the model didn't have, such as a glance out of the window that revealed gathering storm clouds, the new figure would be a **subjective** probability – that is, produced by a person, using her or his best judgement.

Many objective probabilities can be drawn from past experience. For instance, if decades of climate in a given city show that it rains on average about one out of every four days in June, then a summer-solstice festival has a 25% chance of getting dampened in any given year. (Those odds may change as the event nears, much like the example above.)

Scientists can't use purely objective probability when it comes to climate change, because by definition the climate is entering new territory – there's no set of past human-induced warmings on which probabilities could be based. One way around this is for humans to apply expert judgement, as the forecaster above did, and assign probabilities based on the evidence at hand. For example, in 2001 the IPCC stated that doubling carbon dioxide in the atmosphere would make a warming of 1.5–4.5°C "likely". Elsewhere in that report, the IPCC defines "likely" as a probability range of 66% to 90%.

While these ranges can be useful to policymakers, they aren't based on hard-and-fast scientific experiments. Now that climate simulations can be carried out en masse, it's possible to derive probabilities as if those simulations were a set of climates similar to ours. That's how various groups are coming up with percentages and ranges like the ones discussed on p.298. These are more objective than before, but they're still based on models rather than real-world climate, since we can never have an archive of thousands of global warmings to compare this one to.

CO_2 would be somewhere around 1.0–1.2°C (1.8–2.2°F). It's believed that water vapour roughly doubles the impact from the extra CO_2. Together with smaller effects from other greenhouse gases, this brings the total change into the 3.0°C ballpark.

Be that as it may, we can't be assured that larger (or smaller) changes are completely out of the question. The small but real risk of a dangerously high sensitivity has been borne out by a number of studies based on observations, and more recently by the massive experiment in distributed climate modelling known as ClimatePrediction.net (see p.258). In its first study, the project ran more than 2000 variations of a sophisticated climate model from the UK Met Office. The results, published in 2005, showed that the sensitivity to doubled CO_2 could be higher than 8°C (14.4°F). If such a warming came to pass, the consequences could be catastrophic.

Most of the ClimatePrediction.net variations showed much less sensitivity, with a consensus value of 3.4°C (6.1°F). That's well within the range of previous work (and expected by the nature of the model used), but the upper-end estimates raised enough of a red flag to trigger follow-up research. Several of these studies found a 95% chance that sensitivity should lie below a point ranging from 4.5–6.5°C (8.1–11.7°F), depending on the particular study.

Clearly, there's still disagreement about just how sensitive climate might be. But all of the recent studies indicate that the most likely outcome of a doubling of carbon dioxide is a global temperature increase close to the one scientists have been projecting for decades: somewhere not too far from 3°C (5.4°F). None of this rules out a much more sensitive climate. It simply adds to the weight of evidence for an amount of warming that's certainly nothing to be cavalier about. A 3°C rise, after all, would bring Earth to its warmest point in millions of years. Plus, there's no reason that greenhouse gases couldn't increase well beyond the benchmark of doubled CO_2.

Listening to the models

It's up to policymakers and the public to decide how society ought to respond to the onslaught of data that climate models give us. Probabilities like the ones above can help immensely. It may seem confusing to think of a temperature range and a probability at the same time, but that's the direction modellers are heading, and it's the type of output that's most useful in making decisions. Once upon a time, weather forecasts would

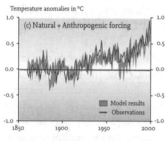

When models look only at natural climate forces (top) or human-produced ones, such as greenhouse gases and sulphate pollution (middle), they fail to accurately reproduce the last century's climate (the red line in each graph). A model that incorporates both kinds of forces (bottom) does a much better job. The grey shading shows the range of each model ensemble.

describe "possible showers" or "a chance of snow" without ever specifying a number. That made it hard to tell exactly how likely it was that your picnic would get rained out. It wasn't until computing became widespread in the 1960s that a new set of statistical weather models enabled forecasters to look at a wide range of outcomes and assign likelihoods. By the 1970s, "probabilistic" forecasts were the norm and people had grown accustomed to phrasings such as "a 30% chance of rain".

In much the same way, climate modelling is now affordable enough that some high-end models can be run a number of times. Each run might use a different rate of CO_2 increase, for example, to show the many ways the climate might unfold depending on how serious we get about reducing emissions. Or the model might simulate the same CO_2 increase a number of times, but with starting conditions that vary slightly, in order to see how natural variations in climate affect the results. These large sets of simulations are called **ensembles**, and they've become increasingly important in climate modelling.

As in the examples cited above, some ensembles include more than one thousand simulations, which was unheard of as recently as the turn of the century. The sheer size of these ensembles allows scientists to calculate statistics and probabilities that a single model run can't provide.

The European Union explored how this technique might help guide policy in a huge multi-year project called ENSEMBLES, one of the largest climate-change research efforts ever conducted. From 2004 to 2009, scientists at 66 institutions in Europe, Australia and the US carried out hundreds of simulations of European climate, looking at changes that manifest globally, regionally and locally over seasons, years and decades. The results will feed directly into assessments of the impact of climate change on various economic sectors, from agriculture to insurance.

Apart from creating ensembles, scientists are hard at work making other improvements to their models. Among the current areas of keen interest are weaving in more interactive chemistry, adding more detailed land cover and ecosystems, and capturing the various effects of aerosols more precisely.

One might ask why we ought to believe anything that global climate models tell us. Some contrarians pose this question often, and it's healthy for everyone to recognize what climate models can and can't tell us. For instance, models will always struggle with uncertainty about the starting-point state of the atmosphere, which affects shorter-term weather and climate outlooks the most. At the same time, our lack of foresight about greenhouse emissions becomes enormous when looking a century out. Thus, the search for "sweet spots" – time frames close enough to be of interest for long-term business and government planning but far enough out so that models can provide insight that other techniques can't. In a 2007 *Science* paper, Peter Cox and David Stephenson (University of Exeter) urged their fellow scientists to focus their research energies on the time frame from about 2040 to 2060.

One good reason to be confident in the general quality of today's climate models is their ability to simulate the past – especially the twentieth century, for which we have by far the most complete climate record. The three graphs opposite, prepared for the 2001 IPCC report, show how one climate model captures the ups and downs in global temperature over the last century. It doesn't reproduce every year-to-year wiggle, but that's not the idea, just as a weather model doesn't aim to tell you the exact temperature in your backyard at 8.27pm. What this model does show accurately is the character of twentieth-century climate. It does so by including the gradual increase in greenhouse gases and Sun-shielding pollutants, as well as the effects of natural features like volcanoes and the Sun.

When such a model tells us that global temperatures could rise by 3°C (5°F) or more in the next century, we'd be foolhardy to ignore it.

PART 4

DEBATES & SOLUTIONS

From spats and spin
to saving the planet

Elk River Wind Project, Butler County, Kansas.

A heated debate

How scientists, activists, sceptics and industry have battled for column inches and the public mind

In recent years, perhaps no realm of science has found itself under as much scrutiny as climate change. That's no surprise: at its heart, the issue counterposes the future of our planet's atmosphere and ecology against the most profitable industries on Earth – those that extract, refine and sell fossil fuels. But there are many other reasons why global warming has been such a contentious topic for more than twenty years. The main cause, carbon dioxide, is literally invisible. The worst effects may occur many thousands of kilometres from where we live, or many decades from now. And unlike environmental problems that can be pinned on a few bad actors, virtually all of us produce greenhouse emissions through the course of our daily lives. All of these factors make it tempting to try to refute, dismiss or ignore climate change rather than confront its profound implications head on.

Global warming politics didn't catch fire at first. For the most part, the topic remained in the scientific background until it became clear that the rise in greenhouse gases was real and serious. Even as research (and the gases themselves) continued to accumulate, it took other events to light the fuse in the late 1980s and make climate change a top-level global concern. That's when industry saw the potential threat to its profits and began to act. Starting around 1990, many oil, coal, gas and car companies and like-minded firms joined forces to sow seeds of doubt about climate change science. Working along similar lines as these companies (and, in some cases, getting funded by them directly or indirectly) was a tiny subset of contrarians – scientists of various stripes who bucked the research mainstream and downplayed the risk of global warming. Facilitated by a media eager to produce "balanced" coverage, this small group of scientists became familiar to anyone reading or hearing about climate change. At the other end of the spectrum, environmental activists argued passionately for the defence of the planet and its ecosystems, often with colourful protests tied to key diplomatic moments.

The landscape of climate-change debate has grown far more complex since the Kyoto Protocol – the first global agreement to reduce greenhouse-gas emissions – was introduced in 1997. While the issue is often portrayed as a one-side-versus-the-other battle, there's plenty of nuance in the mix, even among those who agree on the big picture. Many corporations have taken up the banner of reducing emissions, in order to save on their energy costs as well as burnish their green credentials. And environmentalists have fiercely debated many angles among themselves, including the pace at which emissions ought to be reduced and the merits of nuclear power in a greenhouse-threatened world. The scene grew even more muddled in 2009–10, as a batch of emails leaked from top climate scientists triggered controversy, while international squabbling dimmed the chance of a new global agreement on emissions reduction. All this has unfolded despite the fact that the world's leading scientific societies – including the national academies of more than 30 countries – stand firmly behind the recognition that humans are changing the climate.

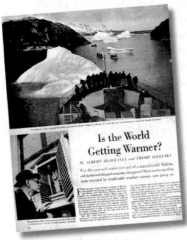

This *Saturday Evening Post* article from 1950 was one of the very first stories on climate change in the popular press.

The early days

As we've seen, the greenhouse effect was discovered right back in the nineteenth century, but serious scientific and media debate about climate change didn't take off until much later. One of the first news stories came in 1950, after global temperatures rose from around 1900 to the 1940s. *The Saturday Evening Post*, then one of the US's biggest magazines, asked a question one might hear today: "Was this past mild winter just part of a natural cycle?" Their article "Is the World Getting Warmer?" rounded up a variety of anecdotal evidence, including "tropical flying fish" sighted off the New Jersey coast. As possible causes for the warm-up, the article cites solar variation and other natural factors. Greenhouse gases aren't even mentioned.

The scientific debate picked up in the 1960s. But at that time, there was plenty else to worry about – nuclear annihilation, for instance

> "Climatological Cassandras are becoming increasingly apprehensive, for the weather aberrations they are studying may be the harbinger of another ice age."
>
> *Time* magazine, 1974

– so few people outside of scientific circles heard much about the risk of climate chaos. Things began to change in the 1970s, when the embryonic environmental movement called out air pollution as an example of humans' soiling of the planet. With early photos from outer space now highlighting Earth's stark aloneness, it was suddenly easier to believe that humans could affect the atmosphere on a global scale.

But what grabbed most of the press in the 1970s wasn't a global warming but a cool-down. Earth's temperature had been gradually slipping for some three decades, mainly in the Northern Hemisphere. A few maverick scientists speculated that dust and sun-blocking sulphate particles emitted from North America and Eurasia could be responsible for the cooling. A British documentary in 1974 called *The Weather Machine* warned that a single brutal winter could be enough to plaster northern latitudes with a "snow blitz" that the next summer couldn't entirely erase, thus leading to continent-encrusting ice sheets within decades. If nothing else, climate had started to seem more fluid and unstable than most people had ever thought possible.

Even as reporters chattered about cold, many scientists were concerned about the long-term outlook for warmth. In a 1972 *Nature* paper entitled "Man-made carbon dioxide and the 'greenhouse effect'", J.S. Sawyer predicted a temperature rise of 0.6°C (1.0°F) for the rest of the twentieth century – a figure that was only slightly off the mark. A landmark 1975 paper in *Science* by

28 April 1973: Newsweek's report on the fear of a forthcoming ice age.

Arguments and counterarguments

Fairly early on in the climate debate, sceptics developed a broad set of criticisms – many of which you still hear today – that were thrown at mainstream climate scientists and at the concept of global warming in general. Taken to the extreme, you could sum up the classic sceptical view like this:

> The atmosphere isn't warming; and if it is, then it's due to natural variation; and even if it's not due to natural variation, then the amount of warming is insignificant; and if it becomes significant, then the benefits will outweigh the problems; and even if they don't, technology will come to the rescue; and even if it doesn't, we shouldn't wreck the economy to fix the problem when many parts of the science are uncertain.

Probably no single sceptic would endorse the whole of that rather convoluted statement. Yet each of the points within it has been argued vigorously over the years by various contrarians. Let's look briefly at each point in turn. For a set of succinct refutations of more than 100 arguments against human-produced climate change, see www.scepticalscience.com.

▶ **"The atmosphere isn't warming"** This one has been put safely to rest, although as recently as the 1990s some sceptics insisted there was no planet-wide warming at all, and the notion still crops up on the Internet. Fuelling this line of argument was the apparent lack of warming in upper-air temperatures as measured by satellites and radiosondes, but it's now clear that globally averaged upper-level temperatures are in fact warming at close to the same rate as the surface (see p.204).

▶ **"The warming is due to natural variation"** This point is still argued often, even though the IPCC has concluded that the warming of the last century, especially since the 1970s, falls outside the bounds of natural variability (see p.8). In any event, the long-term ascent of global warming should outpace the peaks and valleys of natural cyclic processes such as the El Niño/Southern Oscillation or the Atlantic Multidecadal Oscillation in the long run. And variations in the Sun don't appear to explain what's been happening in the last few decades (see p.274).

▶ **"The amount of warming is insignificant"** This claim mingles bona fide uncertainty about the future with a judgement call on how much warming should be labelled as significant. The genuine uncertainty is how much warming we can expect in the coming decades and centuries. As noted in "Circuits of change" (see p.249), the most widely accepted estimate for the rise in global temperature from a doubling of atmospheric carbon dioxide is about 3.0°C (5.4°F) over pre-industrial times, with a range of 2.0–4.5°C (3.6–8.1°F) deemed "likely" in the 2007 IPCC report. A number of sceptics believe that the low end of the range – if even that – is the most likely outcome, but in fact we've already warmed close to 0.8°C (1.44°F) since pre-industrial times. This puts us almost halfway to the low end of the IPCC range, and that's with carbon dioxide up only about 35% from its pre-industrial value.

▶ **"The benefits will outweigh the problems"** Overall, some stimulation of plant growth does appear likely, but it's not at all certain that the benefits will be prolonged or planet-wide or that the nutritive value of crops will be sustained (see p.180). Moreover, while CO_2 may give forests a boost, the changing climate raises the risk of devastating fires and insect attacks (see p.172). Against the potential pluses of CO_2 fertilization, and other benefits such as fewer cold-related illnesses and the possibility of sailing through the Arctic in midsummer, we have to balance the various negatives, including the risk of rising seas, widespread drought and massive species loss.

▶ **"Technology will come to the rescue"** This isn't scepticism about global warming so much as an affirmation of human ingenuity. Some optimists believe that geoengineering might save us from the clutches of global warming (see p.359). Even if such an approach proves feasible, it would face an uphill trek to gain funding, international approval and public confidence. Still, it's important to keep in mind the possibility and promise of technical innovation, while at the same time recognizing the reality of our present situation and the emissions trajectory we're on.

▶ **"We shouldn't wreck the economy"** For sceptics motivated more by economic than scientific considerations, this is the ultimate bottom line. If we don't know with absolute confidence how much it will warm and what the local and regional impacts will be, so the reasoning goes, perhaps we're better off not committing ourselves to costly reductions in greenhouse-gas emissions. However, the eventual costs of environmental remedies have often proved much less than economic models indicated at first. Moreover, it's unclear how much further the scientific uncertainty around specific regional outcomes can be reduced. Perhaps more importantly, it would be foolish to assume that reducing emissions will cost more than coping with a changing climate – a point made emphatically by the Stern Review (see p.326). In the following chapters, we'll discuss some technologies and approaches that could actually produce net savings in the long run.

▶ **Other arguments** Many critiques of climate change that emerged long ago have popped up on talk shows, blog posts and newspaper op-ed pages even after they've been debunked time and again. One example is the oft-cited argument, "If they can't predict the weather for next month, how can they predict the climate a hundred years from now?" Of course, these are two fundamentally different processes. A weather forecast tracks day-to-day changes at a given point. A climate projection looks at longer-term trends that in turn tell you about the type of weather we might expect. If you live in Germany or Minnesota and it's the first day of January, you can say with some confidence that the first day in July ought to be warmer than today, even if you can't predict whether the high will be 20°C (68°F) or 35°C (95°F).

Two other points of contention are the quality of the global models that project future warming and the data that tell us about past climate. The models certainly aren't perfect (see p.249), but they've agreed for years that we can expect a significant warming. Likewise, shortfalls do exist in the records of past weather (which weren't really designed to detect climate shifts in the first place), but they aren't enough to rule out the overwhelming evidence of change already under way. It's hard to debate a world full of melting glaciers.

Wallace Broecker (Lamont-Doherty Earth Observatory) asked if we were "on the brink of a pronounced global warming" (and became the first research paper to use that phrase in its title). Two studies late in the 1970s by the US National Academy of Sciences confirmed that the ever-increasing levels of CO_2 in the air should lead to significant warming. Computer models were improving quickly, and they continued to indicate that warming was on the way. Finally, the atmosphere itself chimed in. By the late 1980s global temperatures had begun an ascent that hasn't abated since, except for a sharp two-year drop after 1991's eruption of **Mount Pinatubo** (see p.218) and a levelling off during much of the millennium's first decade.

The reports and findings continued to pile up through the 1980s, but with little fanfare at first outside of research labs and government hearings. Theories of global warming remained exotic enough to the public that many journalists kept the term "greenhouse effect" in quotes. The stunning discovery of the Antarctic **ozone hole** in 1985 was a turning point. Although it fostered long-lived confusion between ozone depletion and global warming (see p.32), the finding was also a new sign of the atmosphere's fragility, borne out by vivid satellite images.

The other shoe dropped in the United States during its sizzling, drought-ridden **summer of 1988**. Huge tracts of forest burned in Yellowstone National Park; parts of the Mississippi River ran dry; and on a record-hot June day in Washington DC, NASA scientist **James Hansen** delivered memorable testimony before Congress, presenting new model-based evidence and claiming to be "99% sure" that global warming was upon us, most likely induced by humans. Together, *The New York Times* and *Washington Post* ran more than forty stories on climate change in 1988 after fewer than two dozen in the preceding four years, according to Katherine McComas and James Shanahan of Cornell University. *Time* magazine named "Endangered Earth" Planet of the Year, in place of its usual Man of the Year. Even conservative politicians took note. In August, US presidential candidate George Bush (senior) declared, "Those who think we are powerless to do anything about the greenhouse effect forget about the 'White House effect'." And although the meteorological drama of 1988 was focused on North America, the political waves reverberated far and wide. In September, British prime minister Margaret Thatcher warned the Royal Society that "we have unwittingly begun a massive experiment with the system of the planet itself." As Jeremy Leggett recalls in *The Carbon War*, "1988 was the year that broke the mould." Indeed, the events of that year were enough to convince Leggett, who was then teaching at Britain's Royal School of Mines, to join Greenpeace as a science advisor to its climate campaign.

Is the Sun behind climate change?

In an attempt to let carbon dioxide – and human actions – off the hook for climate change, it's often claimed that **solar variations** account for the last century's warmth. Over the very long term, variations in Earth's orbit that shape where and when sunlight reaches the planet are the main cause of ice ages (see p.216), but this doesn't apply to our current situation. It's true that the Sun has produced more sunspots in recent decades than it did in the early 1800s. However, this mainly reflects an increase in the ultraviolet range of sunlight, which is only a tiny part of the solar spectrum. In fact, the total solar energy reaching Earth changes very little over time. Across the eleven-year solar cycle, it varies by less than 0.1%, and even across the period since the Little Ice Age chill of 1750, solar output climbed no more than about 0.12%, according to the 2007 IPCC report. Subsequent estimates by Judith Lean (Naval Research Laboratory) and others have pegged the solar contribution to twentieth-century warming at 10% or less.

Extreme ultraviolet image of the Sun showing the chromosphere, hotspots and a huge solar flare.

There's still a question mark or two when it comes to ultraviolet radiation, where the lion's share of solar variability occurs. It's possible that UV rays interact with ozone in the stratosphere to change circulation patterns, though more work is needed to clarify how this might occur. UV light also helps shield Earth from **cosmic rays** that bombard and ionize the atmosphere – a point much discussed by sceptics in recent years, thanks to work by Henrik Svensmark (Danish National Space Centre) and others. This concept got major play on UK and Australian TV in the 2007 documentary *The Great Global Warming Swindle* and in the popular book *The Chilling Stars*, co-written by Nigel Calder and Svensmark. The idea is that highly reflective low-level clouds might form more easily when tiny particles that serve as cloud nuclei are ionized, helping them to clump together more readily. Should this be true, then an active Sun would inhibit low-level clouds, thus allowing more sunlight to reach Earth and fostering warming. In lab work, Svensmark and colleagues found some evidence for the clumping effect, but it's an open question whether these particles actually make low-level clouds more prevalent in the real world. Vast numbers of potential cloud nuclei are normally present anyway, and studies comparing satellite-derived cloud patterns to cosmic-ray counts have shown conflicting results. Moreover, there's no clear evidence that more cosmic rays have actually made it into Earth's lower atmosphere over the last several decades. *The Great Global Warming Swindle* was criticized by many experts not only for downplaying these unknowns but also for using discredited data and inaccurate graphs.

From progress to roadblocks

In the aftermath of 1988, governments began to pour money into global warming research. In 1989 the Intergovernmental Panel on Climate Change was established (see p.308) as a means of channelling research from dozens of nations and many hundreds of scientists into an internationally recognized consensus. The IPCC produced its first report in 1990, underlining the risks of global warming, and environmental activists did their best to alert journalists and the public to the problem.

By this time, the leading **environmental groups** in North America and Europe were well established, most with fifteen years under their belts. No longer a fringe movement, their cause was now part of the fabric of public life. Years of activism had helped slow the growth of nuclear power to a crawl in many countries (with no small assistance from the Chernobyl debacle in 1986). Governments and politicians had become so attuned to environmental risk that it took fewer than three years – lightning speed by diplomatic standards – from the time the ozone hole over Antarctica was discovered in 1985 to the signing of the 1987 Montreal Protocol that's now guiding the planet towards eventual ozone recovery.

For a while, it looked as if the same success might be seen with global warming. In 1992, thousands of activists joined similar numbers of journalists and diplomats in Rio de Janeiro for the United Nations-sponsored **Earth Summit**. The meeting, and the global climate treaty that emerged, kicked off years of negotiations that led to the historic **Kyoto Protocol**. (For more on Kyoto, see the following chapter.)

"Are existing environmental institutions up to the task of imagining the post-global warming world? Or do we now need a set of new institutions founded around a more expansive vision and set of values?"

Michael Shellenberger and Ted Nordhaus, *The Death of Environmentalism*

As media attention grew and as the scientific evidence strengthened, people in many countries became more aware about the risks of climate change and the possible solutions. Yet something wasn't quite clicking. Although public support for climate-change action was growing steadily, the support was "wide but shallow". In other words, people were concerned, but not sufficiently concerned to force the issue up the political agenda (nor to take personal action to reduce their own greenhouse gas emissions).

The sheer scope of the problem was one factor. Fossil fuels are used in virtually every aspect of modern society, and climate

Debates among the campaigners

Climate change is now a favoured cause of most environmental groups, but key differences in strategy and ideology remain. One question is whether to embrace or reject certain **aspects of capitalism**. With consumerism driving the world economy, some eco groups are going with the flow – encouraging people to use their spending power to make climate-healthy choices, such as buying a hybrid car or using low-energy light bulbs. Other groups, especially those with roots in 1970s counterculture, retain an abiding suspicion of the corporate world and the governments that support it, but many of these groups are working within the system to achieve results. Friends of the Earth stresses its anti-globalization work in its climate change publicity materials, yet FoE's "Big Ask" campaign played a pivotal role in securing the UK's Climate Change Act of 2008.

For environmentalists who *do* accept the idea of green commerce, the struggle to isolate genuinely "ethical" choices can be tricky. For instance, Toyota makes the world's most popular hybrid – the Prius – yet it was one of the plaintiffs in a lawsuit that aimed to block California's strict new emissions standards. Hence some groups encourage consumers to push big companies in a climate-friendly direction through shareholder actions or socially responsible investing (see p.367).

Another difference between climate activists is **how to approach fossil fuels**. Some take a pragmatic approach, figuring that oil, gas and coal aren't going away tomorrow. The US Natural Resources Defense Council has thrown its considerable weight behind "cleaner coal", on the condition that it includes CO_2 sequestration. NRDC notes the large number of coal plants already on order in China and elsewhere and stresses the need to develop cleaner technology. Other groups keep the focus on renewables. "Make no mistake: coal is dirty", says Greenpeace International, which is pushing for a global phase-out of coal as an energy source.

Looming in the background is **nuclear energy**, and the question of whether countries should turn to it to bridge the potential gap between fossil fuels and large-scale deployment of renewables. Some key environmental thinkers have lent support to the idea of using nuclear as a stop-gap, including Gaia theorist James Lovelock, *Whole Earth Catalog* founder Stewart Brand and an interdisciplinary panel of MIT scientists (see p.348). However, mainstream environmental groups remain vocally opposed to nuclear power.

A final point of difference is **picking goals**. A concrete target is the best way to motivate volunteers and supporters. The most commonly cited benchmark, consistent with the European Union's goal, is to stabilize the climate at 2°C (3.6°F) above the pre-industrial global temperature (see p.300). A more ambitious target was set by the group 350.org, which aims to bring the global concentration of carbon dioxide down to 350 parts per million. Yet global emissions would have to be cut drastically in order to meet either goal. With this in mind, some environmental groups use other types of targets as well, including legislative ones. Short-term goals are still another approach, as employed by the 10:10 campaign, which encourages individuals and organizations to aim for a 10% reduction in a year – a challenge taken up by new prime minister David Cameron for the UK's entire central government.

サヨナラ！化石燃料

GREENPEACE

DINOSAUR DIPLOMACY 1; CLIMATE 0

GREENPEACE

Greenpeace protesting against "Dinosaur Diplomacy" at the Climate Conference in Kyoto, Japan, 1997.

change threatens to affect every country on Earth in one way or another. It's hard to motivate people to grapple with such an immense and seemingly intractable issue, and the many options for political and personal action could be too much to process. Moreover, even more than smog or acid rain, human-induced climate change is a classic "tragedy of the commons" – the benefits of burning fossil fuels accrue to individuals, companies and nations, while the costs accrue to the planet as a whole. And while the activists urged concrete action, the benefits – avoiding a global meltdown – were intangible as well.

Another difficulty for climate campaigners was the fact that global warming hit it big just when many people were getting tired of fretting about the state of the world. From its earliest days, the environmental movement had relied on stark, pseudo-apocalyptic imagery to motivate people. In her 1962 book *Silent Spring*, which set the template for environmental wake-up calls, Rachel Carson labelled pesticides and similar agents "the last and greatest danger to our civilization". In her footsteps came a series of similarly dire scenarios, from Paul Ehrlich's *The Population Bomb* to the notion of nuclear winter. Global warming lends itself especially well to this type of rhetoric. It's no exaggeration to talk about the risk of coastal settlements vanishing and the Arctic's summer ice pack disappearing. Yet when activists do hit these points, it sometimes brings to mind other predictions of environmental doomsday that didn't come to pass (partly, of course, because society *did* respond to those earlier threats). Some observers have argued that fear-based rhetoric on climate change makes people defensive and resistant to recognizing the problem, and that it can stoke the politicization of the issue.

Even the most painless ways to reduce global warming – such as improved **energy efficiency** – came with cultural baggage in some countries. Efficiency measures had swept the US during the oil shocks of the 1970s – the top interstate speed limit was dropped to 55mph (89kph), for instance – but the practice of saving energy never lost its taint of deprivation. As oil prices plummeted in the go-go 1980s, efficiency quickly fell by the wayside and speed limits went back up. By the time global warming pushed energy efficiency back onto the national agenda, it was a loaded topic. Because so many activists had been proposing sensible energy-saving steps for years, it was easy for critics to paint them as opportunists, happy to use climate change or any other issue in order to advance their ulterior goals.

To top it all, climate activists were up against some very tricky adversaries, who were doing their best to *stop* the public from getting too worried about global warming.

Sceptics and industry fight back

Especially in the United States, a group of scientists with sceptical views about climate change – perhaps no more than several dozen – have wielded far more influence than their numbers might indicate. Until very recently, many if not most news articles about climate change included a comment from one or more of these contrarians. Their voices have been backed up in many cases by the immense money and influence of the **oil, coal and car industries**, largely through conservative think tanks (see overleaf). This support has enabled them to exert much sway on the US Congress, the media and, by extension, the global fight against climate change.

Many scientist-sceptics have been portrayed as mavericks or outsiders, and with good reason. Although a few are active in climate research, many aren't. Some of the most vocal have backgrounds in subjects like solid-state physics or mathematics, sometimes with impressive résumés in their fields but little if any

> "I am convinced that in fifteen to twenty years, we will look back on this period of global warming hysteria as we now look back on so many other popular, and trendy, scientific ideas."
>
> William Gray, Colorado State University, testifying before the US Congress, September 2005

Industry lobby groups

Like their peers, the few climate sceptics active in research are employed mainly by universities and private labs. Although a few have received grants from oil and coal companies, most rely largely on public funds to carry out their work. However, that work gets an extra dose of clout, especially in the US, thanks to a number of conservative think tanks and lobby groups which cite their findings widely and use them in an attempt to convince legislators that climate change science is full of unknowns. Such centres are often influential, and many are buoyed by funding from corporations with a lot to lose from carbon restrictions.

One highly visible group throughout the 1990s was the opaquely titled **Global Climate Coalition**, which formed in 1989 as the prospect of global diplomatic action on climate change appeared on the horizon. Based at the US National Association of Manufacturers, the GCC included some of the biggest oil, car and coal companies in the world, including General Motors, Ford, BP, Shell and Exxon (aka Esso). Along with lobbying at UN meetings, the coalition angled its way into becoming an oft-quoted presence in the media. They also financed Kyoto-related commercials warning that "Americans would pay the price" for the treaty.

The GCC began to fracture with the departure of BP in 1997, Royal Dutch Shell in 1998 and Ford in 1999. By 2001, it was history, though arguably it had served its purpose and was no longer necessary. A 2001 memo written to Exxon by the US under-secretary of state, Paula Dobriansky, and later obtained by Greenpeace, states that George Bush rejected Kyoto "partly based on input from you [the GCC]". In the group's own words, "The industry voice on climate change has served its purpose by contributing to a new national approach to global warming."

Since the days of the GCC, most of the world's major oil companies have shifted towards public acknowledgement of climate change, but Exxon – the largest of

experience specifically in climate-change science. For example, physicist Frederick Seitz worked on atomic research in the 1940s, presided over the US National Academy of Sciences in the 1960s, and then served as a consultant to the medical research programme of the R.J. Reynolds tobacco company before becoming an outspoken climate-change sceptic. Other dissenters are trained in atmospheric science but have published few peer-reviewed papers on climate change. Many are retirees, affording them the time and freedom to act as consultants, writers and speakers without having to conduct scientific studies of their own. Of course, there have also been plenty of spokespeople for sceptical positions who aren't scientists at all (see box above), just as plenty of non-scientists speak out for climate-change action.

them all – continued to cultivate doubt. From 2000 to 2003, according to an exposé by Chris Mooney in *Mother Jones* magazine, the company poured more than $8 million into more than forty organizations aligned with climate-change scepticism. In a rare move, the UK's Royal Society wrote to Exxon in 2006 asking it to stop funding sceptically aligned groups. In 2008, the company made a high-profile announcement that it would stop contributing to such groups, but *The Times* of London reported that Exxon gave $1.3 million in 2009 to groups affiliated with sceptical positions (albeit a smaller set of groups removed from the most heated climate-change denial).

Working largely below the media radar is Koch Industries, a Kansas-based conglomerate with major oil holdings. Its founding brothers were ranked by *Forbes* in 2010 as two of the six richest Americans. Since the Koch Industries name doesn't appear on consumer products, the firm is virtually unknown to the public. However, it gave more than twice as much as Exxon to sceptic-oriented groups from 2005 to 2008, according to a 2010 Greenpeace report that dubbed the company "a financial kingpin of climate science denial and clean energy opposition."

The global warming wing of the Washington-based **Competitive Enterprise Institute** became the leading institutional voice of climate scepticism as other entities pulled back or lost interest. Its leader, Myron Ebell, was censured by the British House of Commons "in the strongest possible terms" in 2004 after he told BBC's Radio 4 that Sir David King, the chief science advisor to prime minister Tony Blair, "knows nothing about climate science". On the release of the Al Gore documentary *An Inconvenient Truth* (see p.291), the CEI issued a pair of glossy TV advertisements that noted how fossil fuels have made life more comfortable and convenient. They ended with the tag-line "Carbon dioxide: They call it pollution. We call it life."

Right from the beginning, uncertainty has been the overriding theme in the arguments of climate-change contrarians. The core of greenhouse science – such as the consensus estimates on how much global temperature rise to expect from a doubling of CO_2 – has held firm for decades. But climate change is such a multifaceted and complicated enterprise that it's easy enough to find minor weaknesses in one study or another. Furthermore, there are always exceptions that prove the rule, such as an expanding glacier or a region that's cooled in recent decades. When contrarians point to a single event or process as a disproof of global climate change, or when advocates tout a particular heat wave as iron-clad proof that humans are meddling with climate, they're often accused of **cherry-picking**: selecting a few bits of evidence that seem to prove their point

Bjorn Lomborg's sceptical environmentalism

Danish political scientist Bjorn Lomborg marshalled a slew of statistics and nearly 3000 footnotes to make his case that, overall, the environment is in better shape than we might think. In his influential 2001 book *The Sceptical Environmentalist*, Lomborg employed the climate-and-economy models used by the IPCC assessments to argue that major emissions reductions in the short term (*à la* Kyoto) are not only enormously costly but will have little impact on the longer-term climate outcome.

Lomborg's book got rave reviews in *The Economist*, *Rolling Stone* and elsewhere, but it was panned in other publications and pilloried by some leading scientists. The dust-up got to the point where the official Danish Committee on Scientific Dishonesty labelled Lomborg's book "objectively dishonest" (they later withdrew the finding). The magazine *Scientific American* published "Misleading Math about the Earth", an eleven-page critique of *The Sceptical Environmentalist* by four top researchers, eventually followed by a rebuttal from Lomborg himself and then a re-rebuttal from one of the four critics. Another of the four, climate scientist Stephen Schneider from Stanford University, blasted Lomborg in *Grist* magazine for "selective inattention to inconvenient literature and overemphasis of work that supports his lopsided views."

In his discussion of climate change, Lomborg glided past sea-level rise with little concern for the high-end possibilities. Moreover, Lomborg's economic focus failed to take into account the intrinsic, non-monetary value of protecting particular species and ecosystems. Even so, the book's sunny-side-up view of economic and ecological progress and its critique of environmental doom and gloom drew many fans, especially from the sceptical side of the global-warming aisle. Lomborg, in fact, ended his climate-change discussion by claiming that society has the money to control greenhouse emissions if we deemed it a high enough priority. However, he argued that many other issues – such as preventable diseases – deserve to take precedence. In his 2007 follow-up *Cool It: The Sceptical Environmentalist's Guide to Global Warming*, Lomborg staked out his turf even more firmly, remaining sanguine about such concerns as Antarctic ice (it's growing) and polar bears (their real nemesis is hunting, not warming).

In 2010, Lomborg raised eyebrows with what seemed to be an about-face, declaring that the world should dedicate $100 billion per year to addressing climate change. That position emerged from the Copenhagen Consensus, a panel of economists organized by Lomborg. In 2004, the panel had ranked three climate measures (meeting the Kyoto Protocol goals, and assessing two types of carbon tax) at the bottom of a list of sixteen actions on which governments might best spend $50 billion to advance global welfare. But in 2009, apparently more concerned about the risks posed by global warming, Lomberg assigned a new question to the panel: given a set of potential responses to climate change, which ones would provide the best bang for the buck? Not surprisingly, taxes were the least favoured approach, whereas geoengineering and R&D investment topped the list.

while omitting counter-examples. It's a classic rhetorical technique, one well known to skilled lawyers and politicians.

Starting in the early 1990s, sceptics seized such uncertainty and exceptions and – amplified by the PR budgets of corporations heavily invested in fossil-fuel use – gave the false impression that the entire edifice of knowledge about climate change might crumble at any moment (or even that the whole thing was a colossal scam, a claim voiced more than a few times). The tactic of stressing uncertainty wasn't a new one: it had been used to delay tobacco regulation for decades and to stall environmental action in other areas, as laid out in detail by science historians Naomi Oreskes and Erik Conway in their 2010 book *Merchants of Doubt.*

Sceptic attacks have put some climate scientists directly in the line of fire. In the 1970s, the late **Stephen Schneider** studied the powerful effects of polluting aerosols and, in his book *The Genesis Strategy*, stressed the need for society to prepare for intense climate shifts. With the media focused on global cooling during this period, Schneider was often quoted, but his book emphasized the risks of greenhouse warming as well as aerosol cooling. In 1988 Schneider wrote *Global Warming*, one of the first lay-oriented books on the topic. His visibility and prolific output soon made him a prime target for critics who pointed to Schneider's earlier work on aerosol cooling, took his quotes out of context and derided his research on climate risks and assessments. Schneider later noted, "A scientist's likelihood of having her/his meaning turned on its head is pretty high – especially with highly politicized topics such as global warming."

The release of the IPCC's second assessment in 1995 drew one of the decade's sharpest rounds of sceptic vitriol. The report's beefed-up assertion that "the balance of evidence suggests a discernible human influence on global climate" was based in part on detecting climate-change **fingerprints**, the 3-D temperature patterns one would expect from the observed increase in greenhouse gases (as opposed to different patterns of warming that might be produced by other factors, such as strengthening solar input). A pro-

Bjorn Lomborg: climate sceptic?

They're not climate scientists, but...

It's not often that a novelist gets the opportunity to speak to members of a science-oriented committee in the US Congress. The late Michael Crichton, king of the techno-thriller, got his day on Capitol Hill on 28 September 2005, not long after his novel *State of Fear* stormed the bestseller lists. Crichton, a Harvard-trained physician, was invited to discuss the role of science in environmental policymaking. He also scored a private hour-long meeting with US president George Bush, according to the 2006 book *Rebel in Chief* by Fred Barnes.

Crichton is just one of many people from outside the realm of climate research who've been anointed as experts in the field. Many of them are smart, articulate people with strong beliefs about how science should operate or how society should respond to its research. Those qualities can help give non-experts a voice in media, government hearings and the blogosphere that's well

Michael Crichton

out of proportion to their actual knowledge about climate change. In a 2003 speech at Caltech entitled "Aliens Cause Global Warming", Crichton mocked the projections from global climate models, ignoring the differences between how weather and climate models work: "Nobody believes a weather prediction twelve hours ahead. Now we're asked to believe a prediction that goes out 100 years into the future". In *State of Fear*, Crichton alternated between nail-biting action sequences and tutorials in which the hero (and by extension the reader) discovers that the consensus on global warming is full of holes. It's surely the first blockbuster novel to include more than 25 actual graphs of long-term temperature trends at stations

cedural delay in incorporating review and revisions to one chapter following the IPCC's plenary meeting, and confusion involving text that had been leaked to the media, led critics to accuse the IPCC – and, directly or indirectly, convening lead author Benjamin Santer (Lawrence Livermore National Laboratory) – of "doctoring the documents" and "scientific cleansing", Seitz castigated the IPCC in a widely cited *Wall Street Journal* editorial, while Santer and colleagues emphasized that the edits were a required response to review comments, per IPCC rules, and that the chapter's conclusions were virtually unchanged. Eventually, the battle subsided, but it made a lasting impact on Santer. "Nothing in my university training prepared me for what I faced in the aftermath of that report", he later said. "You are prepared as a scientist to defend your research. But I was not prepared to defend my personal integrity. I never imagined I'd have to do that."

Lord Christopher Monckton

across the world. Many show cooling trends; others, such as New York, show dramatic long-term warming, presumably due to the heat-island effect. In real life, however, no climate scientist expects that every station on Earth should warm in lockstep, and urban biases have already been exhaustively studied and corrected (see p.194).

Across the Atlantic, the UK's most prominent climate sceptic may be Christopher Monckton, Third Viscount Monckton of Brenchley. Schooled in classics and journalism, Monckton has a résumé that includes serving as an advisor to Margaret Thatcher, writing and editing for several newspapers, and creating Eternity, a 209-piece puzzle that tantalized thousands of Brits and won two mathematicians a £1 million prize. Monckton later took up the cause of climate change scepticism, which dovetailed with his longstanding Euroscepticism. He addressed the US Congress in 2009 and 2010, claiming in his written 2010 testimony that "there is no consensus on how much warming a given increase in CO_2 will cause" and "today's CO_2 concentration, though perhaps the highest in 20 million years, is by no means exceptional or damaging." Nine of Monckton's assertions were contested in a rebuttal sent to Congress by more than two dozen climate scientists from NASA, NOAA, NCAR, and other labs and universities. "For those without some familiarity with climate science, his testimony may appear to have scientific validity", said the response. However, it added, "in all cases, Mr Monckton's assertions are shown to be without merit – they are based on a thorough misunderstanding of the science of climate change."

Two sides to *every* story

In spite of the flare-up over the second IPCC report, the media had largely lost interest in global warming by the mid-1990s. According to the Cornell study cited above, the number of climate-change articles in *The New York Times* and *Washington Post* dropped from more than 70 in 1989 to fewer than 20 in 1994. In part, the drop-off was typical of how news stories come and go – alarms couldn't ring forever, at least not without some major disaster to make climate chaos seem like an imminent threat. But the sceptics and lobby groups undoubtedly played a role, having successfully convinced many journalists – and large swathes of the public – that global warming was at best an unknown quantity and at worst "ideological propaganda ... a global fraud" (in the words of UK *Daily Mail* journalist Melanie Phillips).

Even when climate change *did* appear in the news, it suffered from the media paradigm that seeks to give equal weight to both sides of every story, with an advocate from each side of the fence battling it out in a fair fight. It's a time-honoured form of reporting, honed to perfection in political coverage, but in the case of climate change it often conveys a misleading sense of symmetry, as if the sceptic camp represented half of the world's climate scientists rather than a small group of contrarians.

Despite its faults, the "duelling scientist" mode of coverage soon became the norm, especially in the US. One study of seven leading US papers showed that a group of five top climate scientists from the US and Sweden were quoted in 33 articles in 1990 but in only five articles in 1994. The use of sceptics changed little over the period, so that what were once minority opinions were soon getting roughly equal time.

In a later study for the journal *Global Environmental Change*, Maxwell Boykoff and Jules Boykoff detailed the sudden shift in US reporting styles at four major newspapers. Most articles in 1988 and 1989 focused on the evidence that human-induced global change is a real concern. But in 1990 these were eclipsed by "balanced" articles that gave more or less equal weight to natural variations and human factors as causes of climate change. This became the standard format into the 2000s, according to Boykoff and Boykoff. As they noted, "through the filter of balanced reporting, popular discourse has significantly diverged from the scientific discourse."

How did the sceptics gain such a high profile? One factor was the powerful public-relations machine discussed above. With the blogosphere not yet a major force, several campaigns were designed to highlight uncertainties in the science through news releases, press conferences and direct vehicles such as TV advertisements. In 1998, for example, just after the Kyoto Protocol was drafted, *The New York Times* learned of a proposal by an industry faction to spend $5 million campaigning against the treaty. "If we can show that science does not support the Kyoto treaty – which most true climate scientists believe to be the case – this puts the United States in a stronger moral position", noted the industry document, which called for "an action plan to inform the American public that science

> "Although the scientific community has known since 1995 that we are changing our climate, the US press has done a deplorable job in disseminating that information, and all its implications, to the public."
>
> Ross Gelbspan, *Boiling Point*

When global warming meets TV weather

For a variety of reasons, most TV forecasters keep remarkably quiet about global warming. With only a couple of minutes available on the daily weather segment, there's little time to explain the greenhouse effect or other global-warming science. What's striking is that TV producers so seldom turn to their resident weather experts for coverage that might tie global warming to local concerns and conditions. "The last thing any station wants is an activist weatherman", Matthew Felling, a US consultant, told the *Salon* website in 2006.

There are far more weathercasters in the expansive US television market than anywhere else – more than 1000 of them. Perhaps half are trained meteorologists, but few of them produce science or environmental stories on top of their regular weather-reporting duties. The American Meteorological Society (AMS) has embarked on a campaign to train weathercasters as "staff scientists", with an eye towards giving them a higher profile in environmental coverage at their stations. Yet many forecasters remain dubious. A 2010 survey of more than 500 US weathercasters found that 26% agreed with the statement, "Global warming is a scam." That's the position of John Coleman, a weathercaster for more than 50 years and the founding president of The Weather Channel. Though he left the channel in 1983, a year after its debut, Coleman has traded on his pioneering role and his recent on-camera position in San Diego to air his sceptical views.

The Weather Channel itself, which reaches most American subscribers to cable TV, is a reliable source of news about global warming. In 2003 the network hired Heidi Cullen as its first-ever climate expert. Cullen – a former research scientist – produced numerous segments on climate-change science and impacts before moving to the nonprofit Climate Central. The network continues to cover global warming research under its "Forecast: Earth" banner. The Weather Channel's 2007 position statement on climate change acknowledges "strong evidence that the majority of the warming over the past century is a result of human activities." Not all viewers are happy with the coverage, though. When Cullen suggested in 2006 that TV weatherpeople who can't "speak to the fundamental science of climate change" shouldn't be certified by the AMS, her website received more than 1700 email protests.

The UK doesn't have an exact counterpart to Cullen, but the BBC has encountered its own challenges in covering climate change. In 2007, the network cancelled a day-long special, *Planet Relief*, that would have focused on the topic. Comments from producers and commentators reflected a fierce internal struggle at the BBC over where reporting stops and advocacy begins. "It's absolutely not the BBC's job to save the planet", said *Newsnight* editor Peter Barron, while writer and activist Mark Lynas told *The Independent* that climate change was now "too hot for the BBC to handle".

Heidi Cullen

Climate change and the church

Most environmental activists operate from a secular viewpoint, but that's not always the case. In the US, there's a small but growing faction of what one headline writer dubbed "earthy evangelists". They made the news in 2006, when nearly a hundred of them signed a statement in support of the fight against climate change. It was the first salvo in the Evangelical Climate Initiative, along with TV ads that include the tag line, "With God's help, we can stop global warming."

This was hardly the first faith-based action on global warming. Christian charity Tearfund is a founding member of the Stop Climate Chaos coalition, and environmental groups with Jewish and other religious ties have entered the fray over the years (many of them as part of the US National Religious Partnership for the Environment). But for those accustomed to thinking of US evangelicals as moving in lockstep with the nation's far-right wing, the 2006 statement was a startling move.

There are a lot of evangelical Christians in the US – at least thirty million – and not all are on the same wavelength as the earthy evangelists. The activists have spun off from the National Association of Evangelicals, which declined to endorse their project. A rival group quickly sprang up – the Interfaith Stewardship Alliance – featuring some of the nation's best-known conservative Christians, including James Dobson and Charles Colson. They wrote their own statement, claiming "global warming is not a consensus issue", and their positions align much more closely with traditional sceptic fare. A paper by one of their founders, E. Calvin Beisner, went so far as to draw an analogy between coal and Jesus: "Vegetation is sown a natural body. Then, raised from the dead as coal and burned to enhance and safeguard our lives, it becomes a spiritual body – carbon dioxide gas – that gives life to vegetation and, through that, to every other living thing." The group has since evolved into the Cornwall Alliance for the Stewardship of Creation, whose 2009 "evangelical declaration on global warming" rails against mandatory reductions of greenhouse gases.

The earthy evangelicals and the conservatives come to their vastly different perspectives from a similar starting point. Both subscribe to the biblical view of humans as stewards of Earth. Both express concern over the fate of Earth's poorest residents, especially in the developing world. But where the activists point to climate change as "the latest evidence of our failure to exercise proper stewardship", Beisner says that "a truly biblical ethic of creation care simply cannot ignore the Biblical mandate for man to fill, subdue and rule the Earth."

The activists note that "millions of people could die in this century because of climate change, most of them our poorest global neighbours", while the rival group claims that the money presumably required to reduce greenhouse emissions could lift millions out of poverty – a point also made by Bjorn Lomborg (see p.282). Of course, this presumes that a giant pot of money is at hand, ready to be spent on either climate-change protection or poverty relief. In truth, of course, policymakers seldom put such big goals side by side in setting budgets. Indeed, it could end up being the risk to the world's biggest economies, rather than to its poorest people, that motivates real action on climate change.

does not support the precipitous actions Kyoto would dictate." Victory will be achieved, it said, "when media 'understands' (recognizes) uncertainties in climate science."

It wasn't only lobby groups and think tanks working to calm the public mind about climate change. After Kyoto, some government officials appeared to be out to stop – or at least slow – the research and views flowing from their agencies' own scientists to the media and the public. In June 2005, *The New York Times* reported that a political appointee of US president George W. Bush (himself a former oil man) had reviewed several climate reports, including the annual overview *Our Changing Planet*, with a heavy hand. For example, in the statement "The Earth is undergoing a period of rapid change", the term "is" had been changed to "may be". Within days the Bush appointee – who had earlier been a staffer with the American Petroleum Institute and a lobbyist against the Kyoto Protocol – left the White House and took a job with ExxonMobil. Similarly, several researchers at NASA and the National Oceanic and Atmospheric Administration claimed they were stymied by higher-ups in their attempts to speak with journalists on the latest findings in controversial areas, such as the effect of climate change on hurricanes. NASA's James Hansen told *The New York Times* and the BBC in 2006 that he was warned of "dire consequences" should he fail to clear all media requests for interviews with NASA headquarters. One of the political appointees charged with such clearance was a 24-year-old who resigned after it turned out he'd falsely claimed a degree in journalism on his résumé. Both NASA and the NOAA reiterated their public commitment to openness and transparency after these stories broke, and the US National Science Board called for uniform federal guidelines to prevent "the intentional or unintentional suppression or distortion of research findings."

On the other side of the Pacific, a media storm erupted in 2006 with an investigative report from the Australian Broadcasting Corporation. According to the ABC report, at least three scientists at Australia's Commonwealth Scientific and Industrial Research Organization (CSIRO) were dissuaded from airing their views on climate policy in various reports. The CSIRO discourages its scientists from commenting on policy, as is typical in the US. However, at least one CSIRO scientist claimed that his comments on policy-relevant science – such as sea-level rise, which could affect immigration from Pacific islands to Australia – were getting quashed as well. Among the casualties in this turmoil was a short-lived communications director at CSIRO who had joined the science agency after serving as a spokesperson for Australia's tobacco industry.

"Conspicuous by its absence has been any sense of urgency in the British media ... the public has been left uninformed about a serious issue."

The British Medical Journal, 1996

Even the UK government, a participant in the Kyoto Protocol, found itself under scrutiny. Sir David King, the UK's chief science advisor, stated in January 2004 that "climate change is the most severe problem we are facing today – more serious even than the threat of terrorism." A few weeks later, US reporter Mike Martin discovered a floppy disk left in the press room at a Seattle science meeting attended by King. On the disk was a memo from Ivan Rogers, the chief private secretary to Tony Blair, asking King to avoid media interviews. The discovery led to an article by Martin in the journal *Science* and a follow-up from *The Independent* in London that was headlined, "Scientists 'gagged' by No 10 after warning of global warming threat". In the eyes of Martin – the only reporter who saw the actual wording – the memo was more of a suggestion than a gag order. But as US journalism professor George Kennedy opined, "It doesn't seem unreasonable to read a 'request' made by one's superior as an 'order' to be followed on pain of consequences."

The tide turns?

Despite the naysayers and lobby groups, coverage of climate change was on the increase as the new millennium unfolded. In her study of three UK newspapers – the *Times, Guardian* and *Independent* – communications scholar Anabela Carvalho found that the number of stories published on climate change per year jumped from about fifty per paper in 1996 to around 200–300 each by 2001. Carvalho noted that the articles were increasingly laced with a sense of urgency and more talk of extreme events, including such contemporary happenings as England's severe flooding in 2000 and the European heat wave of 2003. By 2006, front-page climate stories were a frequent sight in Britain. The environment also shot up the political priority list, with the new leader of the opposition – David Cameron – aiming to entice voters to the Conservative Party with the slogan "Vote Blue, Go Green". In 2007 the three major parties were all jockeying to be seen as the most environmentally friendly, and by the end of the following year the Climate Change Act – which obliged the UK to cut its emissions by 80% by 2050 – had passed into law, a landmark in political responses to global warming.

A shift toward mainstream acceptance of climate change was slower to arrive in the US, where contrarians were highly visible in the editorials and opinion pages of right-wing papers such as *The Wall Street Journal*. But things changed in a big way around the midpoint of the millennium's first decade. One of the first concrete signs was a front-page story in *USA Today* on 13 June 2005, entitled, "The debate's over: Globe is warming". In the sceptic-friendly US press, this was a banner development. The next year, Democrats took control of both houses of the US Congress, a shift that pushed sceptical voices on climate toward the margins of American politics.

Climate change also started featuring more frequently on those most mainstream of media outlets: TV and film. On TV, Alanis Morrissette gave viewers in the UK, US and Canada an overview of the "signs and science" of global warming. A few months later, BBC Television announced a series of sixteen programmes on climate impacts, themed to the massive Stop Climate Chaos campaign organized by a group of British non-governmental organizations. Soon after, Al Gore's long-time mission to raise awareness of global warming suddenly went mainstream with the launch of *An Inconvenient Truth*, a feature-length documentary film (and accompanying book) which smashed per-screen box-office records on its launch in 2006. Magazines also took note, with titles from *Time* and *Time Out* to *Vanity Fair* releasing special environmental editions. By 2007, climate change was a prominent presence in most media – a situation bolstered in October, when the IPCC and Al Gore were co-awarded the Nobel Peace Prize.

While Gore himself was no longer a political figure by this point, other world leaders inclined to act on climate change were coming onto the scene. On his first day as Australian prime minister in December 2007, Kevin Rudd signed the Kyoto Protocol, leaving the US as the only major industrialized country that hadn't ratified the treaty. If anything, the US had become a pariah in the world of climate-change diplomacy. A chorus of boos greeted US delegate Paula Dobriansky at the UN's 2007 climate summit in Bali when she conveyed the nation's refusal to sign up to a

Climate change and the cinema

Movie theatres aren't exactly packed with films about climate change, but the topic has made a few appearances on screen, both major and minor. Filmmakers have used the backdrop of particular climates to great effect ever since the silent-film days. But the abstraction of climate *change* – the evolution of weather over time – is a much trickier concept to put into cinematic terms. Plus, as author and activist Bill McKibben has pointed out, the instigators of global warming are typically far removed from the consequences of their actions. In other words, you can't exactly resolve the plot with a thrilling chase.

Some of the first films to address the dystopian prospects of human-induced climate change were in the realm of science fiction. *Zombies of the Stratosphere* (1952) included a young Leonard Nimoy as part of a gang of Martians bent on exploding Earth from its orbit so that Mars can move Sunward and benefit from a milder climate. The James Bond spoof *In Like Flint* (1965) featured hero Derek Flint, played by James Coburn, facing off against a sinister organization called Galaxy that plans to flood valleys and send icebergs into the Mediterranean. Then there's *Soylent Green* (1973), the first of the lot to present a scenario based in part on global-warming science. It's set in an overcrowded New York City in the year 2022, with pollution and other environmental ills run amok. ("A heat wave all year long", grouses Charlton Heston. "Greenhouse effect", replies his partner, played by Edward G. Robinson.) With food scarce, residents turn to the concoction that provides the film's name – ostensibly a blend of soya and lentils ("soylent"), but actually something far more gruesome.

Waterworld (1995), with Kevin Costner as co-producer and star, was the first major US film to draw on modern concerns about global warming. The most expensive movie ever made up to that point, at $175 million, it was considered a commercial and critical flop, even though it ended up making a profit overseas. In *Waterworld*, hard-scrabble camps of people are left to fight and adapt long after the world has been entirely flooded by the melting of icecaps. (Such a sea-level rise goes far beyond anything possible in the real world, even if every inch of Antarctic and Arctic ice melted.)

Climate change crept into other films of the period as well. In *The Arrival* (1996), aliens intent on occupying Earth pollute the atmosphere in order to boost the greenhouse effect and melt polar ice. *The March* (1990) depicted Africans migrating to a Europe stressed by global warming, and *Prey* (1997) featured a bioanthropologist who fears that a new über-species, nurtured by global warming, wants to kill off *Homo sapiens*.

Steven Spielberg set his futuristic *A.I. Artificial Intelligence* (2001) in a world where New York and other coastal cities have been abandoned due to rising sea levels. High water also figures in *Split Second* (1992), set in London c. 2008. Global warming coupled with "forty days and nights of torrential rain" (as noted in the prologue) has pushed the ocean into the city streets, where a policeman played by Rutger Hauer – who lives on "anxiety, coffee and chocolate" – hunts for a mysterious, lethal creature.

It took until 2004 for Hollywood to produce a big-budget extravaganza with climate change at its centre. The hero of *The Day After Tomorrow* is Jack Hall (Dennis Quaid), a paleoclimatologist who discovers that global warming has triggered a shutdown of North Atlantic currents. Within days, the Northern Hemisphere is plunged into an ice-age-scale deep freeze, thanks to cold air descending from the stratosphere (and disobeying the laws of physics, which dictate that the descending air would become warmer than the surface air it replaced). Before long, Hall is tramping through snowfields up the East Coast to rescue his son from the icebound New York Public Library.

Well before its release, *The Day After Tomorrow* got the attention of people on both sides of the global-warming debate. The activist group MoveOn.org distributed fliers to people outside theatres showing the film, while sceptics bashed it as a propaganda tool. But whatever potential the movie had to sway real-world debate was compromised by its absurdly telescoped view of how fast climate change might unfold. Although the Atlantic currents that help keep Europe warm for its latitude have shut down before, and they could diminish in the future (see p.132), the process should take years to decades to unfold, rather than mere days. Still, many saw the film's success as a sign of growing public interest in climate change.

Al Gore became an unlikely movie star in 2006 with the release of *An Inconvenient Truth*. The Oscar-winning documentary is a glossed-up version of Gore's stump speech, which he claims to have presented over a thousand times. *An Inconvenient Truth* rode the wave of two big trends – interest in global warming and a surge of hugely popular US documentaries that includes Michael Moore's *Fahrenheit 9/11*. Outside of a few nuggets on Gore's life story, the film is what it claims to be: the story of global warming, told in a style that's sober and thoughtful, yet visually rich and emotionally compelling, enhanced by Gore's 30-year storehouse of knowledge on the topic. Expanding on Gore's thesis, *The 11th Hour*, a powerful 2007 documentary narrated by Leonardo DiCaprio, linked climate change with other environmental woes.

An even sharper take on our global predicament emerged from the UK with 2009's *The Age of Stupid*. This hybrid creation involves a fictional archivist (Pete Postlethwaite) looking back dolefully from the warming-ravaged world of 2055. To find out what was and wasn't being done to fight climate change early in the twenty-first century, he reviews clips from the era – which are, in fact, current documentary-style profiles of people around the globe. *The Age of Stupid* was unveiled in a solar-power tent at London's Leicester Square and simultaneously at 62 satellite-linked screens across the UK, making for the world's biggest and greenest film debut. An even larger global premiere was beamed to more than 600 cinemas worldwide, and the film was shown to UN delegates aboard a train en route to the 2009 Copenhagen summit.

Global warming even penetrated the children's film market in 2006 with *Ice Age: The Meltdown*. This sequel to the animated feature *Ice Age* – featuring the tag line "the chill is gone" – follows a batch of prehistoric animals in a newly warmed world, with species going extinct and floods threatening. Though set in the great meltdown after the last ice age, the film subtly tips its hat to modern-day worries.

"road map" for a new global treaty (see p.316). Winds of change would soon reach the US, though, as Barack Obama swept into the White House in 2008. As part of his campaign platform, Obama promised to make global warming a top priority. Given that US politics had been arguably the main obstacle to a truly global climate agreement for years, it seemed a new era just might be at hand.

Copenhagen, Climategate and beyond

In early 2009, with climate-friendly leadership now in place among most of the world's industrialized powers, hopes were high among activists and many leaders that an ambitious post-Kyoto agreement might take shape at the major UN summit in Copenhagen scheduled for that December. But political realities soon dashed those hopes, and a newly revived sense of doubt was nourished by climate sceptics. The global economic meltdown that began in 2007 didn't help matters. Anything that might be construed as an extra cost, or an expense with benefits years away, was suddenly suspect. The US cap-and-trade proposal was successfully recast as "cap and tax" by its opponents, which helped drag it into limbo in the US Senate after it passed the House in 2009. With many Americans fearful that such laws might hit them in the wallet, US support of climate change legislation – along with acceptance of the physical reality of climate change – dipped markedly. By late 2009, only 36% of Americans believed humans played a role in global warming, down from 47% in early 2008, according to the Pew Research Center for the People & the Press.

The lack of a US climate bill, together with other global obstacles, cast a pall over the UN's much-anticipated Copenhagen climate summit in December 2009. Within days, the meeting descended into chaos, with little real progress (see p.320), and it was laden with unfortunate symbolism. Not only did the summit unfold during a long stretch of cold and snow – hardly the setting that brings to mind a warmer world – but the sight of hundreds of world leaders travelling to Copenhagen in carbon-spewing private jets and limousines made the gathering an easy target for ridicule.

To make matters even worse, there was the unauthorized release in November 2009 of more than 1000 emails that had been exchanged among climate scientists over more than a decade. These messages were pulled from a computer at the University of East Anglia's Climatic Research Unit (CRU) and placed online for all to see. Whether the emails were actually stolen or merely leaked (no charges had been filed as of late 2010), the redistribution was no doubt meant to disparage the reputation of the scientists involved and, by extension, the validity of climate science. The media quickly dubbed the event Climategate – a widely used label that implied the emails contained scandalous content.

In fact, the vast bulk of the emails depicted the mundane process of climate researchers at work. Reporters and bloggers zeroed in on a tiny fraction of the emails that appeared to reveal something more sinister. Perhaps the most widely quoted snippet was from CRU's Phil Jones to Michael Mann (then at the University of Virginia), in which Jones cited a technique used by Mann that employed air temperature data in order to "hide the decline" in temperatures derived from tree rings over the latter part of the twentieth century. It isn't well understood why tree-ring data have diverged from air temperature in recent decades, but there's no doubt that the globe was warming during that time frame. Thus, the "hiding" actually produced a more accurate portrait of recent temperatures within the multicentury temperature record. The same email referred to this technique as a "trick", which some observers interpreted as a scheme to obscure global cooling. In context, it was clear that "trick" meant a satisfying approach to solving a problem, as opposed to some nefarious strategy.

A few other emails raised thornier issues, especially with regard to data and software access. Many climate scientists, including Mann and Jones, faced an ever-growing tide of requests for their data sets and computer code. Some of these requests came from parties outside the climate-science mainstream, which meant that techniques might need to be documented and explained more thoroughly than they would be for peers – a task that took time away from research. Moreover, some records had been provided to CRU by national meteorological services without the right of redistribution. Because of these and other factors, responding to requests had become an increasingly burdensome task, and some of the released emails implied that scientists were trying to avoid at least a few of the requests.

A series of investigations in 2010 exonerated the researchers from allegations of scientific wrongdoing, but the reports emphasized the need for greater transparency and responsiveness when it comes to making data and software available to the world at large. Those points were echoed in massive coverage by *The Guardian*, which entrained readers and scientists in an effort to produce a self-correcting, annotated online narrative of the Climategate saga. The debate over open access isn't limited to climate research: science as a whole is struggling with how best to document extremely complex software and make it more widely available in useful forms.

It's hard not to see the timing of Climategate, just one month before Copenhagen, as an attempt to torpedo the summit. However, the ill-fated meeting had so many other headwinds that the email release may not have been a make-or-break factor. The scene was different in the media, the blogosphere and the court of public opinion, where the emails drew intense scrutiny and criticism – even from some of the greenest activists and writers around. Once the dust had settled, it became evident that many of the items that seemed outrageous at first weren't so shocking in context. And as many researchers took pains to emphasize, Climategate had no effect on the rock-solid scientific consensus that greenhouse gases were warming Earth's climate and would continue to do so. In that sense, the whole episode was a colossal distraction – albeit a very effective one.

Another set of controversies that erupted around the same time involved errors in the 2007 IPCC assessment. One of the most prominent of these dust-ups, inevitably dubbed Glaciergate, centred on the report's claim that "the likelihood of [glaciers in the Himalayas] disappearing by the year 2035 and perhaps sooner is very high if the Earth keeps warming at the current rate." For this assertion, the IPCC cited a WWF report from 2005, which in turn drew on a 1999 interview in *New Scientist* magazine with Indian glaciologist Syed Iqbal Hasnian. There was no peer-reviewed work behind the factoid, and it's generally accepted that Himalayan glaciers would take far longer to disappear than several decades (see p.110). Another part of the same IPCC paragraph contained a similar error involving the rate of loss of glacier-covered areas. These errors attracted little notice at first. Buried within the full IPCC assessment, they didn't make it into the summary for policymakers, which serves as the assessment's highest-profile document. But by early 2010, the errors had drawn flak from glaciologist Vijay Raina in an Indian government report, and the nation's environment minister was calling the IPCC "alarmist". Finally, the IPCC commented on the brouhaha: "In drafting the paragraph in question, the clear and well-established standards of evidence, required by the IPCC procedures, were not applied properly."

Although these and a few other unearthed errors involved only a minuscule part of the IPCC's exhaustive 2007 report, they did put a few dents in the panel's Nobel-backed reputation and raised questions about its review process and its leadership. A blue-riband group assembled in 2010 by the world's national science academies, acting through the InterAcademy Council, praised the IPCC's overall success while calling on the panel to "fundamentally reform its management structure" to better handle ever-more-complex assessments as well as ever-increasing public scrutiny. As the council noted, "The world has changed considerably since the creation of the IPCC."

The various dramas swirling around climate change in 2009–10 extended to the atmosphere itself. That winter the US East Coast and much of Europe were entrenched in a prolonged stretch of bitter cold and frequent snow, related to record-low values of the North Atlantic Oscillation (see p.128). The onslaught was perfectly timed to build on scepticism fuelled by the revelations above. In February 2010, the family of US senator James Inhofe – for many years the most outspoken climate sceptic in Congress – built an igloo in deep snow near the Capitol, with a sign reading "Al Gore's New Home". The igloo was nowhere in sight a few months later, however, when Washington endured its hottest summer ever observed. Even with the epic snow and cold taken into account, it was one of the warmest years on record for the US mid-Atlantic coast.

Indeed, 2010 brought new monthly peaks in temperature for the globe as a whole, and the seasonally adjusted concentration of carbon dioxide topped 390 parts per million for the first time. Yet the continued reality of a warming planet, and the overall solidity of climate science, couldn't hold a candle to the scepticism fanned by politicians and commentators and bolstered by the emails and IPCC errors. From 2008 to 2010, public opinion polls in the US, Britain, Germany and Australia showed drops of 10–15% in acceptance of, and concern about, human-produced climate change. Developments in 2010 only seemed to accentuate the backslide. David Cameron became UK prime minister that year and immediately promised to lead "the greenest government ever", but six months on there was little tangible progress, with the climate agenda having been somewhat lost amid savage budget cuts in the public sector. In Australia, a defection to the Green Party led to the first hung parliament in 70 years, threatening to complicate climate action there. And the US Congress swung sharply to the right in late 2010, making it all but certain that any major American climate bill would be years away.

It's too soon to tell whether the resurgence in global-warming scepticism will persist through the 2010s. It does seem a safe bet that protecting the economy will be a chief concern for politicians and the public, with the emphasis increasingly on green actions that make both short- and long-term economic sense. Whether sufficent innovative policies and technologies can leap into place quickly enough to avert the worst scenarios of twenty-first-century global warming remains to be seen. In the next two sections, we'll look at some of the leading political and technological solutions being proposed.

The predicament

What would it take to fix global warming?

The global-warming problem isn't going to be solved tomorrow, next week or next year: we're in this one for the long haul. And there clearly isn't one single solution. Any sensible strategy needs to include governments agreeing to and enforcing targets; innovators developing low-carbon energy sources and improving efficiency; and individuals and organizations doing their best to reduce their own carbon footprints. These three approaches are explored in the following chapters. But first, let's take a look at the problem – and what we need to do to solve it – as a whole.

Looking at the global-scale challenge before us, the most obvious worry is the sheer momentum that needs to be overcome. We've already added a great deal of greenhouse gas to the atmosphere, and we're adding more each year than the year before. Even if we can lower emissions, there's enough inertia in the physical drivers of climate to keep us rolling towards an even warmer future for years to come. In short, then, if we're to avoid long-term climate trouble, we need to take real action sooner rather than later.

Understanding the goals

At least three types of goals are commonly discussed by policymakers and campaigners working in the field of climate change:

▶ **Stabilizing emissions** Making sure that each year we emit no more than the year before, and ideally less.

▶ **Stabilizing concentrations** Reducing emissions enough so that the amount of greenhouse gas in the atmosphere levels off at a target and stays there (or falls back down).

▶ **Stabilizing temperature** Keeping the atmosphere from warming beyond a certain point.

Obviously, these three types of goals overlap: our emissions build up in the atmosphere, changing the concentration of greenhouse gases, which in turn alters the temperature. However, the relationship between emissions, concentrations and temperature is complicated by a few factors.

First, as discussed in the Circuits of change chapter (see p.227), there's still some uncertainty about exactly how much hotter Earth will get as greenhouse-gas concentrations rise – the so-called **sensitivity** of the climate.

Second, since some greenhouse gases persist in the atmosphere for many years, even centuries, simply levelling off emissions won't be enough to stabilize concentration – and therefore temperature. After all, if water's flowing into a bathtub faster than it can drain out, you need to *reduce* the flow – not just keep it constant. Likewise, though stabilizing global emissions would be an enormous short-term achievement, we need to go much further and make significant *cuts* in emissions in order to keep the world from getting hotter. As we'll see later, some countries have already managed to stabilize or reduce their emissions, but globally there's a long way to go, despite a flattening off caused by the recession in 2009.

Third, there's the **time-lag** factor to consider. Even if we *do* manage to make significant emissions cuts, it will take decades for the concentration to begin levelling off, and centuries for the temperature to stop rising (see graphs). If there are lasting changes to the Earth system, such as the loss of major ice sheets or the release of trapped methane, then substantial global warming could be locked in for far longer.

To help simplify this picture, some researchers have begun focusing on the **cumulative carbon** that's put into the atmosphere, rather than the year-

Slow progress: how temperature lags behind CO₂ concentration, which lags behind emissions

Even if we reduce our CO_2 emissions like this...

...CO_2 concentration will keep rising for many decades before levelling off...

...and the temperature will climb for even longer

by-year emissions. There's quite a wide range of estimates on how much total carbon we can afford to emit, but surprisingly, the ultimate temperature jump doesn't appear to hinge on the timing of emissions – just on the amount. Still, it stands to reason that the longer we wait to start cutting emissions, the more stringent the later cuts will need to be. According to a 2010 review by the US National Academies, the best guess for cumulative emissions likely to trigger an eventual 2°C of warming is 1.15 trillion tonnes of carbon. We've already emitted roughly half that amount, so to avoid exceeding the 2°C target, we'd need to keep our total future emissions less than our total past ones – a Herculean task.

Selecting a target

Perhaps the ultimate question when it comes to defining goals for tackling climate change is this: how much global warming would be truly dangerous? Many studies suggest that a rise of 3°C (5.4°F) relative to pre-industrial levels would be well past the edge of the comfort zone. It may be enough to trigger unstoppable melting of the Greenland ice sheet, for example, or a net drop in world food production. Any further warming could jeopardize parts of the Antarctic ice sheet and cause other dire consequences.

With these and many other factors in mind, the most commonly cited temperature target for climate stabilization is a **2°C** (3.6°F) warming above pre-industrial levels, or around 1.2°C (2.2°F) above today's global temperature. This value was agreed to by the European Union in 1996 and reconfirmed in 2005, and it's shared by many climate scientists and activist groups.

Some parties are calling for more ambitious goals. At the Copenhagen meeting in 2009, a group of more than 40 small island nations and like-minded activists pushed for a target of **1.5°C** (2.7°F). The idea is to minimize the risk that rising sea levels could swamp low-lying countries. The accord that emerged from Copenhagen included the standard 2.0°C target, but it also called for an assessment to determine whether or not to strengthen the target to 1.5°C.

Equally stringent is the goal proposed by the grassroots effort 350.org, which has organized more than 10,000 demonstrations and events since 2009. The group is pushing to bring global CO_2 concentrations down from their recent 390 parts per million to **350 ppm**, which is deemed by the group (and by NASA's James Hansen, among others) to be the value consistent with a reasonable shot of keeping the eventual temperature rise to 1.5°C. Reaching the 350 target is "a non-negotiable demand from the planet itself", says author and 350.org founder Bill McKibben. Even if it were ultimately unachievable, the 350 number serves as an icon of urgency and the need for action, as well as a reminder that we're already on a path toward much higher concentrations.

2°C: can it be done?

Given physical and political realities, the world may not be able to avoid 1.5°C. After all, we made it halfway there in the twentieth century, which saw more than 0.7°C (1.4°F) of warming. And as noted earlier (see p.18), an additional 0.5°C (0.9°F) can be expected as Earth adjusts to emissions already in the atmosphere.

To have any chance of achieving even the 2°C goal, we can't afford decades of delay. The European Union and Group of Eight (G8) nations have called for global emission cuts of 50% or more by 2050 in order to keep the 2°C target within reach. Those values were confirmed for a midrange IPCC scenario in a 2010 analysis led by Brian O'Neill (US National Center for Atmospheric Research). How might that translate into yearly goals? The UK government's Committee on Climate Change found in a major 2008 report that the world could buy itself a 50-50 chance of reaching the 2°C goal by cutting global emissions by 3% per year from 2014 onward.

Another way to gauge the 2°C challenge is by assessing how much total carbon we can afford to emit between now and some future date. A group led by Kirsten Zickfeld (University of Victoria) found that the total carbon from CO_2 emissions for the period 2000–50 needs to stay below about 590 gigatonnes, which would average about 12 gigatonnes per year. The current annual rate, by comparison, is already more than 8 gigatonnes per year of carbon from CO_2.

At first glance, these are highly discouraging numbers. However, when one considers greenhouse gases other than CO_2, the picture brightens just a bit. Most of those gases are less prevalent and shorter-lived than CO_2 but far more powerful. Together, they account for more than 30% of the global-warming impact from gases we're adding to the atmosphere each year. The good news is that some of them could be reduced more easily and affordably than CO_2, and political agreements that can smooth the way are already in place, such as the Montreal Protocol and regional air pollution laws. In the EU, for instance, the nitrogen oxides emitted from diesel-powered cars and light trucks sold after 2009 have been cut by roughly 20%; they'll be reduced more than 50% beyond that new limit starting in 2014. There is some hope among policy experts and climate scientists that a two-step approach – cutting the emission rates of non-CO_2 greenhouse gases right away, plus reducing CO_2 over a somewhat longer time frame – might prove fruitful. New approaches to keeping carbon in Earth's ecosystem, including stronger **deforestation** limits and **biochar** production, could also have a big impact relatively quickly.

Whether the 2°C goal can be met depends on many other factors as well, such as how fast **new technology** is developed and adopted, and how seriously we take **energy efficiency**, which in turn depends partly on the political will to prioritize climate change. Another important factor is the state of the world **economy**. Global emissions of CO_2 from fossil fuels actually fell more than 4% between 1980 and 1983, a period of high oil prices and widespread recession, and they dropped by about 2% in 1992 and 1999, when the economies of eastern Europe and Russia were struggling. Despite the intensity of the worldwide downturn of 2009, global CO_2 emissions fell by only a little more than 1% that year.

Naturally, all these factors are interrelated in complex ways. For example, high oil prices can simultaneously dampen economic growth, encourage efficiency and stimulate investment in alternative energy sources – all of which are likely to reduce emissions. But if the economy suffers too much, governments may feel pressure from voters to prioritize short-term growth over long-term environmental issues.

Despite all these complications, it seems at least theoretically possible that we could manage global emission cuts of at least 10% within a decade or two. And, though doing so could cause some short-term fiscal pain, the long-term gains from energy efficiency and renewable energy are likely to be enormous – economically as well as environmentally.

The wedge strategy

Since almost all human activity contributes to greenhouse gas emissions on some level, the task of reducing global emissions can seem somewhat overwhelming. What if we thought of it as a series of simultaneous smaller goals? That's the philosophy behind Stephen Pacala and Robert Socolow's "wedge" approach to climate protection. The two Princeton University scientists brought the wedge concept to a wide audience through a 2004 article in *Science*. The idea is to break down the enormous carbon reductions needed by mid-century into more manageable bits, or wedges, each of which can be assigned to a particular technology or strategy.

When it was introduced, the wedge concept triggered widespread interest and excitement, in part because Pacala and Socolow claimed that the needed emission cuts through 2050 could be handled entirely through existing technologies. That may not be the case, according to critics of the plan (see below), who believe it could mislead us into undercutting research on new forms of energy. Still, the concept has proven hugely influential, and it's a handy way to compare and contrast various pieces of the overall emission-reduction puzzle.

The wedge concept originates from Pacala and Socolow's projection of historical CO_2 emissions into the future (see diagram, p.305). Let's assume that emissions could be instantly stabilized – in other words, the yearly increases in CO_2 all went to zero right now – and remained that way until at least the 2050s. This is represented by the flat, black line on the diagram. Pacala and Socolow argue that this would correspond to an eventual CO_2 concentration of about 500 parts per million and, under certain assumptions of climate sensitivity, a global temperature rise somewhere in the ballpark of 2°C.

However, if emissions continue to increase as they have in the last several decades – at more than 1% per year – then we'll be adding twice as much CO_2 to the atmosphere in 2054 as we are doing now (see the red "business as usual" line on the diagram). The result will be a warming far more severe than 2°C.

The triangle between the red lines shows the difference between the desired path of steady emissions and the dangerous uphill path. To get rid of the triangle, we'd need to come up with at least seven wedges, each of which would eliminate a billion tonnes (a gigatonne) of annual carbon emission by 2054. (Further wedges would probably be needed after 2054 to stay below the 2°C target.) According to Pacala and Socolow, we're almost spoilt for choice on wedges that we can implement promptly. They identify fifteen examples (see p.304–305), each of which is already being implemented on a reasonably large scale somewhere in the world.

While the wedge strategy is a helpful way to assess which directions we might want to go in, it doesn't make the job itself any easier. "Each of these wedges represents a staggering amount of effort from the public and private sector", noted Joseph Romm in a 2008 *Nature* analysis. Moreover, in tallying the number of wedges needed to meet the 2°C target, Pacala and Socolow have been accused of undue optimism. Apart from the recent economy-induced dip, global emissions have been rising a bit faster than the 1.5 % annual increase assumed in their graph. Marty Hoffert (New York University) argues that holding emissions constant from now to the 2050s would actually require closer to eighteen wedges than seven. To get those, he adds, we'd need a huge ramp-up in research and development on new modes of alternative energy beyond the existing forms.

The wedges (expressed in terms of the year 2054 versus 2004)

▶ 🔺🔺 **Doubling vehicle fuel economy** and **cutting the distance driven per car in half** (one wedge each).

▶ 🔺 **Installing lights and appliances with state-of-the-art energy efficiency** in all new and existing residential and commercial buildings, thus reducing emissions by about 25%.

▶ 🔺 **Improving the efficiency of coal-fired power plants** from 40% to 60%, plus cutting in half the energy lost when fossil fuels are extracted, processed and delivered to those plants.

▶ 🔺🔺 **A 50-fold expansion in wind energy**, or adding about two million wind turbines of the standard one-megawatt size (there are now more than 30,000 turbines worldwide), replacing coal. These turbines would cover an area about the size of Germany, but the land beneath them could be used in other ways. A second wedge could be obtained by adding another four million turbines to generate hydrogen for fuel-cell vehicles (see p.352).

▶ 🔺 **A 700-fold expansion of photovoltaic (PV) solar energy**, again replacing coal. This would require enough panels to blanket every inch of an area the size of New Jersey. However, many of the panels could be mounted on roofs and walls.

▶ 🔺 **A 50-fold expansion in ethanol**, displacing petrol. Even with plants that yield far more energy per acre than the corn and sugar cane used now for ethanol, these fuel crops would still require a region about the size of India.

▶ 🔺 **A halting of current deforestation**, coupled with plantations on non-forested land, together covering a total area about 40% the size of Australia by 2050 (this would require a doubling of the current rate of reforestation).

▶ 🔺 **Employing conservation tillage** on all cropland, in which farmers avoid ploughing and thus reduce the amount of CO_2 escaping from tilled soil. It's now used on less than 10% of cropland globally.

It's obvious that some of the proposed wedges are far more practical and/or politically palatable than others. Even if alternatives are deployed, there's no guarantee that they will supplant traditional fuels. And if the history of energy development is any guide, the economies of scale tend to push each sector towards a strongly preferred fuel type – petrol for vehicles, coal for electric power plants and so on. This would work against, say, the parallel large-scale growth of electricity generation from wind power, nuclear power and cleaner types of coal.

There's also the obvious fact that we can't stabilize emissions today or tomorrow as if we were flipping a switch. Even using current technologies, such massive change would take years to implement. This is why many scenarios constructed by the IPCC and others include a ramp-up phase of continued growth, followed by substantial cuts in emissions – ultimately bringing us well below today's emission levels – in contrast to the flat-line "stabilization trajectory" shown on the above graph.

And looming in the background, not considered directly in the wedges, is the inexorable rise in global **population**, which threatens to swamp the emission cuts made in efficiency and technology. Because the idea of

After Robert Socolow, Roberta Hotinski, Jeffery Greenblatt and Stephen Pacala

▶ **Tripling the amount of energy now generated by nuclear sources**, adding about 700 one-gigawatt plants as well as maintaining or replacing all nuclear plants now in use.

▶ **Quadrupling the use of natural gas in power plants**, replacing an equal number of coal-fired plants.

▶ **Sequestration** – capturing carbon emitted by large fossil-fuel power plants and storing it underground – with one wedge each coming from standard coal or natural gas plants, synfuel plants (which generate synthetic fuel from coal) and hydrogen plants that draw on coal or natural gas.

stabilizing global population has proven to be such a political minefield, it's sometimes omitted from climate change discussions. But there's no getting around the fact that more people on the planet will – all else being equal – lead to more greenhouse emissions.

Despite all these various qualms, the wedge concept remains useful as a rough framework for examining the total emissions picture. Each wedge is technically feasible – it's a matter of society choosing which ones to emphasize. Government subsidies or other incentives could shape this course, and it's possible that different countries will opt for different technologies, thus providing a well-rounded portfolio on the global scale. The wedges also allow one to work backwards and set even shorter-term goals, taking into account that some wedges might grow more rapidly at first while others would take a while to get going. As Socolow puts it, the wedges "decompose a heroic challenge into a limited set of monumental tasks".

We'll be considering the above options in the "Technological solutions" and "What you can do?" chapters, after we explore what governments are doing – and not doing – to speed us along the path to a low-carbon future.

Political solutions

Kyoto and beyond

Climate change must be dealt with on a global level – and this poses an unparalleled challenge in international diplomacy. For one thing, a whole host of intractable issues, from excessive consumption to South–North inequity, tend to glom onto climate change, as if the latter weren't a daunting enough problem in itself. Over more than twenty years of addressing climate change, the United Nations has made fitful headway. Nearly all countries are on record as taking the problem seriously, and the UN's 1997 Kyoto Protocol – though weaker than many had hoped – was ratified by countries that together represent the majority of people in the developed world (though only a third of global emissions). However, the effort to forge a binding post-Kyoto agreement experienced a major setback in 2009 with the near-collapse of UN negotiations in Copenhagen, delaying progress towards any new plan for global action.

Global agreements aren't the only game in town, though. Some countries that opted out of Kyoto are exploring other mechanisms for grappling with the carbon problem. For instance, many of the most populous US states and cities are volunteering for reductions that match or exceed Kyoto goals. And a whole smorgasbord of new ideas will be on the table over the next few years as diplomats attempt to thrash out a global agreement to extend or replace Kyoto.

The road to Kyoto

In some ways, global warming couldn't have arrived on the diplomatic radar screen at a better point than in 1988. A year earlier the UN had forged the **Montreal Protocol**, which called for the gradual phase-out of the industrial chemicals (chlorofluorocarbons) implicated in the Antarctic ozone hole (an issue separate from global warming, as explained on p.32). The Montreal

Protocol came about only after a thorough assessment of the science behind ozone depletion. Previous work had already shown the connection to chlorofluorocarbons, but it was the assessment that gave an international imprimatur to the findings. It showed that researchers from around the world agreed with the science, which in turn provided the green light for global action. The protocol even had a happy side effect for climate: because CFCs are potent greenhouse gases, their phase-out has been one of the biggest emission-reduction successes to date. Had nations not acted to cut CFCs and other ozone-depleting chemicals from the 1970s onward, the total greenhouse impact of these gases by 2010 would have nearly rivalled that of carbon dioxide, according to an analysis led by Mario Molina (University of California, San Diego). With this in mind, similar gases might be added to the protocol – especially hydrofluorocarbons (HFCs), which are ozone-friendlier than CFCs but just as dangerous from a greenhouse standpoint.

The Montreal Protocol's success made it natural for the UN to ask for another high-level scientific assessment when global climate change reared its head. In response, the World Meteorological Organization and the UN Environment Programme established a group called the **Intergovernmental Panel on Climate Change** (IPCC). Few if any people at the time could have guessed how influential, and at times controversial, the IPCC would turn out to be. Its findings have been featured in countless news stories and cited in thousands of reports. The IPCC's four major assessments – released in 1990, 1995, 2001 and 2007 – have become the definitive reference tools on the state of climate science, and the panel shared a Nobel Prize with Al Gore in 2007 for its cumulative efforts. See overleaf for more about how the IPCC works. Several nations have taken a cue from the IPCC and produced their own comprehensive assessments; some of these are listed in Resources (see p.391).

The IPCC's first assessment in 1990 underscored the seriousness of the climate-change threat, so there was plenty of momentum driving world leaders to the 1992 UN **Earth Summit** in Rio de Janeiro. Global warming was only one of a host of environmental issues on the table at this Woodstock of green politics, but it played a starring role. An agreement called the UN Framework Convention on Climate Change (**UNFCCC**) was introduced at Rio and signed by 166 nations that summer (and, by 2010, a total of 192 nations). Because the US and several other countries were steadfast in their opposition to binding emissions reductions,

> "I told the world I thought that Kyoto was a lousy deal for America."
>
> US president George W. Bush, March 2006

the UNFCCC doesn't include any targets or timelines, much to the dismay of climate activists. Yet it was, and is still, a landmark agreement. Among its key points:

▶ **The ultimate objective** is to stabilize climate "at a level that would prevent dangerous anthropogenic interference with the climate system" and in a time frame "sufficient to allow ecosystems to adapt naturally to climate change, to ensure that food production is not threatened and to enable economic development to proceed in a sustainable manner."

Inside the IPCC

The IPCC has only a few permanent staff, but it's a far larger enterprise than the term "panel" might suggest. Indeed, it's one of the biggest science-related endeavours in history. That said, the IPCC doesn't conduct any science of its own. Its role is to evaluate studies carried out by thousands of researchers around the world, then to synthesize the results in a form that helps policymakers decide how to respond to climate change. Each assessment is a bit different, but typically each of several IPCC working groups generates an exhaustive report as well as a summary for policymakers. All of the reports are available online (see p.391). For their 2001 and 2007 reports, the three working groups dealt with: **the basis in physical science** (how climate change works), **impacts, adaptation and vulnerability** (options for dealing with climate change) and **mitigation** (options for minimizing it).

Every IPCC assessment involves around 100–200 researchers from many dozens of countries, generally nominated by their governments or by a nongovernmental organization. Each working group is headed by a pair of scientists, one each from a developed and a developing country. By and large, these scientists volunteer their time to be involved with the IPCC with the blessing of their employers. They survey peer-reviewed science studies and other pertinent materials; meet with peers to gather input as needed; and draft, revise and finalize reports. Several hundred other experts then review each report. Finally, each document is scrutinized by technical reviewers within each government and accepted at a plenary meeting. The policymaker summaries take shape on a parallel track; they're approved by a panel of governments on a word-by-word basis.

With so much riding on its conclusions, it's no wonder that the IPCC has drawn scrutiny from those aiming to discount the risk of climate change. The second assessment provoked an attack from the Global Climate Coalition (p.280) and other sceptics, who criticized the process by which the wording of one of the summaries for policymakers was prepared (including the fateful statement "the balance of evidence suggests that there is a discernible human influence on global climate"). The process was revised for the next assessment in response to this and other

▶ **International equity** is given a nod with a reference to "common but differentiated responsibilities and respective capabilities", a theme that ended up in the Kyoto Protocol.

▶ **Uncertainty in climate science** – one of the favourite criticisms from sceptics – was acknowledged but put in its place: "Where there are threats of serious or irreversible damage, lack of full scientific certainty should not be used as a reason for postponing such measures."

▶ **Countries agreed to inventory their greenhouse emissions** and publish yearly summaries.

concerns, although no IPCC rules had been violated and the panel's earlier conclusions were unchanged. After the third assessment came out, sceptics' focus turned to its "hockey stick" graphic depicting climate over the last thousand years, including the sharp upturn of the twentieth century (the "head" of the stick – see p.246). Controversy aside, the IPCC's pronouncements on the state of global climate resonate worldwide. As noted earlier in this book, the IPCC has grown increasingly emphatic about its conclusions on human-induced climate change.

Although the IPCC doesn't commission research, the world's major climate modelling activities are timed to coordinate with IPCC schedules. The fifth major assessment, scheduled for release in 2013–14, will draw on a series of decadal- and regional-scale simulations that together will provide substantially more detail than those used for previous assessments.

Putting together an IPCC report is unlike any other job in the science world. Susan Solomon, the NOAA scientist who unravelled the role of polar stratospheric clouds in creating the ozone hole (see p.32), served as co-chair of Working Group I for the 2007 assessment. "It's a very intense activity", says Solomon, who estimates she went through more than 17,000 comments from more than 500 reviewers on the first draft of her group's report. "I've learned a tremendous amount about climate, but it is demanding, both personally and professionally." More than anything, she stresses the community aspect of the panel: "It's very important for people to understand that the IPCC is not one scientist's voice."

Susan Solomon with a draft of the 2007 IPCC assessment.

With the UNFCCC in place, the world's industrialized countries plus "economies in transition" (mostly in central and eastern Europe) – together referred to as **Annex I** – agreed to reduce their emissions to 1990 levels by the year 2000. This was fairly easy for Russia and the former Soviet republics due to their lacklustre economic performance and the resulting drop in energy use, but many other countries missed the goal. Britain managed it only because of the phase-out of subsidies for coal-fired power plants and a resulting shift to natural gas. As a whole, the Annex I emissions were 6% above 1990 levels by 2002.

It was already apparent by the mid-1990s that some type of **mandatory reductions** were needed to make real progress on emissions. Setting targets, though, was far easier said than done. At a 1995 meeting in Berlin, it was agreed that industrialized countries (the ones that had caused the lion's share of the greenhouse problem thus far) would bear the full brunt of the first round of agreed-upon emissions cuts. This was in keeping with the principles of the UNFCCC. The idea of mandated cuts eventually gained support from US president Bill Clinton (albeit weaker cuts than those promoted by the EU). But in 1997 the US Congress voted 95–0 against any treaty that did not specify "meaningful" emission cuts for developing as well as developed countries.

The standoff roiled the diplomatic world right up to the fateful meeting in Kyoto, Japan, late in 1997, where targets would be set. In the end – after virtually nonstop negotiations and some last-minute assistance from US vice president Al Gore – the parties agreed to a Kyoto Protocol that would exempt developing countries and produce an average 5.2% cut among Annex I countries by 2008–12, compared to 1990 levels.

The US and Australia signed the treaty, but neither country proceeded to ratify it right away. (Australia eventually did, as we'll see.) However, the Kyoto Protocol was crafted so that it could become international law even without US participation – but just barely. The rules required ratification by at least 55 Annex I countries that together account for at least 55% of the total Annex I emissions in 1990. Thus, if every other major developed nation apart from the US and Australia ratified the protocol, it could still take effect. Most of the industrialized world quickly came on board, but it wasn't clear whether or not Russia would. Years of suspense passed before Russia, after stating its opposition to Kyoto in 2003, ratified the protocol in late 2004. The protocol became international law 90 days later, on 16 February 2005. Australia belatedly joined the treaty in December 2007, with incoming prime minister Kevin Rudd signing the ratification document on his first day in office – which happened to coincide with the kickoff of that year's UN summit in Bali (see below).

How Kyoto works

The meat of the Kyoto Protocol (which is technically an amendment to the UNFCCC) was the requirement for developed nations to cut their emissions of six greenhouse gases: CO_2, **nitrogen oxides**, **methane**, **sulphur hexafluoride**, **hydrofluorocarbons** and **perfluorocarbons**. Different countries have different reductions targets and, for various reasons, some countries are actually allowed an *increase* in their emissions over the period:

▶ **Bulgaria, Czech Republic, Estonia, Latvia, Liechtenstein, Lithuania, Monaco, Romania, Slovakia, Slovenia, Switzerland:** 8% reduction
▶ **US:** 7% reduction
▶ **Canada, Hungary, Japan, Poland:** 6% reduction
▶ **Croatia:** 5% reduction
▶ **New Zealand, Russia, Ukraine:** no change
▶ **Norway:** 1% increase
▶ **Australia:** 8% increase
▶ **Iceland:** 10% increase
▶ **All other countries:** no mandated targets

These targets applied to emissions in the period 2008–12 – the "commitment period" – and were expressed in CO_2 equivalent, which takes into account the varying power of each of the six gases (see p.38). With a few exceptions, the targets were relative to 1990. Those for the US were hypothetical, since it didn't ratify the treaty.

As you might guess, the targets that emerged at Kyoto were political creatures. Australia is a good case in point: how did this prosperous country win the right to increase its emissions? As in the US, Australia had come up with voluntary reduction plans in the 1990s, hoping to avoid mandated cuts, but these failed to do the trick. Its negotiators then argued that Australia was a special case because it was a net energy producer with a growing population and a huge, transport-dependent land area. They also pointed to reductions in Australian deforestation after 1990. Kyoto diplomats were keen to get all of the industrialized world on board, so in the end they acceded to Australian wishes, but the country still opted out of the treaty for nearly a decade.

Another wrinkle is the so-called "EU bubble". The EU was assigned an 8% target but was allowed the freedom to reallocate targets country by country within its overall goal, given that some EU economies would find the 8% limit easier to meet than others would. The resulting targets range from reductions as big as 29% for Denmark and Germany to increases as high as 19% and 20% for Portugal and Luxembourg, respectively.

The emissions market

Because Kyoto didn't become international law until 2005, many countries found themselves playing catch-up in order to try and meet their targets. They were assisted by several market-based mechanisms woven into the protocol, designed to help countries meet their targets at the lowest possible cost. The basic idea is that countries or companies finding it tough to cut their emissions directly could meet their targets by purchasing reductions achieved more cheaply elsewhere. A secondary goal is to promote the export of clean technologies to emerging economies. The mechanisms included the following:

▶ **Emissions trading** is just what its name implies. Annex I countries that exceed their emissions targets can buy allowances from another Annex I country that's doing better than its goal. This idea was inspired by the successful US drive to reduce sulphur dioxide pollution by allowing corporations to buy and sell emission credits from each other under a federally imposed cap on total emissions. Such systems are often labelled "cap and trade" because they operate by capping the total emissions from a group of companies or countries (which assures the overall target will be met) and by allowing the various participants to trade emission credits (which helps prioritize the lowest-cost reduction strategies). Ironically, it was the US that lobbied hardest for emissions trading in the early days of Kyoto negotiations, while Europe balked. But the US ended up out of the Kyoto picture – at least for the initial period – while the EU has not only participated in the Kyoto trading scheme, which began in 2008, but also created an internal trading system of its own (see p.323).

▶ **The Clean Development Mechanism (CDM)**, first proposed by Brazil, allows developed countries to get credit for bankrolling projects such as reforestation or wind farms that reduce emissions in developing countries. The scheme was slow to get off the ground, with many

The pros and cons of clean development

The Clean Development Mechanism seems like a win-win part of the Kyoto Protocol. It allows nations that come up against their emissions targets to pour money into emission-reducing projects in developing countries. But not everyone likes the CDM. Some oppose it on principle, claiming it's simply a tool that legitimizes the polluting legacy of Annex I nations by allowing them to buy their way out of any commitment to actually reduce emissions. Others are more concerned with how the CDM is implemented. The parties who arrange a CDM project have to demonstrate that it produces a reduction in developing-country emissions that wouldn't have happened otherwise. But many of the initial projects got started before they came under the CDM umbrella, thus complicating the task of deciding what would have happened without the CDM.

Some campaign groups, including the WWF, have lobbied to see certain projects ruled off-limits for CDM, such as those involving coal, large-scale hydropower and forest-based carbon sinks. A particular sore spot is the Plantar project in southeast Brazil, funded by the World Bank. Grassland at the Plantar site is being replaced by a eucalyptus plantation spanning about 230 square km (89 square miles). The trees are being harvested and converted to charcoal that goes to local pig-iron smelters, providing them with a less carbon-intensive form of fuel than coal. But activists say that such monoculture plantations have been a problem in Brazil for decades, displacing local residents as well as indigenous ecosystems. As of 2010, activists were still pushing to keep Plantar from CDM certification.

In a more distributed approach, a solar-energy project in Bangladesh is now being expanded through CDM support. Since 1999 the Bangladeshi bank Grameen (winner of the 2006 Nobel Peace Prize, with founder Muhammad Yunus) has provided small, no-collateral loans to more than 70,000 householders – most of them women – which allow them to install solar photovoltaic (PV) systems. Fewer than 30% of homes in Bangladesh are on the national electric grid. The others typically use kerosene for lighting and cooking and perhaps a lead-acid battery, recharged at a regional centre, to power a TV. Adding a solar system not only reduces greenhouse emissions but relieves women from the drudgery of maintaining kerosene lamps and provides cleaner indoor air and better light for reading. Revenue from the CDM will cut the cost of the household loans, allowing the project to expand to as many as a million homes by 2015.

Children at a solar house in Bangladesh.

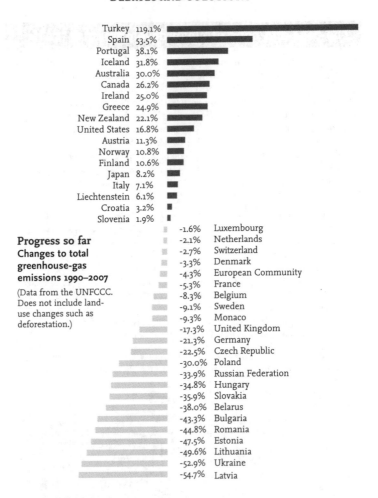

Country	Change
Turkey	119.1%
Spain	53.5%
Portugal	38.1%
Iceland	31.8%
Australia	30.0%
Canada	26.2%
Ireland	25.0%
Greece	24.9%
New Zealand	22.1%
United States	16.8%
Austria	11.3%
Norway	10.8%
Finland	10.6%
Japan	8.2%
Italy	7.1%
Liechtenstein	6.1%
Croatia	3.2%
Slovenia	1.9%

Progress so far

Changes to total greenhouse-gas emissions 1990–2007

(Data from the UNFCCC. Does not include land-use changes such as deforestation.)

Change	Country
-1.6%	Luxembourg
-2.1%	Netherlands
-2.7%	Switzerland
-3.3%	Denmark
-4.3%	European Community
-5.3%	France
-8.3%	Belgium
-9.1%	Sweden
-9.3%	Monaco
-17.3%	United Kingdom
-21.3%	Germany
-22.5%	Czech Republic
-30.0%	Poland
-33.9%	Russian Federation
-34.8%	Hungary
-35.9%	Slovakia
-38.0%	Belarus
-43.3%	Bulgaria
-44.8%	Romania
-47.5%	Estonia
-49.6%	Lithuania
-52.9%	Ukraine
-54.7%	Latvia

investors waiting to see if Kyoto would become official. By late 2010, though, more than 2400 CDM projects had been registered, with the majority of the emissions savings taking place in China and India. Together, these projects were expected to keep a total of more than 1.8 billion tonnes of CO_2 equivalent out of the atmosphere between 2001 and 2012. That's a substantial amount – close to 6% of the CO_2 emitted globally per year. However, the CDM is not without its detractors, some of whom object on philosophical and environmental grounds while others criticize the methodology (see box on p.313). For example, due to a loophole that was present in the CDM until 2007, more than

£3 billion was allocated for reduced emissions of trifluoromethane (HFC-23). That's more than ten times the actual cost needed to remove HFCs at the factory, according to US researcher Michael Ware, who argued in a 2007 *Nature* article that CDMs would be far more effective if they were devoted purely to CO_2 than to all six Kyoto gases.

▶ **Joint Implementation projects** allow an Annex I country to earn emission credits by subsidizing an emissions-reduction project in another Annex I country. As with the CDM, these projects officially began in 2005 and took a while to gather steam. As of late 2010, there were more than 200 projects under way in 14 host countries. The favoured locations have been Russia and the former Eastern Bloc, where there are plenty of inefficient technologies ripe for upgrading.

Towards the end of the initial Kyoto commitment period (2008–12), it looked as if Annex I countries would manage to meet their collective target of a 5.2% cut over 1990 levels, even before considering any help from the mechanisms above. However, this collective success hid a great deal of variation among countries, as can be seen in the graphic opposite. As of 2007, several of the wealthiest Annex I members, including Canada, Australia and New Zealand, had racked up double-digit emission *increases*. Meanwhile, the UK's emissions were down by more than 17%, well below its Kyoto target of a 12.5% decline – though even here, things look far less positive if you consider that the UK has massively increased its reliance on imported goods from countries with rising emissions (see p.45). Moreover, most of the carbon cuts achieved within the UK's own borders were in place before Kyoto was even agreed to, thanks to the 1990s transition from coal to gas as the nation's principal fuel for generating electricity.

What enabled Annex I to meet its Kyoto goal were the "economies in transition" that emerged from the fall of the Soviet Union. These nations experienced big drops in emissions during the 1990s simply due to the economic pain they endured in the initial post-Soviet years. As a result, many of them, including Russia, are now emitting anywhere from 30% to 50% less than they did in 1990. This comfortable margin has been derisively termed "hot air", since it arose from the peculiarities of the Kyoto target-setting process and the downfall of the Eastern Bloc economies rather than from any concerted effort to cut emissions. Were it

> **"In effect, Kyoto is learning by doing, even if what is agreed initially is minimal."**
>
> Dieter Helm, *Climate-Change Policy*

not for these countries, the Annex I group would have missed its Kyoto target by a long shot – ending up with overall emission gains instead of cuts. Some Annex I nations were on track to miss their goals despite the option of taking advantage of the CDM and/or buying emissions reductions through the European trading system.

Even if all countries were set to meet their 2012 Kyoto targets through bona fide emission reductions, it wouldn't have made a great deal of difference to the atmosphere. Together, all of the Annex I parties to the protocol represented only about 30% of the world's 1990 greenhouse-gas emissions, and only about 20% of 2005's global emissions. This is largely due to the absence of the US from the protocol and the lack of targets for China and other large developing countries. Even if the Kyoto-prescribed reductions were maintained for a full century – rather than until 2012 – they would only bring down the year-2100 temperature increase by a few percent, according to a 1998 analysis by Tom Wigley of the US National Center for Atmospheric Research. Critics have cited the weakness of even the potential best-case outcome in order to bash the protocol altogether. But as Wigley noted about his findings, "This does not mean that the actions implied by the protocol are unnecessary." Indeed, the larger goal of Kyoto was to establish a template through which the world could work for even greater reductions in subsequent agreements.

Beyond Kyoto: what's next?

There's no expiration date woven into the Kyoto Protocol itself: participants are free to choose a new commitment period with new targets. Still, with the world's biggest emitters not included in the protocol, the focus was eventually bound to shift to a possible successor agreement. Momentum began to build with a "road map" that emerged from the UN's 2007 climate summit in Bali. Two years of planning culminated in the mammoth meeting that took place in Copenhagen in December 2009 (see box on p.320). However, the results were a stunning disappointment to almost all parties involved. The resulting Copenhagen Accord, drafted by a handful of leaders at the last minute, was merely "noted" by participants rather than ratified by them. Nations were asked to submit their own voluntary targets for 2020, which were later added to the agreement – but unlike Kyoto, the accord was not legally binding. Weak as the deal was, it did manage to codify unprecedented pledges from China (a 40–45% reduction in carbon intensity – see p.44) and the United States (a 17% reduction over 2005 emission levels by 2020). In addition, participants agreed to beef up efforts to protect rainforests (see p.321) and pledged to establish a $100-billion-per-year green climate fund from 2020 onward that would help developing countries mitigate and adapt to climate change. These efforts gained more ground at the 2010 UN meetings in Cancún, though much difficult work still lay ahead.

If US participation is critical to the next binding global treaty, it may be years in coming. In the aftermath of Climategate and the resurgence of conservative forces, the US Congress found itself returning to scepticism on the value of cap and trade and other policy mechanisms for reducing carbon emissions. After the 2010 elections, US president Barack Obama acknowledged the near-impossibility of passing cap-and-trade legislation for at least several years.

All of this adds to the urgency of mitigation efforts being undertaken on local and regional scales, as discussed below. It also highlights the need for serious work on adaptation (see box, p.322), given that some degree of climate change is inevitable. For many years, adaptation was considered to be a flag of surrender to climate change, with policymakers encouraged to focus purely on reducing emissions. But governments are increasingly recognizing that there's no sense remaining unprepared for impacts that are increasingly likely.

As for what an eventual successor to Kyoto might look like, there's been no shortage of suggestions on the table. A report from the US Pew Center on Global Climate Change summarized more than 40 ideas, with names that range from the grandiose ("Orchestra of Treaties" and "Climate Marshall Plan") through the humble ("Broad but Shallow Beginning" and "Soft Landing in Emissions Growth") to the droll ("Keep it Simple, Stupid"). The main points in question include:

▶ **What's the best time frame to consider?** Some plans have been focused on the second Kyoto commitment period (2013–17), while others extend all the way out to 2100.

▶ **What type of commitments should be specified?** There is a whole array of possibilities, from emission targets by nation or region to non-emission approaches such as technology standards or financial transfers.

▶ **How should the burden of climate protection be shared among developed and developing countries?** This remains a key sticking point, as was the case from the very beginning.

▶ **How does the world make sure that commitments are enforced?** It's an issue that many say Kyoto hasn't fully addressed.

▶ **Should there be a single global plan or an array of decentralized alliances?** From the ashes of Copenhagen emerged a stronger interest in new kinds of partnerships outside the Kyoto-style global model, such as the impromptu group of nations that pulled together the summit's last-minute accord. Developing countries are often highly sceptical about such groupings, since they've traditionally been excluded from them. However, according to former UK secretary for energy and climate Ed Miliband, "The old order of developed versus developing has been replaced by more interesting alliances."

Certainly, the European Union has made real progress in emissions reductions in a variety of ways, thanks in large part to its rule-making power. However, regional alliances that don't have such teeth may not be able to accomplish so much. In 2005, the US teamed up with Australia, China, India, Japan and South Korea (with Canada joining in 2007) to form the Asia-Pacific Partnership on Clean Development and Climate. The group was hailed by US president George W. Bush as a "results-oriented partnership" preferable to the Kyoto process. The group's members are heavy hitters indeed – they account for more than half the world's carbon emissions, GDP and population – but the partnership itself has focused on voluntary, non-binding actions such as "aspirational" emission goals and the sharing of energy technologies. This strategy has been slammed as "Kyoto-lite", and as of 2010, none of the partnership members had actually managed to reduce its own emissions. Even US conservative John McCain blasted the formation of the group, calling it "nothing more than a nice little public relations ploy". According to UK representative John Ashton, "I think that the argument that we can do this through voluntary approaches is now pretty much discredited internationally."

Among the most intriguing plans offered to date for a post-Kyoto agreement is the **contraction and convergence** (C&C) model developed by the Global Commons Institute, a British group headed by Aubrey Meyer. It was introduced by the Indian government in 1995 and adopted by the Africa Group of Nations in 1997 during the run-up to Kyoto. The plan has also received votes of support from the European Parliament and several UK and German advisory groups. The two principles at the heart of C&C are:

▶ **Contraction:** the need to reduce overall global emissions in order to reach a target concentration of CO_2, with a commonly cited goal of 450 parts per million.

▶ **Convergence:** the idea that global per-capita emissions, which now vary greatly from country to country, should converge towards a common amount in a process more or less parallel to contraction.

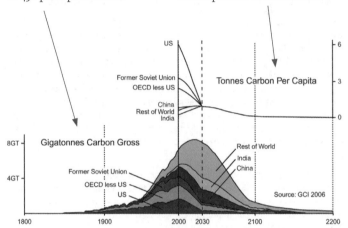

In short, C&C calculates how much carbon the world can emit in order to reach its target, then apportions that total equally among the world's residents based on population. It's an elegant concept that moves the process towards a climate-protected future that virtually everyone recognizes as fair.

Some critics of C&C point out that it could provide developing countries with an anti-incentive for birth control, since it allocates emissions rights based on population. And many activists feel that the C&C plan lets developed nations off the hook for their hundred-plus years of creating our current predicament. A competing plan, the **Brazilian Proposal**, uses the historical pattern of emissions as a starting point. On behalf of a wide range of nongovernmental organizations, the Climate Action Network (CAN) proposes a three-tier **Global Framework** in which developing countries move into tighter emissions brackets as they prosper. Focusing on the root of the

> "In the politics of climate change, the Kyoto Protocol is the equivalent of kerb-crawling. It is utterly inadequate and doesn't provide the legal framework we need."
>
> Aubrey Meyer, Global Commons Institute

Copenhagen and its aftermath

The 2009 Conference of Parties to the UN treaty – more commonly referred to as the Copenhagen meeting – was the world's biggest-ever climate change summit. It was long expected to be the venue where a post-Kyoto treaty would be hammered out. The US arrived at the summit backed by a new president and new Congress, both far more amenable to climate-change policy than their predecessors.

Despite the hoopla, Copenhagen ended up as a bitter disappointment for activists and policymakers alike. The dank, chilly weather reflected the evolving mood inside the cavernous Bella Center, as delegates were torn between the demands of poorer countries, including African nations and island states, and the mutual reluctance of the United States and China to commit to emission cuts that would go beyond Kyoto levels, which neither nation had agreed to in the first place. A round of spirited protests midway through the meeting was quashed by Danish police, sending hundreds into detention. Afterward, the conference went into virtual lockdown for its final week. Most of the 22,000 registrants from nongovernmental organizations were shut out in the final two days to make way for heads of state and their vast security teams.

On the final day, negotiations culminated in an offsite meeting among the leaders of Brazil, China, India, South Africa and the United States. From behind those closed doors emerged the bare bones of the Copenhagen Accord. Instead of a step beyond the Kyoto Protocol, the document was seen by many as a step backwards. Other countries were left out of the drafting process, and Kyoto-style binding targets were omitted. Dissatisfaction among delegates was so widespread that, rather than formally adopting the document, the assembled parties managed only to "take note" of it.

Although a few participants and observers sounded notes of optimism – "We have come a long way", said British prime minister Gordon Brown – these voices competed with a chorus of critics. The accord was "nothing short of climate change scepticism in action", said Sudan's Lumumba Di-Aping, chief negotiator for the

problem, the **Kyoto2** plan would combine market-based mechanisms with regulation of emissions at or near their source, including a global cap on the total amount of carbon extracted in the form of fossil fuels.

There's no disguising the fact that many of these plans would require a massive explicit and/or implicit transfer of wealth from rich to poor countries, at least in the beginning, in order to get results. One might argue that such a transfer is unavoidable if the developed world is to take full responsibility for its outsized contribution to the greenhouse effect. However, there's sure to be fierce political resistance in some quarters.

It's also argued that global emission cuts won't happen – especially in fast-growing countries like China and India – until technological break-

G77 group of developing countries. Even the conference's own news service acknowledged that "enthusiasm for the Copenhagen Accord is scarce". Dubbing Copenhagen a "fiasco" and a "foundering forum", the *Financial Times* claimed that the big winners in the process were climate sceptics, China and the Danish economy. The losers, according to the *FT*: carbon traders, renewable energy companies and the planet.

One of the few bright outcomes of the Copenhagen process was the creation of a "green climate fund" as well as new mechanisms for transferring technology and reducing deforestation. The Kyoto Protocol had assigned credits for renewing destroyed forests and planting new ones, but it had no mechanism for preventing deforestation in the first place. Without a global agreement, any nation's preservation work might simply push deforestation to the next country. But in Copenhagen, at the urging of a coalition of activist groups – as well as such luminaries as Jane Goodall and Richard Branson – billions of dollars were pledged to help cash-poor, rainforest-rich nations gain more value from preserving their forests than from exploiting them. The agreement emerged after years of research in the area known as REDD (reducing emissions from deforestation and degradation). Newly precise monitoring by satellite will help keep track of progress.

Protesters at the Copenhagen summit and world leaders as marionettes.

throughs make low-carbon energy sources so cheap that countries can reduce emissions while improving their standard of living. Otherwise, when individuals and countries juxtapose the promised long-term benefits to climate against their own near-term economic future, the climate can easily lose out. Still, there's widespread acknowledgement that a price of some sort must be put on carbon. Roger Pielke Jr (University of Colorado) suggests that a new global agreement might start by setting a relatively low, but slowly rising, global tax on carbon, levied at the fuel source, with the revenues flowing directly into intensified research and development on decarbonized energy sources. That way, the link between sacrifice and benefit would be clearer and shorter-term. Scientist

Adaptation vs mitigation

What might seem like a straightforward response to climate change – adapting to it – is actually fraught with politics. There's nothing especially novel about being prepared for what the atmosphere may bring. As the IPCC notes in its 2007 assessment, "Societies have a long record of adapting to the impacts of weather and climate." There are plenty of sensible steps that vulnerable nations and regions could take right now to reduce their risk of climate-change trauma. Just as energy efficiency makes sense for other reasons than climate protection, many forms of adaptation – such as moving inwards from coastlines – will help keep people and their property safer regardless of the extent of global warming. Yet there's a tension between adaptation and mitigation: to some, the former implies a disregard of the latter, as if society were giving up on trying to reduce greenhouse-gas (GHG) emissions. "There's one way to directly address climate change, and that's reducing the GHG emissions that drive it", writes David Roberts in the online magazine *Grist*. "In the context of the climate-change debate, advocating for adaptation means advocating for a non-response."

Others argue that the problem is so vast that neither adaptation nor mitigation alone can do the trick – we need both. "The UK government is very clear on a twin-track approach", says Chris West, director of the UK Climate Impacts Programme. The UKCIP was founded specifically to help cities, business and other entities determine their climate-change risk so they can better prepare for it. "Regardless of mitigation efforts to reduce greenhouse gas emissions, there are unavoidable impacts already in the system", says West. With the North Sea rising against its dykes, the Netherlands is also considering adaptation in a big way. Among the long-range options being explored are large-scale relocation to the eastern Netherlands and the creation of a "hydrometropole" – in essence, a floating city of fifteen million people.

and activist James Hansen, who is adamantly opposed to cap-and-trade plans, also believes a global carbon tax is the way to go, but he calls for a fairly large one – to help drive immediate emission cuts – that's revenue-neutral, with benefits returning to citizens. The latter idea, often dubbed a "feebate" or fee-and-dividend plan, was pioneered by Amory Lovins of the Rocky Mountain Institute. According to Hansen, "As long as fossil fuels are the cheapest form of energy their use will continue and even increase."

> "What is needed above all right now is US leadership, for no country bears greater responsibility for climate change, nor has greater capacity to catalyse a global response."
>
> Eileen Claussen and Elliot Diringer, Pew Center on Global Climate Change, 2007

The next few years should continue to make for fascinating and hugely important politics, as various plans vie for consideration and leaders from around the world jostle for position. In the words of the C&C position statement – as true now as when C&C was first proposed in the 1990s – "the global community continues to generate dangerous climate change faster than it organizes to avoid it. The international diplomatic challenge is to reverse this."

National, regional and local schemes

Governments of all sizes are taking a variety of actions independent of Kyoto and its potential successors. Some of these actions involve setting reduction targets based on much longer time frames than Kyoto specifies, while others focus on specific sectors or market mechanisms.

A big part of the EU's strategy for climate protection is its **emissions trading scheme** (ETS), launched in 2005. It's similar in spirit to, but separate from, the Kyoto emissions trading. When the EU's scheme began, each nation assigned each of its largest industries the right to emit a certain amount of carbon. The scheme covers a variety of activities, including oil refining, cement production, iron and steel manufacture, glass and ceramics works and paper and pulp production, as well as fossil-fuel-based power generators of more than twenty megawatts. Together, these companies emit almost half of the EU's carbon – more than two billion tonnes of CO_2 per year. In order to meet its own targets, each of these big EU companies can either cut its emissions or "pay to pollute" by purchasing credits from another company that's done better than targeted. The credits can even be purchased from a country outside Annex I, via the Kyoto mechanisms discussed previously.

Largely because the initial targets assigned by EU governments to their biggest companies were so lenient, there were far more sellers than buyers for carbon credits in the first ETS phase (2005–07). The price plummeted from a peak of nearly €30 per tonne of CO_2 in 2006 to only a few cents per tonne by late 2007. Prices held more firmly in the second phase (2008–12), which expanded from CO_2 to include all six of the greenhouse gases covered by Kyoto as well as clean-development and joint-implementation credits from the Kyoto process above. As of late 2009, the equivalent of more than 200 megatonnes of CO_2 was being traded each month. A third phase, beginning in 2013, will include aviation emissions and other new wrinkles.

It's hard to know exactly how much good the EU's scheme has accomplished, but some studies suggest that a regional emissions decrease of several percent could be attributable to the system thus far. Elsewhere, New Zealand, Tokyo and some US states (see below) have adopted their own emission trading schemes, but Australia failed to gather enough support among legislators to institute such a plan until 2013 at the earliest. Aside from these efforts, the Annex I countries have tackled emission reduction with a smorgasbord of policy tools that, for any one country, could include a blend of taxes, support for low-carbon fuel sources, campaigns to raise public awareness about energy efficiency and environmental aid to developing countries.

The **UK** took a leap forward with its Climate Change Act 2008, which mandates an 80% reduction from 1990 levels of the six Kyoto greenhouse gases by 2050. This is among the most stringent goals set by any government, and a considerable ramp-up from the nation's previous goals of a 60% reduction by 2050 and 80% by 2100. The UK has also put into place a range of directives dealing with everything from co-generation to energy efficiency in buildings (see p.356). However, these measures haven't yet produced the sustained emission cuts that will be needed to reach the UK's newly ambitious goals – especially when international aviation and goods manufactured abroad are taken into account. Indeed, as noted above, the country is on target to meet its Kyoto goals mainly due to the 1990s switch from coal to gas power plants.

The most stunning example of national energy transformation may be **Portugal**, where greenhouse emissions had soared by more than 40% from 1990 to 2004 – the most of any Annex I country except for Turkey. In 2005, Portuguese prime minister José Sócrates led the push to revolutionize the country's use of energy, including a massive increase in wind and solar farms and a carefully designed mix of government-owned transmission lines and privatized energy producers. With only modest increases in utility costs to consumers, Portugal now generates close to half of its electricity from renewable sources.

The **US**, by contrast, has put only a smattering of climate-protection policies in place, and some American laws seem to encourage more greenhouse gas rather than less. For instance, hulking sport utility vehicles (SUVs) were required to meet only minimal fuel-economy standards for years, and the passenger-car standard in 2010 was the same as in 1985. The rules are now tightening, thanks to legislation passed in 2007 for passenger vehicles and 2010 for trucks and buses. The US Environmental Protection Agency (EPA) has made gains with its Energy Star efficiency labelling scheme (see p.372). And in a landmark 2007 ruling, the US Supreme Court ordered the EPA to begin regulating carbon dioxide as a pollutant, signalling the potential for eventual nationwide action on climate change.

For now, though, much of the US action has been happening on the **state** level, where there's more of a chance for a few motivated legislators and policy analysts to make a real difference. **California**, as usual, is a bellwether. Its greenhouse emissions are comparable to those of Australia or Brazil, so its actions could have a real impact in themselves, as well as through their influence on other US states. In 2005, California's governor and erstwhile movie star Arnold Schwarzenegger signed an executive order calling for massive emission reductions in the long term – a reduction to 1990 emission levels by 2020 and a further reduction of 80% by 2050 – as well as a variety of policy moves, including a proposed petrol tax. State voters soundly rejected a move to repeal the law in 2010. Meanwhile, a group of northeastern states extending from Maryland to Maine has put its own emissions trading scheme in place, with a goal of cutting emissions from power plants 10% by the period 2015–20. Similarly, eleven states and provinces from the western US and Canada teamed up for an ETS set for a 2012 launch (though Arizona later backed out of the voluntary scheme).

Given the scope of the US economy, private endeavours can make a big difference. Teaming with the Environmental Defense Fund, **Walmart** – the world's largest company – aims to more than offset the growth of its carbon footprint by reducing 20 million tonnes of emissions from its global chain of suppliers between 2010 and 2015. "Looking deep into the supply chain and across product lifecycles for carbon pollution [reductions] is uncharted territory," said EDF's Elizabeth Sturcken.

Across the world, **cities and towns** are also getting involved. More than 1000 US mayors, representing more than a quarter of Americans, have signed pledges to meet Kyoto targets. In the UK, Newcastle – a cradle of coal mining for centuries – made a splash in 2002 when it set out to become the world's first carbon-neutral city. It's still aiming for that goal, along with a growing number of other municipalities.

A few brand-new settlements are being planned with energy efficiency topmost in mind. Just northeast of Shanghai, **Dongtan, China**, will be taking shape on Chongming, China's third-largest island. The population – expected to reach 500,000 by the year 2050, although the city's 2010 debut was delayed – will drive hybrid cars or pedal bikes along a network of paths. Renewable power for homes and businesses will flow from a centralized energy centre being developed in partnership with the University of East Anglia, drawing in part on an array of wind turbines on the island's west edge. The British firm Arup, which is designing Dongtan, is working with several universities to use Dongtan as a laboratory for green design. Also angling for eco-leadership is **Masdar City**, a project seemingly at odds

An architect's image of Dongtan, soon to be China's greenest city.

with the extravagance of its home turf of Abu Dhabi. The financial crisis has delayed Masdar City's development, but the plans are impressive – a carbon-neutral, car-free landscape powered by wind and solar farms and housing up to 50,000 residents. Although WWF is a supporter of Masdar City, some activists have derided purpose-built green towns as exclusive communities irrelevant to the challenge of making everyday cities more sustainable.

How much will it cost?

No analysis of climate-change politics would be complete without considering economic costs and benefits. As major players appeared to finally coalesce around the need for action on climate change, the presumed economic impact of taking action became a hot topic. Conservative commentators and politicians, especially those in the US, tended to stress that emissions cuts could bring major economic downswings. Green-minded writers and legislators, on the other hand, typically suggested that the cost of *not* acting could be even higher.

The terms of the debate shifted sharply in 2006 with the release of the *Stern Review on the Economics of Climate Change*. Commissioned by Gordon Brown, then the head of the UK treasury, this massive study headed by Nicholas Stern (formerly the chief economist for the World Bank) reverberated in financial and political circles around the globe. Like a mini-IPCC report, Stern's study drew on the expertise of a wide range of physical and social scientists to craft a serious, comprehensive look at the potential fiscal toll of global warming.

Stern's conclusions were stark. Our failure to fully account for the cost of greenhouse emissions, he wrote, "is the greatest and widest-ranging market failure ever seen". He estimated that climate change could sap anywhere from 5% to 20% from the global economy by 2100, and that "the benefits of strong, early action considerably outweigh the cost".

While the Stern Review drew massive publicity and galvanized leaders around the world, it wasn't universally praised by economists, many of whom found it overly pessimistic. One of the biggest sticking points was how Stern handled the discount rate – the gradual decline in value of an economic unit over time. In traditional analysis, a unit of spending today is valued more highly than a unit spent tomorrow, because the present is more tangible and the unit will be worth less in the presumably richer future. But climate change is a far more complex matter because of its global reach and the long-term effects of current emissions. In that light, Stern chose not to "discount" tomorrow the way most economists would: "if a future generation will be present, we suppose that it has the same claim on our ethical attention as the current one".

Stern's report also drew fire for its assumptions about weather extremes. In the journal *Global Environment Change*, for example, Roger Pielke Jr criticized the report for extrapolating a 2%-per-year rise in disaster-related costs observed in recent decades over and above changes in wealth, inflation and demography. As it happens, 2004 and 2005 brought a number of expensive US hurricanes, and Pielke argues that this happenstance skewed the trend used by Stern. At the same time, Pielke grants that the Stern report "helped to redirect attention away from debates over science and toward debates over the costs and benefits of alternative courses of action".

Long before Stern's report, the insurance industry realized that climate change threw an uncertain element into the detailed calculations it uses to gauge risk. As early as 1989, the Lloyd's of London insurance market began incurring massive extra losses. The head of the American reinsurance association said in 1993 that "changes in the number, the frequency and the severity of natural catastrophes are threatening to bankrupt the industry". According to industry giant Swiss Re, the global total of insured losses from natural disasters hit $120 million in 2005 (as expressed in 2009 dollars). That was nearly twice the constant-dollar record set only a year earlier. Part of this rise was due to steep rises in property prices in hurricane-prone regions such as the US Gulf Coast, and the years 2006–09 saw numbers falling back to levels seen earlier in the decade. Still, the company cited global warming as an important cofactor in the long-term rise in damages from natural disasters. Both Swiss Re and Munich Re have been among the strongest corporate voices calling for climate protection, issuing reports and raising public awareness. They're also using their leverage as institutional investors (in the UK, they own around a quarter of the stock market) to persuade other companies to take climate change on board.

By comparison, US insurance companies have been rather mute on the issue, despite suffering massive financial hits. As late as 2010, the Institute

The New Great Game – who owns the Arctic?

Throughout the 1800s, Britain and Russia vied for power in Central Asia in a century-long contest dubbed the Great Game. Today, the geopolitics around the fast-thawing Arctic region have been described as the "New Great Game", as the erosion of the region's summertime ice pack opens the door to high-latitude wheeling and dealing. Later this century, shipping routes that now link Europe, Asia and North America through the Arctic could be open months longer than they are now, and huge vaults of oil and gas that are believed to lie beneath the sea may become accessible for the first time.

The five countries with Arctic coastlines – Canada, Denmark, Russia, Sweden and the US (via Alaska) – long paid little mind to each other's business in a region seemingly devoid of usefulness. Intersecting borders across the Arctic were once drawn informally or not at all. Now they're a topic of intense debate, as countries angle for the best shipping routes and undersea resources.

The competition heated up in 2007, as did the Arctic itself. Russia planted a flag on the seafloor beneath the North Pole, which it claims is an extension of the Lomonosov ridge running northward from Russia's Arctic coast. Canada and Denmark, meanwhile, are attempting to claim the ridge as their own. These and other border disputes will be resolved through the Law of the Sea, the United Nations convention ratified by more than 150 nations, including all of those bordering the Arctic except the United States. Although there's bipartisan US support for the Law of the Sea, a small group of conservative legislators has long blocked a vote on ratification. At the same time, US geologists are sailing across the Arctic each year to measure the outermost extent of the continental shelf claimed by the US.

The Law of the Sea dictates that each nation has an "exclusive economic zone" – including the right to drill for oil and gas – extending 200 nautical miles (about 370km) poleward from its Arctic coastline. That leaves a sizeable area under contention, from near the North Pole westward toward the Siberian coast. As it happens, the unprecedented melt of summertime ice in the Arctic has opened much of this area, with prevailing currents pushing much of the remaining ice toward Greenland.

for Business and Home Safety, a major industry group, had no sections about climate change on its website. A former president of the group told climate reporter Ross Gelbspan in 2003 that US insurers are "burying their heads", dropping customers and abandoning high-risk areas. Indeed, hundreds of thousands of Floridians saw their home insurance cancelled in 2005 and 2006, after the state's string of hurricanes. Over two million people along the US Gulf and Atlantic coasts turned to "insurers of last resort" established by state governments – with the public often paying the tab if major disaster strikes.

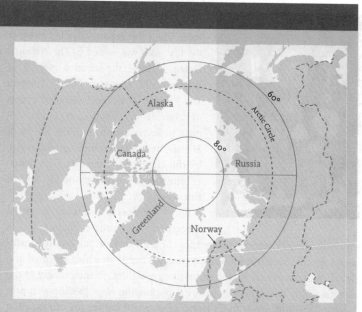

Entrepreneurs and companies are also getting involved in the New Great Game. For example, Pat Bode of Denver snapped up the port of Churchill, Manitoba, from the Canadian government for a token $7 US in 1998, as part of an $11 million purchase of train lines running into Canada's heartland. Though it's only fitfully busy now, Churchill's port – which lies on the west side of Hudson's Bay – could become a major transport hub if climate change thaws the region as much as projected. In his book *The World in 2050*, Laurence Smith draws parallels between this century's Arctic and the wide-open, resource-rich western US. As he puts it, "Imagine the Arctic in 2050 as a frigid version of Nevada – an empty landscape dotted with gleaming boom towns."

The risk of international conflict

Given how difficult it's been to achieve global agreement on even minimal steps to fight climate change, it's not hard to imagine international tensions bubbling over at some point, especially as the climate itself heats up. Defence planners have begun paying more attention to how such conflicts might unfold over the coming decades. Thus far, their analyses have focused on "what if?" scenarios in which the effects of global warming mingle with other environmental and geopolitical pressures in dangerous ways.

CLIMATE CHANGE
TO THE YEAR 2000

Such research has played out in the American military sphere since at least 1978, when the report *Climate Change to the Year 2000* was issued by the National Defense University amid a period of intense cold snaps in North America and Europe. The report included expert opinion from twenty-four climatologists in seven nations on how Northern Hemisphere temperatures might evolve by the close of the century. The actual readings ended up near the top end of the experts' predictions.

One landmark defence-oriented assessment of climate change was a widely publicized report prepared for the US Pentagon in 2003. It considered a specific scenario for abrupt climate change in which regional and global conflicts over land and water use grow through the century. If the planet's ability to support its growing population were to drop abruptly, "humanity would revert to its norm of constant battles for diminishing resources", the report warned. The message was underscored in 2007 with another Pentagon-commissioned report, *National Security and Climate Change*, in which a group of retired generals and admirals examined climate-change research and commented on its implications. "Projected climate change poses a serious threat to America's national security", concluded the team, which envisioned conflicts brewing even in now-stable parts of the world. Starting in 2008, US military planners were required to consider the effects of climate change on future missions. The UK's defence ministry has also been taking global warming into account, as with its 2009 *Climate Change Strategy* report – much of which deals with reducing the military's own greenhouse emissions. It's been estimated that UK defence activities are responsible for 1% of the nation's carbon dioxide emissions, and US forces reportedly account for 1.5% of American oil use.

While some observers interpret military concern about climate change as a sign that the issue has gone truly mainstream, others worry that the notion of global warming itself could become "militarized". In this view, climate refugees and water wars might simply take the place of more traditional conflicts as a rationale for military activity – or by the same token, an emerging climate-related crisis could open the door for intensified efforts in border control, surveillance and the like.

Scholars have also continued to tussle over the longstanding question of how the environment intersects with social factors to help breed warfare. Drought, in particular, is a perennial marker of conflict. Unequal access to water has long been associated with unrest in the Middle East, and the collapse of several ancient cultures has been linked to drought. But the connection isn't a simple one. While early historians often saw drought as a primary trigger of societal ills, many cultures have been able to adapt to even severe water pressures. It appears that drought isn't a prime cause of war so much as a critical backdrop that can help lead to conflict if it persists long enough and if it syncs with an unstable political structure. The horrific violence in Darfur, Sudan, offers one example. It followed decades of strained relations between nomadic herders and farmers, coinciding with a sustained drop in rainfall from the 1970s onward. However, the dramatic onset of warfare in 2003 doesn't appear to be directly linked to short-term dryness.

As the world's subtropics heat up and dry out, many important regions – including parts of the southwest US and northern Mexico, and the nations of Europe and Africa ringing the Mediterranean – will see an intensifying risk of major drought, which would only feed other sources of societal stress and international conflict. These same parts of the world could also be well suited for hosting mammoth solar farms to send energy across international lines, making for another potential wrinkle in the geopolitics of tomorrow.

Technological solutions

Energy, engineering and efficiency

Human ingenuity got us into our greenhouse mess, and we'll need to call on it again in order to find our way out. Even if we can develop affordable ways of storing carbon underground or removing it from the air – both of which are far from certain – the world faces a titanic challenge in reducing its voracious appetite for oil, coal and gas, especially with many developing countries on a growth track.

The good news is that plenty of alternatives exist that produce little or no greenhouse output. However, to make a real dent in carbon emissions, they'll need to be deployed and coordinated on a far larger scale than they are now. And even if that can be accomplished, it doesn't ensure that the remaining fossil fuels will stay in the ground – which is one reason that technologies for capturing CO_2 from power stations and storing it underground could end up being crucial.

Of course, it's not just where we get our energy that counts. Equally important is how *efficiently* we use it. Hence this chapter also covers technologies and policies designed to raise efficiency and reduce energy waste. Finally, we look at geoengineering plans – schemes to pull greenhouse gases directly from the atmosphere or reduce the amount of sunlight reaching Earth. No matter what technologies come down the pipeline, the economics and politics discussed in the previous chapter will play a huge role in shaping their cost-effectiveness and thus how quickly they're adopted.

The future of fossil fuels

For the time being, fossil fuels remain at the core of the global energy supply. Of course, these fuels are all finite resources, which, if we continue to

When will the oil run dry?

One key question in predicting the energy mix of the future is the amount of oil left in the ground. A flurry of recent books has publicized the concept of **"peak oil"** – the point at which we've extracted half of the world's accessible oil reserves. The claim isn't that oil will suddenly become scarce, but that production will start to drop, demand will outstrip supply and prices will rocket, causing massive economic dislocations. There's intense disagreement on whether the peak-oil point is at hand or decades away, but one thing is certain: since the 1980s, the world has consumed more oil each year than it discovers. In fact, the ratio has been around two to one since the late 1990s.

The US offers an example of how an oil peak can sneak up on people who want to deny its existence. In 1956, M. King Hubbert, a geologist at Shell, told colleagues that American oil production would peak in 1971. Few people took his warning seriously – after all, production was climbing year upon year through the 1960s – but sure enough, the peak arrived in 1970, just one year from Hubbert's forecast. Annual US oil production has fallen ever since and now stands at about half of that 1970 peak. Pundits vary in their predictions about the global oil peak, but some believe that it may arrive within the next few years (or even that we're already there but don't realize it yet). It's impossible to know for sure, because the data on the vast bulk of oil supplies are held by members of the Organization of the Petroleum Exporting Companies (OPEC), which keeps a tight lid on numbers and methods. Oil companies tend to put an optimistic face on the situation.

So-called unconventional oil sources, such as **tar sands** in Canada and **oil shale** in the US, could provide as much oil as the world's current reserves, but the present methods of extraction are hugely greenhouse-intensive and environmentally problematic – not to mention expensive. Thus, it seems quite possible that global oil supplies *will* gradually tighten in the coming decades, assuming that long-term economic woes or major efficiency efforts don't cut demand dramatically.

Natural gas supplies don't appear to be quite as imperilled as oil reserves, but they're already beginning to dwindle across some countries that are traditional exporters, including Canada. The world's largest stocks of natural gas are in many of the same locations as oil reserves, including Russia and the Middle East, and transport is a big issue. Liquefied natural gas is fiercely explosive, and few coastal cities are eager to play host to LNG tankers.

Any drop in supplies of either oil or natural gas would have massive effects on the economics of other energy sources. Oil is used above all as a portable fuel, while much gas is burnt for domestic heating and cooking. The alternatives – from hydrogen cars to electric heating – would require massive increases in electricity generation. For this, the world would need to turn to a mix of coal (clean or dirty), renewables and/or nuclear. The relative investment in each of these options could be the single greatest factor in our success or failure in combating climate change.

use them, will cease to be available at some stage. Indeed, many commentators believe that our currently abundant supplies of **oil and gas** will start to dry up relatively soon (see box on previous page). But even if they're correct, that won't be enough to stop climate change. There's certainly enough remaining oil and gas to push us deep into the greenhouse danger zone – especially when you include carbon-intensive "unconventional" oil reserves – and we'll consume extra amounts of energy retrieving some of those hard-to-get supplies. The vast tar sands of western Canada now provide the US with twice as much crude oil as does Saudi Arabia. This oil has to be stripped from the land or cooked out of the soil, and the environmentally destructive process takes several times more energy than regular oil production. In turn, this boosts the total carbon footprint of each gallon of gas by close to 20%.

Even if oil and gas ran out tomorrow, there would still be enormous reserves of **coal** to consider. Currently, around 40% of the world's electricity comes from coal-fired power plants, and coal is responsible for more than a quarter of global CO_2 emissions. What's scary from a climate perspective is that there are virtually no brakes on greenhouse emissions from coal. The world's most extensive coal reserves are in countries that fall outside the Kyoto Protocol's emissions mandates. This includes China, India and the US, which happen to be the world's three most populous countries as well as the three nations with the greatest coal reserves (more than half of the global total).

There's a lot of momentum driving the world's reliance on coal. The typical coal-fired power plant is in operation for sixty years or more, and the number of new coal-fired plants in the works is astonishing. The equivalent of some 3000 of them will be built between 2005 and 2030 at a cost of more than $5 trillion, according to forecasts from the International Energy Agency. Once built, these would generate more power than all of the coal-fired plants in use globally in 2005. Thus, even with improved efficiency, we can expect more carbon dioxide (perhaps as much as 60% more) to emerge from the world's coal plants in 2030 than from the smaller number now in place unless some of that CO_2 can be captured and buried (see below) or unless the rate of construction slows down, which is something many commentators are calling for.

China and India are currently building the most new plants by far – dozens each year. While some of these include emission-reducing technologies, the sheer volume points toward a huge spike in future emissions. In the US, new plants are also coming online, but some roadblocks have emerged. In Kansas, a state agency denied permits for a pair of new plants in 2007 based on their anticipated greenhouse emissions; it marked the first time climate change had been invoked to block a US coal plant. British activists have fought coal vigorously, including plans by a German utility to build the UK's first coal plant in decades, at Kingsnorth in Kent. This proposal was scrapped in 2009, and any future coal plants built in the UK will be required to capture and store nearly all of their emissions (see below).

How much can coal plants be cleaned up? Incremental improvements in technology over the years have raised the efficiency with which coal-fired plants can pulverize and burn their fuel by a percent or so each year – no small achievement. Yet even now, less than 30% of the energy in the coal burned at a typical plant across the world is translated into electricity. (The most modern coal-fired plants have efficiencies closer to 45%.)

Another approach to cleaning up coal is **gasification**. Instead of burning the coal directly, a gasification plant heats the fuel to a high temperature with just a dash of oxygen. The resulting gases are either burned or further refined into **synfuels** – synthetic versions of oil-based fuels such as petrol or diesel. This isn't a new idea; Hitler was driven to synfuels when Germany's oil supplies ran low during World War II. On the face of it, gasification might seem like an inefficient way to use an already problematic fuel, but there are benefits to the process. Chief among them, in terms of greenhouse gases, is that CO_2 can be easily separated out of the stew of gases after combustion. There's even the option of underground gasification, another old idea now being re-tested. In this case, the coal isn't mined at all – instead, it's heated and burned in place below the surface, where the resulting carbon dioxide could theoretically be stored.

Carbon capture and storage

The benefit of isolating CO_2 created in a coal gasification plant – or any other power station – is the chance to capture the gas and keep it from entering the atmosphere. This process is known as **carbon capture and storage** (CCS), or **sequestration**. There was no economic or regulatory motive for most of the world to explore sequestration until the Kyoto Protocol kicked in. Now there are dozens of pilot projects scattered around the world, especially in Europe and North America. Most involve large power plants, since these are the biggest point sources of greenhouse gases. Many campaigners view CCS as an extension of the throwaway culture that's led to climate trouble in the first place, but it may become an important tool nevertheless – not least because it's hard to imagine a situation where all the existing and planned fossil-fuel power plants get switched off.

The idea behind CCS is to put the captured CO_2 (which could represent as much as 90% of a power plant's carbon footprint) in some sort of underground location where it will remain stable for a long time. Some proposals have explored injecting CO_2 into the sea – at shallow or intermediate depths, where it would dissolve into the water, or at the ocean bottom, where the high pressure would keep it liquid for long periods before it eventually dissolved. These options have yet to overcome a serious environmental objection: we know that CO_2 is making the oceans more acidic (see p.136).

A more promising solution is to put the carbon underground. If it's pumped to depths of more than 800m (2600ft) below ground, CO_2 takes on a liquid-like form that greatly reduces the chance of it rising to the surface. Given the right geological formations, the gas could remain underground almost indefinitely. Among the possible locations being explored are:

▶ **Depleted oil and gas wells** This technique has been used on a limited basis for decades in the practice called enhanced oil recovery, which has the economic – if not climatic – benefit of prolonging production from ageing reserves (as the CO_2 is pumped down into the wells, it helps push out the remaining oil or gas).

▶ **Saline aquifers** Found in basaltic formations and other types of rock, these typically hold salty water that's separate from the fresh water found in other aquifers. There may be enough saline aquifers to hold centuries' worth of CO_2.

In a special 2005 report on carbon capture and storage, the IPCC found that the technique holds real potential as part of the world's climate-protection package. They estimate that many of the largest point sources of CO_2 probably lie within 300km (180 miles) of potential underground storage sites. According to the IPCC, these sites would likely hold 99% or more of the CO_2 added to them for more than a thousand years. The IPCC also found that the most expensive part of CCS isn't likely to be the storage but the capture and compression of CO_2 – although these costs could drop by 20–30% by the mid-2010s.

There are two big pieces to the CCS puzzle that are only now starting to align. One is the industrial-scale testing of sequestration itself. To date, there have been only a few such projects, including one in the North Sea, where millions of tonnes of CO_2 have been injected into an undersea aquifer from a natural-gas drilling platform. The Barents Sea is home base for a similar project. In North America, CO_2 from a synfuel plant in North Dakota is being piped to Saskatchewan and used to enhance recovery at an ageing oil field. Germany, in 2008, unveiled the first coal plant to use sequestration. Meanwhile, the UK plans to spend more than £1 billion on CCS demonstration projects during the 2010s. These are likely to involve at least two electricity plants, one of them coal-based and the other fired by natural gas. The latter test would be especially significant given Britain's shift toward generating power from gas rather than coal.

The other puzzle piece is to develop the technologies to reliably and inexpensively separate out the CO_2 within power plants, ready for burial. This can be done in various ways. The **post-combustion** approach uses chemical processes to remove the CO_2 from the exhaust stream of a regular power plant. **Pre-combustion** CCS, by contrast, involves using a plant with gasification capability to turn the fuel into a mixture of CO_2 and hydrogen before any burning takes place. The CO_2 is then separated off and the hydrogen is used as the carbon-free energy source. Finally, the **oxyfuel** approach uses special equipment to extract large quantities of pure oxygen from the air. The fuel is then burned in the pure oxygen (rather than the usual mixture of oxygen and nitrogen). That means the exhaust contains only CO_2 and steam, which can be easily separated.

Activists block the construction of a new coal-fired power plant in Rotterdam, 2008.

Making sense of power

As you read through this chapter, you may find the following rough comparisons useful. Watts and their derivative terms (kilowatts, megawatts) are measures of the power being delivered or consumed; a watt is one joule of energy per second.

▶ **100 watts** = the power used by a strong incandescent (non-energy-saving) light bulb

▶ **1000 watts** = **one kilowatt** = close to the average electricity demand of a typical household in the developed world (the actual demand goes up and down depending on the time of day)

▶ **1,000,000 watts** = **one megawatt** = the amount of power delivered by a medium-sized commercial wind turbine (some of the newest ones provide 6MW or more) = a thousand households' worth of electricity

▶ **1,000,000,000 watts** = **one gigawatt** = the amount of power generated by a large coal-fired power plant = a million households' worth of electricity

Kilowatt-hours (kWh) and similar terms describe the amount of energy delivered or consumed over a period of time. The world's total electricity usage in 2007, including both residential and commercial needs, was close to 19 trillion kWh. If consumed at a steady rate, that would represent about 2200 gigawatts of power at any one time.

Each of these approaches has its pros and cons. Post-combustion and oxyfuel are attractive principally because, at least in theory, they could be "retrofitted" to existing power plants – which could be crucial considering the long operating lifespan of the world's existing and planned power stations. A small-scale oxyfuel plant is already up and running in Germany, and many other trials of both these technologies are being developed around the world.

Pre-combustion CCS, on the other hand, involves using specialist modern power stations, such as those based on **integrated gasification and combined cycle** (IGCC) technology, which was first explored in the 1980s. IGCC plants employ gasification but also use the waste heat to make steam, which drives a second turbine to increase efficiency. As of 2010 there were a handful of such plants in operation across the world, mainly in Europe, the US and China, but these numbers should increase substantially through the decade.

Although it would be a fairly simple matter to capture the CO_2 from an IGCC plant, it's taken a long time for the first such facility to emerge. The trailblazer is China's GreenGen plant, set to ramp up between 2011 and

2014. GreenGen aims to hit 55–60% efficiency (about double that of today's typical coal plant) and to sequester more than 80% of its emissions. The US had originally planned to bring a similar plant online in 2011, but its FutureGen project was cancelled in 2008. However, one of the sites considered for FutureGen is now playing host to the Texas Clean Energy Project, an IGCC–CCS plant set to go online in 2014 that aims to keep 90% of its CO_2 emissions out of the atmosphere.

At this point, it's hard to tell how quickly the various CCS technologies will develop. Some experts doubt that any of them can be refined and deployed globally soon enough to avoid major climatic trouble. After all, even if we did manage to store 90% of *all* fossil-fuel plant emissions underground, the remaining 10% going into the atmosphere is no small matter. Still, the coal industry is likely to face intensified pressure to step up its CCS efforts. James Hansen, the eminent NASA scientist who helped trigger the first wave of climate-change awareness in 1988, is among many who see coal at the heart of the greenhouse threat. Hansen caused a stir in 2007 when he called for a moratorium on all new coal plants that do not include functioning CCS systems.

Beyond fossil fuels: renewables and nuclear

Cleaning up fossil-fuel-based electricity sources is all well and good, but it's equally important to consider the alternatives. These fall into two camps: **renewables**, such as wind, solar, hydro, tidal, geothermal and biomass, and **nuclear**. The degree to which each of these sources can meet our demand for low-carbon energy is among the most important and fiercely debated issues in climate-change policy. It's an immensely complex area – an intimidating mix of economics, technology and politics.

For one thing, it's not just the *amount* of energy each source can generate that counts. Equally important are various other factors, including:

▶ **Generating time** Demand for electricity varies throughout the day and year, and storage is difficult, which can be a problem for solar, for example.

▶ **Reliability** Wind power, for instance, has enormous potential but needs to be backed up by other sources during calm periods.

▶ **Public safety** Nuclear being a case in point.

▶ **Roll-out time** In an era of proliferating coal power, every year wasted is bad news for the climate.

As for **cost**, this is complicated by subsidies and tax regimes and the fact that each source becomes cheaper and more efficient as it becomes more widespread. Things are further clouded by the lack of definitive statistics on the total greenhouse emissions or maximum potential contributions of each energy source. As UK commentator Simon Jenkins put it, "Nobody agrees about figures ... energy policy is like Victorian medicine, at the mercy of quack remedies and snake-oil salesmen." To scale up renewables in a serious way, major new infrastructure would be critical in order to transmit and/or store the vast amount of energy generated at different locations and times. And there's also the fact that some amount of fossil fuel will be needed to build, deploy and maintain renewables such as wind and solar on the mammoth scales needed.

Despite these complexities and challenges, it appears that an intersecting web of renewables could go a long way toward meeting the world's total energy needs, especially if cars go electric in a big way. One of the most detailed and optimistic analyses of how the UK and the planet as a whole might go carbon neutral was produced by energy guru David MacKay (University of Cambridge) in his 2009 book *Sustainable Energy – Without the Hot Air*. The task won't be simple, MacKay says: "To complete a plan that adds up, we must rely on one or more forms of solar power. Or use nuclear power. Or both."

With so much to consider, a full assessment of non-fossil energy is beyond the scope of this book, but following is a brief description of the pros and cons of each major source.

Solar power

Nearly all our energy ultimately comes from the Sun. Wind, for example, is kicked up by the temperature contrasts that result when sunlight heats Earth unevenly. And you can't have biofuels (or even fossil fuels) without the sunshine that makes plants grow. However, when people refer to **solar power**, they're usually talking about devices that convert sunlight directly into electricity or heat.

Solar energy took off in the oil-shocked 1970s, but its momentum sagged as the limits of that era's solar technology became clear and as oil prices dropped. Today, with the risks of climate change looming, the Sun is rising once more on solar power. As of 2009, more than 20 gigawatts of grid-based electricity was being derived from solar systems. That's a five-fold increase from 2005, with no end in sight to the growth.

Overall, solar power currently costs several times more than electricity from a standard coal-fired power plant, mainly because of the expensive raw materials that go into solar cells and concentrating systems (see below). In some locations, though, tax structures and other factors make solar power roughly competitive with other sources. And for people who want to go off the grid – or who have no grid to draw from – solar power ranks near the top of the list. The costs are likely to go down as solar power continues to grow.

Some solar technologies have been used for centuries, such as orienting windows towards the south (in the Northern Hemisphere) to allow warming sunlight in. These and other **passive** solar techniques are now commonly used in home and office construction. The higher-tech, "active" approaches fall into two camps:

▶ **Solar cells (also known as photovoltaic or PV systems)** use sunlight to split off electrons from atoms and thus generate electricity. A typical PV cell is about 10cm x 10cm (4in x 4in). Dozens or hundreds of these are combined into each solar "panel", and panels can be added together to make an "array". In small numbers, PV cells are ideal for small-scale needs such as patio lights. With a small array and a sunny climate, you can power an entire household; a large array can serve as a power plant. Despite its northerly latitude, Canada boasted the world's largest PV plant as of late 2010, with over a million panels providing a capacity of 80 megawatts. Europe has also embraced PV power in a big way: a 60MW plant opened in 2008 in Spain, and Germany now has at least five plants capable of producing at least 40MW.

▶ **Solar thermal systems** use sunlight to create heat rather than converting it directly into electricity. On a small scale, this usually involves using flat, roof-mounted "collector" panels to provide hot water in homes or commercial buildings. Tubes that carry water (or, in some cold climates, antifreeze) are sandwiched between a glass top and a back plate, which is painted black to maximize heat absorption. The fluid heats up as it flows through the collector and then proceeds to the hot-water tank. The whole process is typically driven by a pump, though there are also passive systems in which the heated water produces its own circulation.

▶ **On a much larger scale, concentrating solar power plants** use massive arrays of mirrors to direct huge quantities of sunlight onto tubes filled with a fluid, which becomes extremely hot and can be used to drive a steam turbine or Stirling engine. (Mirrors can similarly be used to increase the efficiency of PV panels.) The world's largest solar facilities draw on concentrating systems, including the top two, both in California: the vast 354MW Solar Energy Generating Systems complex in California, and the even larger Ivanpah facility (370MW) being built in the early 2010s.

Solar power is both limited and underused – limited because of the fixed distribution of sunlight around the world and around the clock, and underused because far more buildings could benefit from strategically placed solar systems than now have them. The developing world is a particular point of interest, because even a small PV array can power an entire household if the demand for electricity is modest (see p.371). At the other extreme, the world's most advanced showcases for energy efficiency make good use of passive and active solar systems.

The world's subtropical deserts are potential solar gold mines. David MacKay estimates that a square in the Sahara 1000km (600 miles) on each side, covered with solar concentrating systems, would produce as much power as the world's present consumption of all types of energy, including oil and coal. According to a 2010 IEA study, government and market incentives could enable solar power to provide 22% of global electricity supply by the year 2050. And a consortium of European

An artist's impression of BrightSource Energy's 370MW Ivanpah Solar Electric Generating System in California's Mojave Desert.

companies is already backing a plan called DESERTEC that would develop huge solar power installations in North Africa and the Middle East to serve those regions plus Europe. The main challenge is storing and transporting the resulting energy. This can be done using molten salt tanks for storage and long-distance high-voltage power lines for transmission, though the upfront costs are high – especially when the transmission distances are large. Hydrogen may also help with large-scale solar storage and distribution at some stage, but there are hurdles to overcome (see p.352).

Wind power

Although it's growing at a somewhat less rapid pace than solar power, **wind power** represents a bigger slice of the planet's current renewable energy profile. As of 2009, wind accounted for about 60GW of electric capacity, a doubling in just three years. Factoring in the variability of wind speeds, this represents about 2% of the electricity generated globally. About half of this capacity is in Europe: wind power is especially popular in Germany, and it's grown quickly in Spain as well. After lagging other parts of the world for years, the US now produces more wind power than any other nation, about 25% of the global total. The growth potential for American wind power is impressive, especially across the perpetually windy Great Plains from Texas to North Dakota and on to Saskatchewan in Canada. It's been estimated that the US could conceivably generate up to eleven trillion kilowatt-hours of wind power per year, about three times the total now generated by all US power plants. The UK also has truly massive potential for wind energy – especially when it comes to offshore wind farms. According to the *Offshore Valuation* report by the Public Interest Research Centre, British offshore wind, bolstered somewhat by tidal and wave energy, could produce the electricity equivalent of a billion barrels of oil each year.

Once in place, wind turbines produce no greenhouse gases at all – the only petrochemicals involved are the small amounts used to manufacture and lubricate the units. Another bonus is that the giant wind turbines are typically centred at heights of 80m (260ft) or more, leaving the land below free for other uses. Many farmers and ranchers rake in substantial royalties for allowing turbines to be placed atop fields where crops grow and cattle graze.

Based on all this, wind power seems like a no-brainer, yet wind farms have come up against local opposition, especially in the UK and US, where some activists have fought against them with a vehemence approaching that of the anti-nuclear movement that peaked in the 1970s and 80s. Mostly, the opposition isn't based on ecological or economic concerns but on aesthetics (some Brits have referred to wind turbines as "lavatory brushes in the sky") and noise (the distinctive whoosh-whoosh of the blades and the creaking that sometimes results when the blades are moved to take advantage of a wind shift). Another allegation is that wind farms are a danger to bird life.

These are all contested claims. Many people enjoy the motion and symbolic cleanliness of the structures, and wind-power proponents claim that you can hold a normal conversation while standing right underneath a modern turbine. A low-frequency hum may disturb some people within a kilometre or two, depending on the level of other ambient noise, but if wind farms are properly sited and built, these effects can be minimized. As for bird deaths, it's hard to measure accurately, but a 2005 report by the UK's Royal Society for the Protection of Birds claims that, as long as the farms are positioned appropriately, they "do not pose a significant hazard for birds". Adding fuel to the argument is the suggestion that some "local" opposition has been orchestrated behind the scenes by those with vested interests in other energy sources. In fact, for each unit of energy generated, fossil-fuel plants appear to kill many more birds than wind farms do – and few would consider them any more appealing to look at.

Placing wind farms offshore is one way to minimize objections. Denmark has built several major installations at sea, and 2010 saw the opening of the world's biggest offshore wind farm at Thanet, off the coast of Kent in the UK. The London Array, not far from Thanet, is due to be even bigger. However, going offshore adds substantially to the upfront cost – and it isn't always enough to placate local protesters. People living near Nantucket Sound in New England fought plans for the first US offshore wind farm for nearly a decade before the facility was finally approved in 2010. The Cape Wind project will include 130 turbines, each extending about 80m (260ft) from sea surface to turbine centre.

Outside of the local controversies, there is strong overall public support for wind power – typically running around 80% in UK and US polls, with young people the most supportive – so at least for now, the tugs-of-war at specific sites appear unlikely to throw wind power off its course of vigorous expansion. In countries with lots of wind potential, a bigger long-term challenge could be intermittency. Despite myth-making by some protesters, dealing with the inconsistency of wind power isn't generally a big problem when only a small proportion of a country's total electricity is produced from wind turbines. But above a certain threshold,

The UK's Scroby Sands wind farm includes 30 offshore wind turbines in the North Sea off the coast of Great Yarmouth.

intermittency can become a constraint. Hence the huge amount of energy currently focused on so-called **smart grid** technologies designed to more dynamically deal with electricity supply and demand. In particular, many commentators are excited by the prospect of the batteries in millions of parked electric cars (see p.376) becoming a crucial buffer for coping with increasing amounts of intermittent wind power.

Geothermal energy

The same forces that can blow the top off a volcanic mountainside can also generate electricity. Most **geothermal power plants** tap into the hot water produced where rain and snowmelt percolate underground to form pools that are heated by adjacent magma and hot rocks. Some facilities draw hot water and steam from more than a mile below ground, while others harvest steam that emerges directly at the surface. Either way, a small amount of naturally produced carbon dioxide escapes into the air – but it's less than 10% of the amount emitted by a standard coal-fired power plant of the same capacity. **Binary geothermal** plants avoid even these minimal emissions of CO_2 by using water from underground to heat pipes that carry a separate fluid; the geothermal water returns to Earth without ever being exposed to the atmosphere.

Iceland's low-carbon lifestyle

Thanks largely to its fortuitous location atop a geological hot spot, Iceland may point the way towards a cooler and cleaner future for the rest of the world. This island nation of about 300,000 people draws on a unique portfolio of energy that's already low in carbon emissions. By 2050 Iceland could be the world's first hydrogen-based, carbon-neutral economy.

The two continental plates that play host to North America and Europe are separating beneath Iceland, which leads to the world's most concentrated zone of **geothermal** energy. The very name of the capital city, Reykjavik, derives from *reyka* – the plumes of steam visible from many points around the country. Steam and hot water from underground furnishes heat for some 90% of Iceland's buildings. Although depletion of geothermal energy is a concern in some other parts of the world, it's less so here: the magma, rocks and steam beneath Iceland are estimated to hold many centuries' worth of energy, even without factoring in the potential for their recharge through future volcanism. Although some of Iceland's electricity comes from geothermal power plants, most of it is produced by **hydroelectric** plants. They're a reliable source of power in the island's wet climate, but also a source of friction from environmentalists who want to limit the intrusion of dams and reservoirs into wilderness areas.

There's far more potential for geothermal power in Iceland than the country can use, but at the same time the nation still imports oil for its cars and ships. These intertwining factors make Iceland a perfect test bed for **hydrogen,** as evidenced by a long-term project being carried out by a public–private partnership that includes DaimlerChrysler, Shell and NorskHydro. The first step was a mini-fleet of three fuel-cell-powered municipal buses in Reykjavik, which drew power from the world's first public hydrogen filling station. Passenger vehicles were tested beginning in 2007, though progress was slowed by the nation's sharp economic downturn. Still, by mid-century, Iceland hopes to be exporting hydrogen to the rest of Europe.

About 10GW of electric capacity was available from geothermal sources as of 2010. Geothermal plants can run close to their capacity almost 24/7, since their power source doesn't vary like sunshine and winds do. The best sites for geothermal plants are where continental plates separate or grind against each other, especially around the "ring of fire" that surrounds the Pacific Ocean. Around a third of global capacity is located in the US, where the world's biggest plant – The Geysers, in northern California – provides more than 700MW of capacity. In Kenya and the Philippines, geothermal sources provide about 20% of all electricity. They're also part of the palette of carbon-free energy in Iceland (see box above), where geothermal energy provides direct heating as well as electricity.

A completely different approach to geothermal that's recently been attracting interest and investment is known as **enhanced geothermal**

systems (EGS). In an EGS plant, dry hot rocks deep under the surface are fractured artificially using explosives, and water is piped down from the surface. This approach could, its advocates claim, open up vast new areas for geothermal exploitation, as it removes the requirement for particular under-surface water systems. There is already a plant up and running in Landau, Germany, with huge potential for similar projects in the US, Turkey, Australia and various other countries.

On a much smaller scale, geothermal energy can also be employed in buildings through the use of heat pumps (see p.358).

Hydroelectric, tidal and wave power
Hydropower plants use dams to channel water through electricity-generating turbines. This approach currently provides far more energy than any other renewable source – more than 850GW of electric-generating capacity – while generally producing minimal greenhouse emissions (although in the tropics, vegetation decomposing beneath a reservoir can be a substantial emissions source). In theory, hydropower capacity could be scaled up several-fold, especially in Africa and Asia, but actual growth is expected to be much more modest. One reason is the high start-up costs: many of the more easy-to-engineer sites have already been used up. Another is that the construction of dams can wreak havoc with river-based ecosystems far upstream and downstream. There are also social issues, as large dams can flood huge areas and displace many residents (China's Three Gorges Dam forced more than a million people to resettle).

Perhaps more promising for the long term are other ways to generate clean electricity from the movement of water. Turbines driven by the **tide** and **waves** are both being developed, but various technological and economic barriers need to be overcome before either source could meet a significant proportion of the world's energy demands. Tidal-stream power shows particular promise in certain countries (including the UK and Canada, for example), as its timing is predictable and it could be harnessed offshore and out of sight using relatively simple technology – the equivalent of underwater wind turbines. The main challenge is building turbines that can withstand the enormous forces of fast tidal streams, although first-generation devices, such as the Seamen turbine installed in the Strangford Lough off Northern Ireland, are already operational.

Biomass
While the practice of using plant-derived fuels as a substitute for oil – as with ethanol – has engendered both enthusiasm and controversy (see

pp.351–6), there's been comparatively little buzz about using biomass as a source of electricity, where it can displace carbon-belching coal. The world currently derives about 40GW of electricity from biomass. About 25% of that is generated in the US, where the biomass sources range from pulp and paper to landfill gas and even solid waste from municipal sewage systems. Though biomass is often burned directly, many of the same practices used to make coal plants more efficient, such as gasification via IGCC technology, can be applied to yield greater efficiencies. And since the root sources of biomass pull carbon dioxide from the air as they grow, there's the tantalizing possibility that a biomass power plant using IGCC and sequestering its carbon might actually exert a net reduction on atmospheric CO_2.

Nuclear power

Just when it looked as if it might keel over from the weight of negative public opinion and high start-up costs, nuclear power has got a new lease of life thanks to climate change. It's hard to dismiss a proven technology that produces massive amounts of power with hardly any carbon emissions. Yet it's also impossible to ignore the political, economic and environmental concerns that made nuclear power so contentious in the first place.

Though it's taken a somewhat lower profile since the 1990s, nuclear power never went away. The world's 400-plus nuclear plants provide close to 15% of electricity globally, and a much higher fraction than that in some countries – particularly France, which embraced nuclear power in the 1950s and never looked back. France now gets around 80% of its electricity from more than 50 nuclear plants operated through the state-owned Electricité de France and a private partner, Areva. The US has even more nuclear plants – just over 100 – with construction under way in 2010 on the nation's first reactor since 1996. And China has more than 170 nuclear plants in the construction or planning stages, which together would make it the world's leader in nuclear energy.

Global warming has already changed the nuclear equation in tangible ways. In 2003, Finland's legislature voted narrowly in favour of building its fifth nuclear reactor – one of the world's largest, to be completed in 2012 after extensive delays and huge budget overruns – partly because of the hope that it might help Finland meet its Kyoto goals. In 2006, France and the US agreed to give nuclear technology to India, which as of 2010 had 19 reactors online and more than 60 on order. Also in 2006, Tony Blair announced that he'd "changed his mind" on nuclear, paving the way for likely approval of a new generation of British atomic-energy plants, with ten such plants proposed or planned as of 2010.

Though a certain amount of CO_2 is released in the process of building a nuclear plant (plus mining and processing its fuel), there's no question that nuclear power is far kinder to the climate than fossil fuels. That simple fact has engendered some powerful crosscurrents. Environmentalists who oppose an expansion of nuclear power run the risk of appearing callous about climate change. Thus, the anti-nuclear arguments have shifted somewhat in recent years. The unsolved problem of disposing of nuclear **waste** remains a key point, as does the risk of **accidents** like Chernobyl in 1986. And recent heat waves have underscored how nuclear plants themselves can be vulnerable to a warming climate. Plants were closed in Germany in 2003, Michigan and Spain in 2006 and Tennessee in 2007, in each case because lake or river water had warmed too much to cool down the reactors. On the other hand, industry advocates claim that the risks of accidents are overstated, pointing out that Chernobyl was a low-budget, Soviet-era construction quite unlike modern reactors, which have never resulted in a single death. It's also true that coal, oil and gas production lead to many deaths each year, and their waste products – especially greenhouse gases – also pose unsolved risks to ecosystems and society (not to mention accidents like BP's massive 2010 oil spill in the Gulf of Mexico).

Nuclear power opponents have also turned to **economic arguments**, claiming that the massive costs of building and decommissioning nuclear plants – which are typically subsidized and insured in whole or in part by governments – drain money that's urgently needed to fund research on renewables and energy efficiency. According to Amory Lovins of the Rocky Mountain Institute, nuclear's "higher cost … per unit of net CO_2 displaced, means that every dollar invested in nuclear expansion will worsen climate change by buying less solution per dollar." Terrorism also figures in the mix: the more nuclear plants we have, say opponents, the more risk there is of radioactive material slipping into the wrong hands, or of an attack on a power plant that could kill many thousands. And uranium

> "The only technology ready to fill the gap and stop the carbon dioxide loading of the atmosphere is nuclear power."
>
> Stewart Brand, "Environmental Heresies", *Technology Review*

> "Nuclear power has already died of an incurable attack of market forces, with no credible prospect of revival."
>
> Amory Lovins, Rocky Mountain Institute

supplies are in question beyond the next few decades, unless the world turns to breeder reactors, which pose their own set of problems. (One option would be to expand the use of thorium, which is more abundant, less radioactive and more resistant to proliferation.)

Some influential figures have spoken out in favour of increasing the world's nuclear power capacity in the light of the climate-change threat. These include prominent greens such as Gaia theorist James Lovelock and Stewart Brand, founder of the *Whole Earth Catalog*. Along with other advocates, they argue that nuclear is the *only* low-carbon energy source that's sufficiently developed and reliable that it could be rolled out on a big enough scale to take a meaningful chunk out of the world's green-house emissions in the short and medium term. In particular, nuclear plants could replace coal in providing a 24/7 power supply to pair with time-varying sources such as wind and solar power.

A blue-riband panel assembled in 2003 by the Massachusetts Institute of Technology gave ammunition to both sides of the nuclear debate. The report recommends keeping open the option of a tripling of current global nuclear capacity by 2050 – which alone could reduce greenhouse emissions by a massive 25% – but it also cites several major obstacles, with the risk of nuclear-weapon proliferation perhaps the most daunting. "Nuclear power should not expand", the report warns, "unless the risk of proliferation from operation of the commercial nuclear fuel cycle is made acceptably small." Whether the risk of proliferation can be compared to the risk of climate change is a debate that's bound to run and run.

Making transport greener

Transporting people and goods by road, sea and air is responsible for around a quarter of global greenhouse emissions, and this percentage is rising. So if we're to tackle global warming, we'll have to develop lower-emissions vehicles – not to mention promoting public transport and, some would argue, reducing the flow of people and goods around the world.

The easiest gains could come from simply improving the **efficiency** of conventional vehicles, something that is already happening but which could go much further, especially in the US, where the average vehicle achieves roughly half the efficiency of those in Europe and Japan. One positive development is the growing popularity of cars with highly effi-cient **hybrid** engines (see p.376), which pair a standard engine with an electric one and a battery and can manage as much as 70 miles per gallon.

Ironically, the hybrid family also includes hulking SUVs and executive cars, which improve upon their non-hybrid counterparts but still require plenty of fuel per mile. For a car to be truly efficient, it also has to be light-weight. Plug-in hybrids and fully electric cars, which draw power from the grid or from renewable home energy systems, provide even greater efficiency, and an array of new models began coming onto the scene early in the 2010s (see p.377).

Within a decade or so, some new cars, trucks and even planes may be powered by energy stored in **hydrogen** (see box overleaf). In the mean time, engineers are refining ways to power vehicles through **biofuels** – such as ethanol and biodiesel – which are made from plants. The pri-mary attraction of biofuels is simple: though they release plenty of carbon when burned, much of this will be soaked up by the next round of fuel crops (after all, plants grow by absorbing CO_2 from the atmosphere). How climate-friendly a biofuel is, then, depends on factors such as the amount of fossil-based energy used to fertilize, harvest, process and distribute the crops.

In recent years, many environmentalists have soured on biofuels as a tactic for fighting climate change. Perhaps the most contentious issue is land. It would take *huge* areas of cropland to turn biofuels into a major part of the global energy picture. This could squeeze food production or require the conversion of virgin land into farms, something which has implications not just for wildlife habitat but for the greenhouse effect itself, depending on where these land changes occur. If forests are chopped down to make room for biofuel crops, the shift could actually produce a net rise in carbon emissions. Advocates of James Lovelock's Gaia theory (see p.226) see further changes in land use as one of the great-est threats to Earth's ability to maintain a stable temperature.

The food-versus-fuel issue rose to prominence in 2007–08, when world food prices spiked dramatically just as ethanol production was gaining momentum in the US and, on a smaller scale, in Europe. Corn prices tripled in just three years, and many farmers who grew other crops began shifting to corn. Food riots rocked Mexico, Haiti, Bangladesh and other emerging nations as the price of tortillas and other staples jumped. However, analysts disagreed on how much of the price spike was driven by biofuels as opposed to other factors, including market speculation and high prices for oil (which goes into making fertilizer and other agri-cultural components). Whatever the reasons, as of 2010 the commodity prices for corn – as well as for other key crops, including wheat, soybeans and rice – were still running substantially higher than before the 2007–08

How much will hydrogen help?

Hydrogen, the simplest and most abundant element in the universe, has been hyped as a solution to the greenhouse crisis in recent years, especially in the US. In truth, hydrogen is no panacea for global warming. Indeed, hydrogen isn't even an energy *source*, as we have to use power from other sources to produce the pure hydrogen gas that's useful as a fuel. Hydrogen does hold potential as a clean and portable energy *carrier* – but with an array of long-term challenges to its use, it's looking increasingly likely that electric cars will leapfrog over hydrogen vehicles in the race for climate-friendly motoring.

On our planet, hydrogen is usually bound to other molecules, as in water (H_2O). But in its pure form, H_2, hydrogen is a lightweight gas (or a liquid at extremely low temperatures) that contains a great deal of energy. Hydrogen gas can be burned directly as a fuel, but the preferred means of exploiting it is through fuel cells, which convert hydrogen into electricity. This process produces zero emissions of carbon dioxide and other pollutants – the only by-product is steam, and though water vapour is a greenhouse gas, the net amount emitted is insignificant.

A few years ago, the potential for zero-emissions vehicles led to a good deal of buzz and some big-time investment, including the EU's European Hydrogen and Fuel Cell Technology Platform, which runs through 2015. In the US, the Bush administration launched a programme in 2004 aimed at bringing hydrogen-fuelled vehicles into the marketplace. And Honda began rolling out its first mass-market hydrogen vehicle, the FCX Clarity, in 2008, albeit in small numbers – only 200 in the first three years.

The enthusiasm for hydrogen cars has been waning, though, as various problems have proven harder than expected to crack. There's much more to hydrogen motoring than the vehicle itself. Before it can provide energy to fuel cells, hydrogen first has to be broken off from other molecules, stored

Honda FCX Clarity: a hydrogen-fuelled first.

(by compression or liquefication) and distributed. Not only is each of these steps a daunting technical challenge in itself, but each one also takes energy. One Swiss study found that hauling a hydrogen-filled truck 500km (310 miles) would consume an amount of energy equal to almost half of the cargo. On top of that, the hydrogen itself is substantially more costly per unit of energy than petrol. Such issues are partly why the Obama administration cancelled Bush's hydrogen-car initiative in 2009.

Hydrogen could conceivably be produced in a home unit, but even that technology (which is far from ready) would signal a continued reliance on fossil fuels. Critics such as long-time researcher Joseph Romm, author of *The Hype about Hydrogen*, argue that the natural gas or renewables used to produce hydrogen would make a bigger and more immediate dent in the global greenhouse picture if they were used to replace coal-fired power.

spike. In addition, the EU was coming under increasing fire from environmental groups for its goal that all liquid fuels in Europe be composed of 10% biofuel by 2020.

In response to these pressures, there's a growing push for **second generation biofuels**, which are derived from cellulose (see below) and other biological materials outside the food stream. In addition, hundreds of facilities in Europe and a few in the US are using animal waste to generate natural gas – or, in this case, **biogas**. Bacteria feed on the animal waste and generate methane that can be burned directly or used to generate electricity. (In theory, human waste could be used in exactly the same way, though it's hard to envisage the idea taking off.) Even **algae** can be used as biofuel – and once the economics are sorted out, that could become the most useful biofuel of all, thanks to algae's ultra-speedy growth habits.

Ethanol

Ethanol (ethyl alcohol) is a plant-based substitute for petrol that's already quite common in some parts of the world. The total global supply in 2008 stood at around 64 billion litres (17 billion US gallons), of which nearly half was produced in the US and more than a third in Brazil. Across the States, ethanol is sold in a standard mixture of 10% ethanol and 90% petrol (E10). If people want to use higher proportions of ethanol, such as 85/15 (E85), they need a flexible fuel vehicle (FFV) that can handle any blend of the two. More than 20 million FFV vehicles have been sold, mostly in the US and Brazil, though many American FFVs never see a drop of biofuel. Ethanol made from sugar cane provides more than 25% of Brazil's vehicular fuel (all petrol there is at least 20% to 25% ethanol) and the country also exports to Asia and elsewhere. A plus for ethanol use in Brazil is that its mild climate enables motorists to avoid the cold-start problems that can otherwise occur with high-ethanol blends.

The US produces more than a third of the world's corn, and almost half of US corn stocks were going to make ethanol as of 2010. Even though the US is the world's leading ethanol producer, its output is only about 8% of the fuel swallowed each year by the nation's cars and trucks. That percentage has doubled over the last few years due to tax breaks and incentives, but further growth is difficult because the 10/90 ratio for standard vehicles is enshrined in federal law. As a result, the US has begun exporting low-cost ethanol to Brazil and even to oil-producing nations such as Nigeria and the UAE.

There's been much debate over whether ethanol really helps protect the climate. Most US ethanol to date has been derived from corn, which is grown through farming that makes fairly extensive use of petrochemicals. All in all, corn-based ethanol seems to provide some climate benefit, but not much. The University of California at Berkeley estimates that corn-based ethanol uses about 74 units of fossil-fuel energy to produce 100 units of ethanol energy. That's somewhat like a charity spending £74 to raise £100. As for emissions, a vehicle running on ethanol does produce greenhouse gases, but when you factor out the CO_2 absorbed by the plants it's made from, corn-based ethanol blends add up to roughly 30% less carbon to the atmosphere than standard petrol does. A variety of factors can affect these carbon ratios, including the wetness or dryness of the source crops and the choice of fuels used for distilling and drying the biomass at the ethanol plant.

A more promising biofuel is on the horizon: **cellulosic ethanol**. Standard ethanol comes from fermenting the sweet starches in corn kernels and sugar, but ethanol can also be brewed from the sturdy material called cellulose (as in celery). Cellulose is found in corn husks as well as a variety of other plants, such as fast-growing grasses, and even in human and animal waste. Since many plants high in cellulose can thrive on land that's marginal for regular farming, they're less likely to displace food crops, and the yield per acre can be twice or more that of corn. Deep-rooted perennials like switchgrass also help stabilize soil and need little if any help from petroleum-based farming techniques. As a result, 100 units of cellulosic ethanol require only about 20 units of fossil energy to produce, according to the US Department of Energy.

In time, cellulosic ethanol could end up being carbon neutral or even a net remover of carbon from the atmosphere, according to the Rocky Mountain Institute. The main catch is that cellulosic materials don't break down very easily, so the cost of producing cellulosic ethanol is still about double that of the corn-based variety. Chemists have been working on methods to bioengineer enzymes that could be produced en masse to break down the cellulosic materials more affordably. That might level the playing field with corn-based ethanol over the next decade, although the economic downturn has slowed the development of cellulosic alternatives somewhat.

Biodiesel

Diesel cars – which are very popular in Europe, though far less common in the US – are typically more climate-friendly than the petrol-based alternatives, since they burn their petroleum-based fuel about 40% more efficiently. (The flip side is higher emissions of nitrogen oxides and particulates, though in both the US and Europe these pollutants are being reduced by increasingly tighter emission standards.) Theoretically, diesel cars can be greener still when powered by **biodiesel** made from sustainably grown crops. This isn't a new concept. The term "diesel" originally described the engine type rather than the fuel, and its inventor, Rudolph Diesel, used peanut oil to drive the prototype that won him top honours at the 1900 World's Fair in Paris.

Biodiesel has received a boost from tax breaks and government mandates on both sides of the Atlantic, and in 2008 there were about 15 billion litres (4 billion US gallons) produced. As with ethanol, biodiesel is often added to fossil fuel in small percentages and sold as a way of helping engines run more smoothly while emitting fewer pollutants. But you can also buy 100% biodiesel in some parts of the UK and US. Country singer Willie Nelson even peddles his own blend, called BioWillie, across several Southern states.

At its most climate friendly, biodiesel can be a convenient product of recycling – as in the UK, where it's often derived from used vegetable oil donated by chip shops and other commercial kitchens. In the US and mainland Europe, however, it's produced commercially from canola, soybean and other inexpensive vegetable oils. This allows for a far greater level of supply, but also raises questions about the wider impacts of the burgeoning international market for oil crops. Huge areas of tropical rainforest have already been cleared to make way for soybean- and palm-oil cultivation, with disastrous impacts on both biodiversity and the climate. For this reason, biodiesel from crops (rather than from waste products) is just as controversial as ethanol, with some environmentalists seeing it as a problem rather than a solution.

Making buildings more efficient

Buildings and the appliances they contain are responsible for around 20% of all global energy consumption, according to the US Energy Information Administration. So no matter what energy source we use to power, heat and cool them, it makes sense to do so as efficiently as possible. The main challenge here isn't developing new technologies – there are plenty to choose from already, with more on the way – but getting consumers, manufacturers, architects and governments to choose (or require) efficiency.

The role of buildings as contributors to the greenhouse effect is largely a legacy of the way offices and homes have been designed over the last few decades. With fuel for heating and cooling assumed to be cheap and plentiful, aesthetics and short-term costs have often trumped energy efficiency and long-term savings. "Commercial and public buildings in the developed world have generally become sealed, artificially lit containers", says architect Richard Rogers of the UK design firm Arup. "The increasingly evident threat to the global environment posed by buildings of this sort cannot be ignored."

Efficiency will be key if the UK is to stand any chance of meeting its overall goal of a 80% emissions cut by 2050. A report by the UK's Environmental Change Institute (ECI) suggested that a 60% reduction in emissions from houses *is* possible, even with an expected 33% increase in the number of households by 2050. Among other measures, this would require:

Low-power towers

Skyscrapers have always been more than mere office buildings. The best rank among the world's greatest architecture, and they often serve as icons for the companies they house. Now a new breed of eco-oriented towers is going up, aiming to impress with their low-carbon credentials as well as with their high-rise aesthetics.

Swiss Re's award-winning, pickle-shaped 2004 tower, located at 30 St Mary Axe in London's financial district, was dubbed the **"Gherkin"** years before it was even built. The design can be seen as an elongated riff on Buckminster Fuller's geodesic dome, with the curvature provided by gentle turns in the frame. Love it or hate it, the building's provocative shape has a point. Wind flows easily around it, reducing the amount of support structure needed. Less obvious is a latticework of tunnels through which air can flow into and through the Gherkin's inner reaches, cutting heating and cooling costs by up to half. The tunnels also bring natural light to the building's interior. Also in London, the 43-storey **Strata** apartment building made history in 2010 as the world's first skyscraper to incorporate wind turbines (a row of three is mounted on top).

In New York there's **Four Times Square**, also known as the Condé Nast Building. The 48-storey structure, completed in 2000, was the first major US skyscraper built with green credentials in mind. The building draws on a wide array of technologies, including fuel cells that kick in at night, state-of-the-art windows, plentiful solar panels and a ventilation system that pulls in 50% more fresh air than comparable towers. New York also hosts the 58-storey **Bank of America Building** (see photo opposite), the city's second tallest building. Completed in 2009, it's been billed as the world's most eco-friendly skyscraper. Its sides are subtly canted to provide some of the streamlining effects of the Gherkin. Windows reflect ultraviolet light while letting in visible rays. The main power source is a 4.6-megawatt co-generator drawing on natural gas, while a heat pump helps keep the building cooler in summer and warmer in winter.

▶ **Fully insulating wall spaces and lofts and installing high-performance windows** in both existing and newly built homes.

▶ **Demolishing about 14% of current houses by 2050** (four times the recent rate) in a targeted strategy to eliminate the worst energy offenders.

▶ **Employing low- or no-carbon technologies** to heat most homes, including heat pumps, solar cells and solar hot water, combined heat and power and wind turbines.

Progress is slow, however. Overall, home electricity consumption in the UK *increased* about 1% a year in the decade after 2000. One factor was the profusion of electricity-guzzling equipment, from computers to big-screen TVs, which cancelled out gains in other areas. An even bigger problem is persuading people to make substantial changes to their homes – such as insulating solid walls and changing doors and windows. These changes save money in the long run but require more upfront investment of time and money than many homeowners are prepared to put in. In 2010, David Cameron's new government announced measures to try to get around this problem that will see companies offering to insulate homes for "free", the cost being ultimately paid off through a premium added to the household's freshly reduced energy bills.

The US hasn't yet adopted stringent goals for building efficiency. That's a serious concern, because American homes and offices are massive consumers of energy. Since the 1970s, the size of the average new US home has grown by more than 50%, and with the population shifting southward, the fraction of homes with air conditioning has risen from a third to more than 60%. Amazingly, per capita energy use in US homes has dropped by more than 15% in this

Bank of America Building, New York – the future of greenscrapers?

357

time, but countering that trend is the nation's relentless growth in population – now just over 300 million, and projected to top 400 million by 2050. This implies a great opportunity to bolster the US's stock of energy-efficient homes. The American Institute of Architects has endorsed a proposal to halve the energy consumption of the typical new US house by 2030. However, the road to such targets is impeded by the fragmented nature of the US construction industry and the lack of a federal mandate. Still another hurdle is the collapse in US construction due to the financial downturn starting in 2008. Progress may be more rapid inside US homes: from 2000 to 2009, Americans bought more than 3 billion products approved by the Energy Star programme, which identifies and certifies the nation's most efficient appliances and electronics. But as in the UK, increasing the energy efficiency of appliances won't make a huge difference if the number of devices in use keeps rising.

The quest to make large buildings – such as schools, factories and offices – more energy efficient differs somewhat from the strategy for homes. In the US, lighting and office equipment make up almost half of the energy consumed by commercial buildings. All those lights and computers produce lots of warmth, which means commercial buildings typically demand less heating but more air conditioning than homes. Given this energy profile, some of the most constructive steps to cut energy use in big buildings involve illumination. Along with **LEDs** – the highly efficient light-emitting diodes used in many traffic lights and stadium signs, and increasingly in homes and offices – large buildings can also make use of **hybrid solar lighting**, in which sunlight is piped into upper floors through optical cables feeding from a roof-mounted solar collector. Green roofs – gardens that spread across much or most of a large building – help keep the building cooler (and, as a bonus, whatever's planted absorbs a bit of carbon dioxide). And several high-profile skyscrapers have become showcases for a variety of energy-saving features (see box on p.356).

The emissions that result from heating and cooling buildings can be slashed by drawing on the steady year-round temperature of soil through **ground-source heat pump** systems (also called geothermal heat pumps). More than a million such units were in place worldwide as of 2009; they're most common in the US and northern Europe, particularly in Sweden. Whether ground-source heat pumps are cost-effective and carbon-efficient depends on climate, household size and other factors such as the fuel used to create the electricity that drives the pump. As a rule, they're quite expensive to install but require 30–60% less energy than systems that burn fossil fuel directly. Ground-source units rely on fluid flowing through pipes buried underground, either in flat arrays a little below the surface or vertical coils that can extend down more than 100m (330ft). The fluid returns at ground temperature, which may run anywhere from 10–16°C (50–60°F) in mid-latitude areas but which changes little through the year in any one location. A compressor is used to concentrate this low-level warmth and the resulting heat is then distributed around the home (often via an underfloor heating system), with the cold fluid returning underground to be warmed up once more. A major bonus in warm climates is that many ground-source heat pumps can also provide air conditioning that's up to 40% more efficient than conventional cooling systems.

Air-source heat pumps are another option in many regions. Although slightly less efficient, these are much cheaper to install and have the advantage of being suitable for existing homes that lack big yards or gardens. Both air- and ground-source systems must be installed and used correctly to provide their promised benefits, as found in a 2010 report by the UK Energy Saving Trust.

Geoengineering: practical solutions or potential disasters?

It's possible, if not probable, that the various strategies discussed so far in this chapter won't be implemented widely enough, or soon enough, to prevent the risk of the worst effects of climate change. But some visionaries think there's another way: using technology to tackle climate change head-on, either by sucking greenhouse gases from the air or changing the amount of sunlight that reaches Earth. These ambitious plans are often lumped under the heading of **geoengineering** or **climate engineering** – global-scale attempts to reshape Earth's environment. Although there are plenty of warning lights around the concept, it's become a topic of increasing research and even some tentative steps toward policymaking – a trend that's perhaps related to the lack of meaningful progress on a post-Kyoto plan.

Humans have long dreamed of controlling weather and climate. In the 1950s, the USSR pondered the notion of damming the Bering Strait that separates Alaska from Russia. The idea was to pump icy water out of the Arctic and make room for warmer Atlantic currents, thus softening the nation's climate and easing the way for ships. Famed computer scientist John von Neumann reflected the era's slightly wild-eyed optimism when he wrote in 1956 of spreading dark material over snow and ice to hasten its melting and warm the climate: "What power over our environment, over all nature, is implied!" Today's geoengineering schemes operate in a more sober realm, the idea being not to create the perfect climate but simply to stop the existing climate from changing too much.

Climate engineering plans fall into two main categories. One of them, **carbon dioxide removal**, is focused on reducing the amount of CO_2 in the atmosphere. As envisioned by David Keith (University of Calgary) in a 2009 *Science* article, "there are important niches where air capture might play a role in reducing emissions in the near term, long before we are able to bring emissions near zero." Keith is studying the option of large facilities at which lye would react with carbon dioxide and remove it from the air. Similarly, Klaus Lackner (Columbia University) has been prototyping "synthetic trees", devices through which CO_2 in the air reacts with an ion-exchange resin.

It remains to be seen how much money – and, crucially, energy – it would take to produce, operate and maintain systems like these, though the theoretical framework looks strong. There's also the issue of how to safely store the vast quantities of captured CO_2, although one of the benefits of capturing CO_2 from the open air is that the collection devices can be placed anywhere in the world (with no need for a pipeline infrastructure), enabling the operator to select the most suitable geological storage sites. For example, some advocates of air-capture technology hold great hopes for the storage potential of basalt rocks underneath regions such as Iceland.

Another variation on this theme is to pull carbon from the air into the oceans. Experiments hint that spreading iron over parts of the ocean where it's limited could produce vast fields of carbon-absorbing phytoplankton. The many geopolitical challenges of geoengineering came to the fore in 2007 when the private US firm Plantos launched plans to dump 100 tonnes of iron near the Galapagos, ten times more than previous iron-fertilization studies. The plan, which could lead to profit-making projects designed to offset emissions elsewhere, drew scrutiny from the International Maritime Organization and the global London Convention. In yet another approach, Gaia theorist James Lovelock and Chris Rapley (director of London's Science Museum) have proposed dotting the ocean with a set of vertical pipes that would allow wave action to pull up rich, deep water and promote algae blooms. Still another marine take on geoengineering is the idea of adding huge volumes of lime to the oceans. This could theoretically pull lots of carbon from the air in a way that avoids increasing ocean acidification – though the plan would only be worth trying if the CO_2 generated in the production of the lime could be captured and stored.

The third approach to removing CO_2 from the air – arguably not "geo-engineering" as such – involves exploiting biological systems such as **plants and soils**. The approach that has generated the most excitement in recent years is **biochar** – the simple technique of creating charcoal (which is a very stable and virtually pure form of carbon) by heating crop wastes, wood or other sources of biomass in a simple kiln that limits the presence of oxygen. The resulting "char" can then be mixed in with agricultural soils. This practice – which appears to go back centuries in some parts of the world – not only locks the carbon into the soil for hundreds or even thousands of years, but also appears to significantly increase crop productivity. This virtuous circle could, its advocates claim, sequester huge amounts of carbon and increase the availability of food in poor areas. Field trials back up these claims, especially in the tropics, where soils tend to have lower carbon content. Some environmentalists are cautious about biochar as a climate solution on the grounds that – if carried out on a large scale – the practice may create extra demand for wood, which in turn could boost deforestation and therefore carbon emissions. But many other commentators pin a lot of hope on this low-tech form of carbon capture and storage.

The other main category of geoengineering concepts – **solar radiation management** – involves trying to reduce the amount of sunlight reaching Earth. This could be done by adding massive amounts of sulphates to the stratosphere, or deploying a colossal (and hugely expensive) array of mirrors or lenses far out in space that would deflect or refract sunlight before it reached our atmosphere. Astronomers Roger Angel and Pete Worden of the University of Arizona proposed an array of ultrathin lenses spanning an area the size of Western Europe that would sit about 1.6 million km (1 million miles) from Earth towards the Sun. A closer-to-Earth idea, first put forth by engineer Stephen Salter and physicist John Latham, is to deploy many hundreds of ships at sea, each with giant turbines that would stir up the ocean and send salt particles airborne. The salt would serve as nuclei to help brighten stratocumulus clouds – which extend across a third of the world's oceans – and thus increase their ability to reflect sunlight.

Even as research gins up, some observers worry that geoengineering is a less-than-ideal way to deal with the root problem. The late climatologist Stephen Schneider likened it to giving methadone to a drug addict. In the case of managing sunlight, especially, there's also the classic risk of unintended consequences. Even if enough sunlight could be reflected to compensate for the extra CO_2 in the atmosphere, for example, those two factors (less sunlight, more carbon) could torque regional weather patterns and produce other effects quite different than our present climate, especially for vegetation and ocean life. And sequestration has yet to be attempted on anything close to the scale needed for effective carbon removal. On top of these concerns, there are political ones. Nations might easily disagree on which geoengineering routes to take, and there's nothing to block a rogue country from embarking on a project unilaterally.

With all these question marks looming, science organizations and policymakers have become increasingly involved in geoengineering guidance. In a 2009 report, the Royal Society called for international development of a code of practice for geoengineering research. The society is also working with the Environmental Defense Fund and other partners on a governance plan for solar radiation management. And the UN Convention on Biological Diversity, which includes nearly all of the world's nations, called in 2010 for a de facto moratorium on geoengineering plans (except for small-scale research projects) until the full array of risks – social, cultural, environmental and economic – has been thoroughly vetted.

PART 5

WHAT CAN YOU DO?

Reducing your
footprint & lobbying
for action

Artist's impression of a
future building powered by
windmills during a blackout.

Getting started

Where to begin

The gravity of our greenhouse predicament is enough to weigh anyone down. But the smart way to deal with climate change is to channel your angst and frustration into constructive action. In this chapter you'll find a wealth of ways to get started. Not all of them will suit your particular situation, but some probably will.

The most obvious way to take individual action on climate change is to reduce the size of your **carbon footprint** – the total amount of greenhouse emissions that result directly and indirectly from your lifestyle. Since this will include increasing your energy efficiency at home and on the road, you may even save some money in the process. The following chapters take a brief look at various ways to reduce your footprint. For more information about low-carbon living, see *The Rough Guide to Green Living* (UK) or *The Rough Guide to Shopping with a Conscience* (US).

While individual actions do make a difference, climate change is such an enormous global challenge that it's also worth **getting involved**: raising the issue with your political representatives, employers and/or local communities, as described on p.369.

Measuring your carbon footprint

Just as new dieters often keep a food diary, an excellent way to start reducing your emissions is by using carbon calculators. These simple tools, available online or in book form, allow you to calculate how much carbon each activity in your life generates and how your total compares to those of the people around you and elsewhere in the world. Carbon calculators vary by country, reflecting differences in the way energy is generated, priced and taxed. They also vary among themselves in how they organize and categorize activities and how they handle uncertainty about the exact greenhouse impact of particular activities, such as flying. Sometimes the calculations in each step are explained in detail; in other cases they're not, which makes the site simpler but not necessarily more user-friendly. US-based calculators often use short tons, about 10% less than a metric tonne.

The carbon-offset companies (see p.384) all offer calculators for specific activities, but to quickly assess your overall carbon footprint, try:

Guardian Quick Carbon Calculator www.guardian.co.uk/carboncalculator
Carbon Footprint www.carbonfootprint.com/calculator.aspx

Both these sites – unlike many others – cover emissions not just from the fuel and power we're each directly responsible for, such as heating and driving, but also less obvious sources such as the construction of our homes, offices and roads, and the manufacture of the goods we buy.

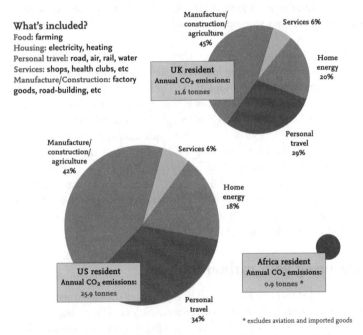

What's included?
Food: farming
Housing: electricity, heating
Personal travel: road, air, rail, water
Services: shops, health clubs, etc
Manufacture/Construction: factory goods, road-building, etc

Manufacture/construction/agriculture 45%

Services 6%

UK resident
Annual CO_2 emissions:
11.6 tonnes

Home energy 20%

Personal travel 29%

Manufacture/construction/agriculture 42%

Services 6%

Home energy 18%

US resident
Annual CO_2 emissions:
25.9 tonnes

Personal travel 34%

Africa resident
Annual CO_2 emissions:
0.9 tonnes *

* excludes aviation and imported goods

These diagrams, based on data from BestFootForward.com, show breakdowns of the typical CO_2 emissions per capita in the US and the UK, with a typical African citizen provided for comparison. The footprint figure for each resident is reached by simply dividing the total for each country or region by the population. Note, however, that these figures don't reflect total carbon footprints as they don't include methane or nitrous oxide. If they did, the figures would shoot up substantially, with the extra emissions allocated mainly to agriculture and manufacture. However, accurate figures for the methane and nitrous oxide emissions caused by the growing of our foods overseas are not easily available.

Making things happen

In addition to reducing your own carbon footprint, you can help move society as a whole towards a lower-carbon future. Joining with other like-minded citizens will enhance your own power.

▶ **Work and school** Find out what policies exist on energy efficiency at your workplace or school. If there aren't any, or they seem half-hearted or inadequate, work for something better by relaying your suggestions to the powers that be. Encourage co-workers or fellow students to do the same.

▶ **Community** Cities, towns and councils vary hugely in how committed they are to solving the greenhouse problem. Contact your local government and ask what climate-change measures they've adopted. Then see what you can do to raise awareness and make change happen – by contacting local media, for example, or attending local-government open meetings.

▶ **Finances** If you've got savings or investments, consider moving them towards a bank or fund that supports action on climate change. Generally this will be one that engages in "socially responsible investing" – that is, it considers global warming as well as other social and environmental issues when deciding which companies to include in its portfolio. If you hold stock directly in a company that's behaving in a less-than-ideal way on climate change, look into introducing or joining a **shareholder resolution**. Typically these require companies to report on their actions and plans regarding climate change and its impact on the company's bottom line.

▶ **Politics** Climate change won't be solved without political will, and there won't be political will without pressure from voters. The Internet makes it easy to contact your political representatives. Simply drop into one of the following websites, enter your postal or zip code and away you go. A good starting point is to ask what emission targets (if any) your legislators support and what they're doing to help ensure these targets are met.

Congress www.congress.org (US)
They Work for You www.theyworkforyou.com (UK)
Write to Them www.writetothem.com (UK)

▶ **Pressure groups** You could also consider joining one of the campaign groups working for political change around the climate issue. The following vary in their policy focus and attitudes towards issues such as globalization, but they're all campaigning for major emissions cuts:

Environmental Defense www.environmentaldefense.org
Friends of the Earth International www.foe.co.uk
Greenpeace www.greenpeace.org
Natural Resources Defense Council www.nrdc.org
Sierra Club www.sierraclub.org
350.org www.350.org
10:10 www.1010global.org

Home energy

How to save money while saving the planet

Saving energy at home will provide a dual satisfaction: less greenhouse gas in the atmosphere and, sooner or later, more money in your pocket. The following tips should help you streamline your household energy use. You may also want to contact your local government office to see whether it offers free home-energy audits or energy-efficiency grants.

The Web is a good place to find out more. For example, Home Energy Saver allows US residents to carry out a simple online audit of their household energy use, while the UK's Energy Saving Trust provides a wealth of useful information to help you make your home less carbon hungry.

Home Energy Saver (US) hes.lbl.gov
Energy Saving Trust (UK) www.est.org.uk/housingbuildings

Of course, not everyone is in a position to make substantial changes to their home on the grounds of carbon and energy savings – even where grants are available. In particular, renters are often left frustrated that they cannot justify investing time or money in a building that doesn't belong to them. Other than appealing to the homeowner, there's often little that can be done in this situation, though legislation announced in the UK in 2010 will soon effectively force all landlords to offer basic efficiency measures such as cavity-wall insulation if requested by the tenant.

Heating and hot water

Heating and cooling is a key area, accounting for the majority of emissions in most homes. Older houses are especially inefficient in this regard, though there's room for improvement in nearly all domestic buildings. (If you're feeling really ambitious, and flush, consider building a Passivhaus, pioneered in Germany. Though expensive to construct, these ultra-efficient homes get by without any central heating or cooling source, thus cutting utility costs by as much as 90%.)

Insulation

Start small with weather-stripping – sealing up cracks around doors and windows – but be sure to consider beefing up your loft and wall insulation. The attic is a good place to start, since it's easier to access and less expensive to insulate (if it's already been done but only to a depth of a few inches, consider topping up). Energy-efficient windows can also make a substantial difference, though they are typically a more expensive way of saving energy than insulating walls and lofts. If you live in a hot climate, you can save on air-con by using bright, reflective window drapes and shades wherever sunlight enters, and by using light colours or a reflective coating on your roof.

Thermostats and heating controls

Reducing your room temperature in the cool season by just a small amount can make a disproportionate difference to your energy consumption. You may find you sleep better, too. Try 16–18°C (61–64°F) and throw on a sweater. Efficient heating controls – especially those that let you specify the temperature of individual rooms, or programme different temperatures for different times of day – can also take a significant chunk out of your energy demands, in return for a comparatively small investment.

If you use air conditioning, shoot for 25°C (77°F). Better yet, if you're in a warm, dry climate, try a whole-house fan to bring in cooler morning and evening air, or investigate swamp coolers, which can take the edge off summer heat while humidifying the air and using far less energy than an air conditioner.

Hot-water heating

An efficient boiler (hot-water heater) would make a sensible long-term investment for many households. Modern condensing boilers can produce up to a third extra usable heat from each unit of fuel than inefficient older models. For even greater savings, consider swapping your hot-water tank – if you have one – for a tankless or "combi" system. These save energy by heating water only as it's needed. If you do stick with a hot-water tank, though, be sure to insulate it. Purpose-built blankets are inexpensive and can save 25–45% of the energy required to heat the water. (Newer hot-water tanks often have insulation built-in.) To save more energy, aim for a temperature setting of no more than 50°C (122°F).

Pick your fuel

In general, natural gas is a more climate-friendly fuel for home heating and hot water than oil, coal or regular grid electricity. If you're not connected to the gas network, wood could be your best bet (unless you can access wind power – see below), as long as the fuel is coming from forests that are being replenished as fast as they're being harvested. The UK government offers grants for efficient wood-burning boilers, though not for regular stoves.

Showers and baths

Everyone knows that showers use less energy than baths, and that shorter showers use less energy than longer ones. Less widely known is that a low-flow shower nozzle, which mixes air with the water flow, can reduce the amount of hot water needed by half. Look for nozzles rated at fewer than 8 litres or 2 gallons per minute.

Electricity supply
Renewable electricity

Many utility companies allow you to specify that some or all of the power you use is generated from renewable sources, most often wind farms. Typically with such plans (labelled "green tariffs" in the UK), you're not literally getting the power generated by the renewables. Instead, the power company agrees to put renewable energy into the grid in an amount equal to your own consumption. One point to note: in countries such as the UK and certain US states, electricity companies are already mandated to buy a certain percentage of their power from renewable sources, and the companies that exceed this quota can sell credits to those which fail to meet it. The upshot is that your fee won't necessarily increase the overall amount of renewable power generated, so it's important not to use a green energy tariff to justify being less careful about minimizing energy waste. Quiz local suppliers or green groups for more information.

Generating your own power

There are various ways to generate energy at home – but don't expect to be self-sufficient without substantial upfront investment. One sensible starting point is a rooftop solar panel to heat your hot water. Unlike the photovoltaic (PV) systems that generate electricity, these **collector** panels funnel heat directly into your hot-water system. Even in cloudy Britain, a collector panel – typically spanning about 2m x 2m (6ft x 6ft) – can provide up to half of the hot water needs of a house with a south-facing roof and a hot-water tank (they won't work with most combination boilers). Such units typically cost around £2000–3000/$3200–4800.

Full-fledged solar **PV systems** can provide homeowners with half or more of their total power needs. The upfront cost is high, though – expect to spend around £10,000/$16,000 for a system capable of delivering 1800 kilowatt hours per year, which is around half of the typical UK home's demand. At current electricity prices, that kind of investment might take decades to pay for itself in terms of savings on energy bills, but government grants and tax breaks have made PV much more attractive financially in many countries and US states than it used to be. In the UK, for

Starting at the top

"Green roof" on top of the Morgan mail processing facility, New York. Covering nearly 2.5 acres, the roof will reduce runoff by up to 75%.

Besides hosting solar panels and micro wind turbines, roofs offer various opportunities for combating global warming. One option is to make your roof more reflective. A study by the Earth Institute at Columbia University estimated that a world full of entirely white roofs could add a full 1% to Earth's reflectivity, and US energy secretary Steven Chu has promoted the concept. But you don't necessarily have to paint your roof white to make it more reflective. Several types of tile, metal and shingle roofs can boost reflectivity while maintaining a splash of colour. For large buildings in urban areas, another good option is a "green roof" with plants that can not only help with cooling but also absorb excess storm water (and CO_2).

Look out for labels

Electrical devices vary enormously in their energy consumption, so when shopping for such items, be sure to look out for the various labels that give you the information you need to make a low-carbon choice. These include:

UK/Europe

▶ **EU Energy Label** Appears by law on a variety of home appliances, light bulbs and air conditioners, rating products from A (efficient) to G (inefficient).

▶ **Energy Efficiency Recommended** This UK label adorns best performers within the EU Energy Label rating system. For example, refrigerators must rank as A+ or A++ and washing machines must be AAA.

▶ **EU Ecolabel** Not yet very common in the UK, the Ecolabel tags products that pass numerous environmental criteria, including energy efficiency as well as expected lifespan and ease of disposability.

North America

▶ **EnergyGuide (US)** Standard on large appliances, this label shows the typical amount of energy consumed per year and how that product compares to the best and worst performers in its category, as well as the expected annual energy cost.

▶ **Energy Star (US)** Identifies products that meet various criteria for efficiency in more than 40 categories – including entire homes.

▶ **EnerGuide (Canada)** Similar to the US EnergyGuide, this label shows the amount of energy consumed per year and a comparison to similar models (but no energy cost estimate).

Australia

▶ **Energy Rating Label** Shows annual energy consumption (or, for air conditioners, the rate of consumption) and an efficiency rating from one to six stars. Even the best appliances only rate three or four stars, to leave room for future improvements.

example, a generous "feed-in tariff" rewards homeowners with a solar roof to the tune of almost £1000 per year.

For people who live in a breezy enough spot, a **wind turbine** that feeds power into batteries or the grid may be another option worth considering. However, building-mounted models tend to be inefficient and expensive, and the towers can run afoul of local building codes – and even jeopardize your home's structural integrity. Standalone turbines are a much better bet if you are lucky enough to have a piece of windy land to put one on.

For much more information about generating your own electricity, see *The Rough Guide to Green Living.*

Appliances and gadgets
Lighting

Close to 95% of the power that drives an old-fashioned incandescent light bulb goes to produce heat instead of light (which explains why those bulbs are so hot to the touch). **Compact fluorescent bulbs** generate far less heat, enabling them to produce about four times more light per unit of energy (the replacement for a standard 100-watt bulb typically uses about 23 watts). These efficient bulbs also last around ten times longer, meaning that they earn back their higher initial cost several times over and save you time and hassle in buying and replacing bulbs (not to mention the emissions involved in making and shipping them).

Though most compact fluorescents take a few seconds to reach full brightness, many are now "instant-on" and even dimmable, and they no longer produce unattractive light. When asked to compare the light without knowing the source, the participants in a study by US magazine *Popular Mechanics* actually preferred the compact fluorescent bulbs. To make sure you get the right bulb for the job, check its colour temperature (usually expressed in degrees Kelvin). Lower readings produce softer whites, while higher readings more closely resemble daylight.

In 2007 Australia became the first nation to ban incandescent bulbs, effective in 2010. Major retailers in the UK agreed to phase out their sale by 2011, with the US phase-out unfolding between 2012 and 2014.

Note that compact fluorescent bulbs do contain tiny amounts of mercury, so when they eventually die, take them to a municipal dump to be disposed of properly.

What about **halogen** lights? These are a subset of incandescents and tend to be middle-range performers in the greenhouse stakes. High-quality halogen bulbs are around twice as efficient as typical incandescents. However, they're still half as efficient as compact fluorescents, and halogen light fittings often take multiple bulbs, raising their overall energy consumption. Moreover, they burn much hotter than other bulbs, making them more likely to damage paintwork. If you have halogen fittings, look out for compact fluorescent or LED replacements, available from specialists such as the UK's **GoGreenLights.co.uk**.

LEDs (light-emitting diodes) may be the bulbs of the future. Already widely used in traffic lights and outdoor displays, LEDs are showing up on consumer shelves as well. They're already a bit more efficient than compact fluorescents, and further gains are expected over the next few years.

Whatever kind of lights you have, turning them off when you leave the room is an obvious energy-saver. Organizing your home to maximize the use of natural light – putting desks next to windows, for instance – can also help.

Invisible power drains

The proliferation of remote controls and consumer gadgets has come at a surprisingly high energy cost. In the typical modern home, around 5–10% of all electricity is consumed needlessly by TV sets, stereos, digital video recorders and other devices that are supposedly turned off but are actually on **standby**. There are huge variations here: cell phone chargers don't use much when they're not charging, while some devices use almost as much energy in standby mode as when they're operating. To stop the waste, unplug items – or switch them off at the socket – when they aren't in use. A power strip with a switch can help if you don't have switches on your sockets. You can root out the worst offenders by plugging your gadgets into an inexpensive electricity usage monitor, such as Kill-a-Watt.

Computers vary widely in terms of the energy they consume in standby and screensaver modes. Turn them off when not in use, or dedicate their downtime to the fight against climate change (see p.258).

Refrigerators and freezers

These are among the biggest users of energy at home – not least because they're running 24/7. Models produced since the 1990s are far more efficient than their predecessors, so if yours is old, consider taking the plunge: a modern replacement could recoup its costs within several years and save plenty of energy from the start. Be sure to dispose of the old one properly, as many older refrigerators contain potent ozone-depleting and planet-warming chemicals. Check with your local government or a home-appliance dealer for more information.

If possible, locate your fridge or freezer as far away as possible from hot-running items such as ovens and dishwashers, and make sure it's set no lower than the recommended temperature, which is typically -18°C (0°F) for freezers and 1–4°C (34–39°F) for fridges.

Doing the dishes

Like fridges, dishwashers vary widely in their energy consumption, so be sure to consider efficiency when purchasing. As for how they compare with washing up by hand, this depends on the individual machine (some newer models use fewer than fifteen litres or four gallons of hot water per load), the efficiency of your hot-water heater and, most importantly, how economical you are when washing up by hand – where it's possible to use huge amounts of hot water. A recent analysis in the book *How Bad Are Bananas?* finds that a dishwasher actually wins out on energy in most cases, assuming you run it only when fully loaded (and assuming your hot water doesn't come from a renewable source). This is true even considering the energy used to make the dishwasher.

Washing clothes

To save energy (and minimize colour fade) keep washing temperatures as low as possible to get the job done well. Even more worthwhile is letting clothes dry on a line or rack rather than using a dryer, as it takes a lot of energy to evaporate water – many times more than it takes to run the washing cycle. In the US, many suburban condominium and housing associations, fearful that clothes-lines will tag a neighbourhood as déclassé, actually prohibit hanging clothes outside to dry. In such situations, an indoor rack may do the trick.

Transport and travel

Trains, planes and automobiles

Personal transport accounts for around a third of the typical person's carbon footprint in countries such as the UK and US. Moreover, transport as a whole is the fastest growing major source of greenhouse gases. Thankfully, in the case of road transport at least, there's much you can do to make your comings and goings less carbon-intensive. As we'll see, air travel is rather more problematic: for most keen travellers, it makes up the single biggest element in their carbon footprint.

The car and what to do about it

Compared to alternatives such as train and bicycle, most cars are very energy inefficient – especially when carrying only one or two passengers. If you can't live without a car, you can at least reduce its greenhouse emissions by changing your driving habits (see p.378). Making fewer and shorter journeys also helps, of course, as can carpools and liftshare schemes. Depending on where you live, you may even be able to run your present vehicle on ethanol or biodiesel (see p.354).

If you're due for a new vehicle, be sure to opt for the most efficient model that fits your needs. This may be a **hybrid**, such as the Toyota Prius or Honda's Civic Hybrid. These look, work and fuel just like regular cars, but they feature an extra electric motor that charges up when you apply the brakes. Hybrids can achieve up to 70 miles per gallon, though SUVs and executive cars with hybrid engines are far less efficient because big, heavy, high-powered cars eat fuel much faster than smaller, medium-powered ones. If you're concerned about the **safety** of a smaller car, check the official safety data for each model – bigger isn't always better in this regard.

Even greener than hybrids are **electric cars**, which you charge up via a standard household power socket. Until recently, the market for electric cars was dominated by tiny, odd-looking vehicles such as the G-Wiz, of

A Reva G-Wiz
charging in London.

which there are more than 1000 on the streets of London (where they're exempt from the congestion charge). But regular-sized electric cars are now proliferating as research models move into mass production. Some are fully electric, like the Nissan Leaf, while others include a backup petrol engine, such as the Chevy Volt (making it a hybrid, but with the electric mode as the default). Many of these cars are limited in terms of battery capacity, which means you need to recharge at least every 160km/100 miles or so, depending on the model. However, they can be ideal for city dwellers, and they're extremely green: even when charged with electricity generated from fossil fuels, they result in less CO_2 per mile than the most efficient hybrids.

Air travel

Aviation is problematic from a climate-change perspective. It takes extra fuel to move people at the altitudes and speeds of jet aircraft, and people typically fly much farther than they would ever travel by train, boat or car. The contrails produced by aircraft also have potentially important impacts (see p.208). This means that a plane's impact on the climate may be two or more times greater than its already significant CO_2 emissions would suggest. All told, two people flying a round trip between Europe and the US produce the equivalent impact of at least four tonnes of CO_2 – more than half of the average UK household's yearly output and comparable to what a typical US car generates each year.

Planes are slowly gaining in efficiency, with improvements of a percent or so each year owing to technological gains as well as fuller flights. The problem is that these gains are mostly translating into cheaper and more popular air travel, so that the sector's overall greenhouse output continues to grow rapidly. More substantial changes are on the drawing

Low-carbon motoring

Even if your vehicle's a gas-guzzler, the following tips could cut its fuel usage by as much as 30%. Some of them are now incorporated in the tests that new drivers take in the UK, the Netherlands and several other countries.

▶ **Drive at the right speed** Most cars are most efficient at speeds of 45–80kph (30–50mph). Above 90kph (55mph), cars gulp more fuel to travel the same distance – as much as 15% more for every additional 16kph (10mph).

▶ **Lighten the load** Keep heavy items out of your car unless you need them – you'll typically lose a percent or two in efficiency for every 45kg (100lb) you haul. While you're at it, check the tyre pressure often: rolling resistance goes up and efficiency goes down by as much as 1% for every PSI (pound per square inch) below the recommended pressure range. However, there's no benefit, and some risk, to driving with over-inflated tyres.

▶ **Avoid idling** Except when it's required (such as in stop-and-go traffic), idling is a wasteful practice, and it doesn't benefit your car, except perhaps in extremely cold conditions. Even five minutes of idling can throw half a kilo (1.1lb) of greenhouse gas into the air. Anything more than about ten seconds of idling generates more global-warming pollution than stopping and restarting would.

▶ **Use the air-con sparingly** As you'd expect, air conditioning normally saps energy and cuts down on vehicle efficiency by a few percent. However, if you're on a long road trip and it's a choice between driving with the windows down and running the A/C, there may be little difference in fuel usage, according to some studies. That's because wide-open windows can increase the car's aerodynamic drag, especially at high speeds. If outside temperatures are comfortable, try using the vents and fan but leaving the A/C off.

▶ **Starting and stopping** Jack-rabbit starts and stops not only put wear and tear on your car, but they also drain fuel economy. Accelerate gradually, and anticipate stops by starting to brake well in advance. If you have a manual transmission, the best time to change gears is between 1500 and 2500rpm.

The rules are a bit different for **hybrids**, which has led to some confusion. With a standard engine, you're best off accelerating lightly no matter what the speed. In a hybrid, you should accelerate briskly until you get to the optimal in-between speed, around 45–60kph (30–40mph), where the car is at its most efficient. A technique called "pulse and glide" driving – hovering in that optimal speed range through small accelerations and decelerations – can boost your hybrid's efficiency.

When it's time to slow down a hybrid, brake slowly at first, then increase the pressure: this ensures that the maximum energy goes into recharging the battery versus creating unusable heat in the brakes themselves.

board, including the use of biofuels instead of aviation fuel. For example, Virgin Atlantic flew a jet from London to Amsterdam in 2008 using a mix of coconut and babassu oil for 20% of its fuel, and in 2009 Air New Zealand carried out tests using 50% jatropha oil. But biofuels raise their own thorny issues (see p.351), and even the most optimistic commentators don't expect developments in fuel or engine technology to offset the massive growth in the numbers of flights (see p.43). Ultimately, putting a plane into the air and keeping it there is an energy-guzzling task, so it'll be hard to make flying dramatically more efficient than it is now.

With all this in mind, cutting back on air travel is one of the most obvious steps for people concerned about global warming. This might mean taking fewer flights and making up for it by staying longer each time. It might mean favouring holiday destinations that are closer to home. Or it might mean considering alternative ways of travelling. With two or more people on the same itinerary, it can even be more climate friendly to drive than to fly – especially for short distances, where planes are most inefficient per mile travelled. Better still are **trains and boats**, which are typically responsible for many times fewer emissions per passenger mile than either cars or planes. To find out how to travel almost anywhere in the world by rail and sea, visit: The Man in Seat 61 **www.seat61.com**.

When you do fly, consider offsetting your emissions (see p.384) and plan your itinerary with as few stops as possible, since takeoffs and landings are energy-intensive. Use the airline's website to see what model of plane you're booking yourself on. Sites such as seatguru.com can help you figure out whether your flight will be on a newer (and thus likely more efficient) aircraft. Especially on long-haul flights, avoid the perks of luxury seating; thanks to the extra room they occupy, business-class passengers are responsible for about 40–50% more emissions than those in economy, and first-class travel may generate up to six times more carbon. Also, try to favour flights in the **daytime** and during the brighter times of the year; this can make a surprisingly big difference. A 2006 study of flights across southeast England, published in the journal *Nature*, points out that plane contrails reflect no sunlight during night-time and wintertime, thus reducing their ability to compensate for the climate-warming gases spewed by aircraft. For the study area, flights between December and February (less than a quarter of the annual total) caused half of the climate-warming effect, and night-time flights – only about a fifth of the total in Britain – produced 60–80% of the effect.

Another easy trick is to travel light. Two heavy suitcases nearly double the on-board weight of a passenger, adding to fuel consumption.

Shopping

Food, drink and other purchases

It's impossible to know the exact carbon footprint of all the items you buy, but a few schemes are working to fix that. The Carbon Trust, in partnership with the UK government, launched its Carbon Reduction Label in 2007. It's now used by hundreds of companies worldwide to show the grams of carbon emitted as a result of producing a given item (typically expressed in a per-serving format for food). And the Swedish fast-food chain Max Burger began providing carbon data for each item on its menu in 2009. Some experts, such as Mike Berners-Lee, claim that these kinds of labels almost always underplay true carbon footprints of the items they appear on, for the simple reason that it's impossible to include all the virtually infinite economic ripples caused by the creation and use of a product. Nonetheless, such labels can still be a handy way for individuals to get a rough sense of how different products compare.

For products that aren't carbon-labelled, there are a few good rules of thumb to follow. First, you can consider the energy used in transporting an item. Heavy or bulky goods manufactured far away inevitably result in substantial carbon emissions, especially if they've travelled long distances by road (shipping is comparatively efficient, though not negligible). On the other hand, smaller items transported by air – such as perishable, high-value fruit and vegetables, and cut flowers – are *far* more greenhouse-intensive than their weight would suggest.

Another thing to consider is the material that something is made from. Steel, aluminium, concrete and precious metals all require large amounts of energy to create. By contrast, wood from sustainable sources – such as local softwoods and anything certified by the Forest Stewardship Council – is practically carbon neutral. The same can't be said for tropical hardwoods such as mahogany or teak, the demand for which is one driver of the tropical deforestation that accounts for a large slice of the world's greenhouse emissions.

Food & drink

Cutting back on beef

Perhaps the best and simplest way to reduce the emissions of your diet is to cut back on meat and dairy – and especially beef. Cattle belch and excrete a substantial fraction (perhaps 20%) of the world's methane emissions, and much of Earth's rainforest destruction is driven by the clearing of land for grazing livestock or growing their feed. In fact, a much-cited 2006 study by the United Nations Food and Agriculture Organization found that factors related to livestock – from deforestation to fertilizer, feed and cow flatulence – represent an astounding 18% of current greenhouse emissions. The same amount of protein can be obtained from grains, beans and other vegan sources with a much lower carbon footprint.

Food miles

In general, food that's both local and in season is best for you and the climate. So if there's a farmers' market or local delivery scheme in your vicinity, try it out, or consider starting your own vegetable garden. Keep in mind, though, that "food miles" alone aren't the whole story. Some foods – bananas, say – look relatively good from a carbon perspective even if they are shipped thousands of miles. A 2006 study at Lincoln University found that, for UK eaters, New Zealand lamb has only a quarter of the total carbon footprint of British-raised lamb, mainly because the free-ranging Kiwi sheep require far less fuel to raise and are transported overseas

The three Rs: reduce, reuse, recycle

The old green mantra holds true for combating climate change. It's heresy in a capitalist society to suggest that downsizing might have its pluses, but a big part of reducing global emissions is taking a hard look at global consumption. So try your best not to make unnecessary purchases, especially of products that take a lot of energy to manufacture and distribute. Second-hand stores are perfect for satisfying a shopping bug without enlarging your carbon footprint.

Equally, try to make use of the recycling facilities offered in your area: almost all recycling helps to reduce energy consumption to some extent. For food waste, use a composter instead of the trash whenever possible. When food is buried in a commercial landfill, it decomposes anaerobically and generates the potent greenhouse gas methane. Landfills are, in fact, the source of a third of US methane emissions.

Weaning yourself from bottled water

Bottled water has made a tremendous splash among health-conscious consumers over the last twenty years. More than 100 billion litres are sold worldwide each year – that's a greater-than-tenfold increase from 1990. All told, the US consumes more bottled water than any other nation. Mexico now leads the world in per capita use; Italy ranks high as well, though much of its consumption is from glass bottles at sit-down meals rather than from plastic.

Drinking more water is certainly a good move for one's health, but it would be hard to imagine a more gratuitously wasteful product than bottled water, in terms of both cost (up to 1900 times more expensive than tap water) and greenhouse emissions. For starters, all those plastic bottles are made from petroleum, most likely at a factory that burns fossil fuels. Then there are the emissions involved in shipping the bottles long distances (water's quite heavy), keeping them refrigerated and, finally, transporting them for recycling or landfill (yet another ecological impact).

Fortunately, there's a marvellous, time-tested alternative: the tap. There's no denying that a bottle of water can be convenient in certain places and at certain times. However, it's practically as easy and far less costly to buy a sturdy, refillable, washable bottle and keep it with you. Avoid reusing bottled-water bottles, though; they may leach out harmful chemicals after only a few refillings.

As for quality, the supposed superiority of bottled water has been vastly overstated. In most developed countries, bottled water is inspected less rigorously than tap water. In fact, many bottled-water brands are simply tap water from city sources accompanied by a label that shows a mountain stream. (The US Food and Drug Administration now requires such brands to indicate their municipal origins, but there's a loophole if the water has been "purified" or otherwise treated.) As for taste, blind tests on the US TV programme *Good Morning America* found New York City tap water beating each of the three bottled alternatives by a clear margin. If water taste and/or quality is a concern in your own community, then try installing a filter on your tap or keeping a filtered jug in your fridge.

In the broader picture, many activists worry that the soaring popularity of bottled water – most of it distributed by a few huge corporations, including Nestlé and Danone – points to an increasing privatization of the world's water supply. In developing countries, the advent of bottled water as an alternative to poor public supplies only exacerbates the divide between haves and have-nots. In his book *Bottled and Sold*, water expert Peter Gleick notes the irony that Western societies pushed for safe municipal water in the 1800s, but now "we're afraid that public fountains, and our tap water in general, are sources of contamination and contagion." With climate change expected to make water supplies more variable in many parts of the globe, there's all the more incentive to make sure that everyone has access to water that's clean as well as climate friendly.

using relatively efficient ships. The study omits trucking within national borders, and there are other hard-to-quantify elements in the comparison (which was, unsurprisingly, carried out by agricultural specialists in New Zealand). Such examples aside, favouring local and seasonal produce where it's available remains a good rule of thumb. When you do buy imports, such as coffee, choosing items marked with a **fair-trade** or rainforest-certified label will reduce the risk that carbon-rich forests are being chopped down to support your tastes.

Organics

Organic food eschews farming techniques that rely on petrochemicals and tends to result in slightly lower emissions per unit of food – though not all studies agree on this point. Whatever the truth, organic produce is *rarely* low-carbon enough to justify its being shipped (not to mention flown) over vast distances, so check the country of origin and favour local food over organics if it's a choice between the two.

Buy basics – and in bulk

Some food-processing tasks require a surprising amount of energy, so buying ingredients, rather than ready-meals, is usually a good idea from a carbon perspective. Buying in bulk is a good idea, too, as it minimizes packaging. For fluids such as shampoo and washing-up liquid, look out for shops that let you bring in and refill your own bottles.

Bring your own bag

Shopkeepers practically force store-branded plastic bags on us, but each of the estimated 500 billion to 1 billion plastic bags that are used and tossed away each year adds to the world's waste load. Each one also carries a tiny carbon price tag, though some analysts have pointed out that a reusable bag can result in more net emissions than a disposable one if it's of poor quality or used only sparingly – an example of how carbon footprints aren't always aligned with other environmental priorities. In any case, finding a decent reusable bag and using it often will put you in the right direction. Aside from anything else, reusable bags are more comfortable to carry and less likely to break. Happily, the number of plastic bags used in the UK dropped by half from 2006 to 2009.

Offsetting

Paying a carbon cleaner

Carbon offsetting involves paying an organization to neutralize the climate impact of your own activities, thereby making those activities "carbon neutral" or "climate neutral". Some offset schemes focus on reducing future emissions by, for example, giving out low-energy light bulbs in the developing world or buying renewable electricity credits. Others focus on sucking CO_2 directly out of the atmosphere – by planting trees, for example.

Offsetting has become very popular in the last decade, not just with individuals but also with global corporations (HSBC and other office-based giants have gone climate-neutral), celebrities (Pink Floyd, Pulp and the Pet Shop Boys have all neutralized their tours) and even publishers (the original production of the book you are reading was offset by Rough Guides). In 2007, Vatican City became the world's first climate-neutral state by planting trees in a Hungarian preserve. Some airlines and other travel-oriented companies now make it possible for consumers to offset their emissions when they buy a ticket.

How offsetting works

Whether you want to cancel the carbon footprint of a single long-haul flight, a year of car journeys or your entire existence, the process is the same. First, you visit the website of an offsetting organization and use their **carbon calculators** to work out the emissions related to whatever activity you want to offset. This will be translated into a fee which the offsetting organization will use to soak up – or remove the demand for – a matching amount of greenhouse gas.

Offsetting fees vary by organization and over time, but they're often calibrated against the price of carbon established in European emissions trading – in the neighbourhood of £13/\$20 for a tonne of CO_2 at the time of going to press. At this price, a seat on a round trip from London to New York would cost around £20/$32 to neutralize, while a typical year of driving in a typically efficient car would clock in at around £55/$88. Currently, though, many popular schemes charge around half that amount.

If you pick a firm that charges an especially high rate per tonne of CO_2, it doesn't necessarily mean your money is doing more good; as the Tufts Climate Initiative notes, "It is more important to invest in high quality offsets than to buy as many offsets as possible."

Tree planting – help or hindrance?

One of the more clichéd approaches to tackling global warming, tree-planting has also become one of the more controversial – especially in the context of carbon offsetting. It's true that trees soak up CO_2 as they grow. Two or three dozen can be enough to absorb an entire household's emissions. However, at snow-prone higher latitudes, trees can actually accelerate climate change, according to a landmark 2006 study led by Govindasamy Bala of the US Lawrence Livermore National Laboratory. That's because their CO_2-absorbing benefit is outweighed by the impact of their dark colour, which in wintertime absorbs sunlight that might otherwise be reflected to space by bright snow cover atop barren ground.

In snow-free warmer climates, and especially in the tropics and subtropics, the dark colour of trees isn't a problem. But there are other catches. First, when a tree dies, much of its stored carbon returns to the atmosphere. So the offset will only be permanent if each tree planted is replaced by another – and so on. Second, there are quicker, cheaper and longer lasting ways to fight climate change, such as distributing low-energy technologies to displace fossil fuels.

For all these reasons, many offset schemes have switched from tree-planting to energy-saving. (The Carbon Neutral Company, for example, was the result of a rebranding of a scheme previously called Future Forests.) That said, unless you live in a snowy region, planting a few trees in your garden is still likely to be beneficial for the climate – at least for the crucial coming decades.

A separate, and altogether more pressing, approach is protecting the forests that are already standing. That's not so much because of the CO_2 that mature forests absorb. Rather, it's because deforestation, and in particular the destruction of tropical rainforest, is one of the very largest sources of greenhouse emissions (see p.11). To help limit these emissions, the charity Cool Earth (see p.388) enables individuals and companies to sponsor areas of critically endangered rainforest in order to "keep the carbon where it belongs". The forest is purchased and given to a local trust, which protects it but allows sustainable harvesting of rubber, nuts and other forest crops. You can even make sure your sponsored area is still tree-covered thanks to satellite photos from Google.

Cool Earth www.coolearth.org

Offset debates

Despite their popularity, carbon-offset schemes are not without their critics. One argument levelled against offsetting is that it's just a plaster on the wound – a guilt-assuaging exercise that hides the inherent unsustainability of carbon-intensive Western lifestyles. Many observers have likened offsets to the medieval Catholic practice of selling indulgences to "offset" sinful behaviour, while the spoof website cheatneutral.com ridicules the offset concept by offering a service for neutralizing infidelity. "When you cheat on your partner, you add to the heartbreak, pain and jealousy in the atmosphere ... CheatNeutral offsets your cheating by paying someone else to be faithful and not cheat."

Even the offset schemes themselves tend to agree that offsetting emissions isn't as good as not causing them in the first place. That said, there's no reason why people can't buy offsets *and* make efforts to reduce their emissions directly. Indeed, the very act of digging into pockets for an offset fee may make consumers or businesses more conscious of carbon's larger cost.

A separate criticism is that offset projects may not make the swift, long-term carbon savings that are claimed of them. It's true that some projects are poorly conceived and/or executed. Tree planting (see box on previous page) may take decades to soak up the carbon you've paid to offset, which is one reason why many offsetting groups are moving towards sustainable energy projects instead of trees. As for whether the carbon savings are real, the better offsetting services are externally audited (see below) to address just this question.

Still another point of contention is whether offsetting services ought to make a profit. According to the Tufts Climate Initiative, some offsetting firms use as much as 85% of their revenue for expenses before actually funding any emission reduction work. Some of the commercial firms that provide offsets don't reveal how much profit they earn, arguing that the good work they do justifies an unspecified return on their labour and risk. Other offsetters are bona fide charities, which means all of their income goes toward operations and the offset projects themselves.

Ultimately, the benefits of offsetting are open to debate. If you pick the right scheme, it's likely that your contribution will help reduce emissions (especially if you pay to offset, say, two or three times the emissions of the activity in question). On the other hand, critics such as George Monbiot argue that, if we're to stand any chance of tackling climate change, then the kinds of projects invested in by offset schemes need to happen anyway, funded by governments, in addition to individuals and companies slashing their carbon footprints directly.

If you do want to offset ...

With dozens of offset schemes out there and no real regulatory framework, it can be hard to know which one to choose. One of the most useful consumer-oriented guides was first published in 2006 by Clean Air – Cool Planet.

Clean Air – Cool Planet www.cleanair-coolplanet.org

More recently, the UK introduced a quality assurance scheme that evaluates the offsetting projects being sponsored by various schemes (but not the offsetting firms themselves). Projects that pass muster on a variety of criteria – including Kyoto-compliant credits and clear, accurate marketing and pricing – carry a mark of approval. In 2009 and 2010, projects were certified for companies that ranged from the offset firms Clear and Carbon Footprint to two airlines (British Airways and TAP Portugal).

The UK Government's Quality Assurance Scheme for Carbon Offsetting offsetting.defra.gov.uk/cms

Alternatives to offsets

The debate about the validity of offsetting is bound to run and run. In the meantime, some companies and individuals are switching from traditional offsets to alternative charity-like schemes that promise big environmental benefits rather than carbon neutrality. These include:

Sandbag www.sandbag.org.uk
This small London-based charity aims to interact with the EU's emissions trading system, which allocates tradable CO_2 permits to carbon-intensive industries. Sandbag uses supporters' money to buy these permits and tear them up, hence reducing the amount of CO_2 that companies are legally allowed to produce.

WHAT CAN YOU DO?

Cool Earth www.coolearth.org Cool Earth uses donations to protect critically endangered rainforest. In some cases, the charity buys the land and donates it to a local trust, with local people employed as forest stewards; in other cases,

when forest-owning tribes have been offered money for logging rights in their land, the charity offers a similar amount in return for forest protection. Cool Earth says that protecting a single acre – at a cost of around £75/$120 – can avoid more than 200 tonnes of CO_2 emissions, close to twenty times the benefit of a traditional offset scheme.

The Converging World www.theconvergingworld.org This charity accepts donations in lieu of offsets and uses them to fund renewable energy projects in India, with the cash generated from these schemes – typically "more than twice what is put in", according to the charity – funding development projects.

PART 6

RESOURCES

Climate change books and websites

Books

Many excellent books touch on climate change as part of more general narratives about the planetary environment. We've reserved this list – which, of course, is only a starting point – for books that have global warming at or near their centre.

General

David Archer, *The Long Thaw* (2010). One of the strongest overviews of climate science for laypeople, presented with clarity and forthrightness by an eminent University of Chicago researcher. While looking toward the future, Archer also surveys the lessons provided by the wrenching climate changes of prehistory.

Tim Flannery, *The Weather Makers* (2005). Flannery, an Australian zoologist, museum director and writer, exploits his gift for pithy explanations and metaphor in this bestselling climate-change overview. The short, cleverly titled chapters walk readers through climate history, global-warming science, political and technological options and individual action.

Ross Gelbspan, *The Heat Is On* (1997) and *Boiling Point* (2004). Gelbspan is the original climate-change muckraker. The sceptics that shaped 1990s climate debate are the main target of *The Heat Is On*, which includes revealing detail from congressional testimony. *Boiling Point* lambasts a broader set of actors, including "criminals against humanity" (big oil and coal), "bad media" and "compromised activists".

Al Gore, *Earth in the Balance* (1992); *An Inconvenient Truth* (2006); and *Our Choice* (2009). Gore's 1992 book established its writer as the US leader most committed to solving climate change. Though the science has evolved greatly since, *Earth in the Balance* remains powerful and surprisingly relevant. *An Inconvenient Truth* is a handsome summary of the documentary of the same name; if you saw the film, you won't learn a great deal more from the book, though it will enable you to linger over the stunning graphics. *Our Choice* takes a more upbeat tack, presenting a range of potential solutions – again with plenty of eye-catching photos and diagrams.

John Houghton, *Global Warming: The Complete Briefing* (Fourth Edition, 2009). Readers who crave scientific meat but don't have the time or inclination to pore over IPCC reports will enjoy this primer. Houghton led two of the IPCC's major assessments and speaks with authority as well as humanity (there's a thoughtful discussion of climate change from ethical and theological points of view).

Mike Hulme, *Why We Disagree about Climate Change* (2009). A brilliantly conceived examination of the many societal, psychological and spiritual variables that torque our ability to fully confront climate change and its implications. With unquestioned expertise as the first director of the Tyndall Centre for Climate Change Research, Hulme takes a hard look at what science can and can't tell us, while recasting climate change as a prism for learning more about ourselves and society.

Elizabeth Kolbert, *Field Notes from a Catastrophe* (2006). Based on a multipart series in *New Yorker* magazine. The influence of Rachel Carson's classic *Silent Spring* (which helped kick-start US environmentalism) is evident in Kolbert's dry, crisp prose and her one-step-back perspective on how climate change is already affecting places like Shishmaref, Alaska, and how it's altered past civilizations.

Eugene Linden, *The Winds of Change* (2006). Focuses on the risk of rapid climate change – how it's apparently happened in the past, how it could happen again, and what that would mean. Linden is a skilled writer who draws on extensive experience reporting on the global environment, which lends a peripatetic feel to the narrative and allows for first-person asides.

James Lovelock, *The Revenge of Gaia* (2006) and *The Vanishing Face of Gaia* (2009). The man who told us Earth is a self-sustaining system now worries that we're pushing that system beyond its capacity to heal itself. In the slender and sometimes gloomy *Revenge of Gaia*, Lovelock argues we may need nuclear power to avoid what he sees as the truly catastrophic implications of "global heating". *The Vanishing Face of Gaia*, subtitled "A Final Warning", brings home Lovelock's concerns with even more urgency.

Mark Lynas, *High Tide* (2004) and *Six Degrees* (2007). One of the best "you are there" books on the impacts of global warming, *High Tide* sees Lynas hopping from Britain to China to Peru, describing how climate change is affecting people and landscapes. An engaging travelogue and an approachable introduction to climate-change science. With *Six Degrees*, Lynas ups the ante, extending his on-the-scene approach to how each added degree of warming might affect the planet. His portrayals of the type of dire events that may be in store, such as a Superhurricane Ophelia ravaging Houston in 2045, manage to be both sober and hair-raising. The last chapters tiptoe toward doomsday, with

methane release triggering mass extinctions, but as always Lynas is clear and well referenced.

Bill McKibben, *The End of Nature* (1989) and *Eaarth* (2010). McKibben staked out his turf with the now-classic *End of Nature*, which connected the environmental dots between ozone depletion, acid rain and climate change. In *Eaarth*, McKibben calls for adapting to the changes he believes are already shaping a new planet (thus the extra "a") while pushing for a concerted effort to reduce future emissions. Packed with statistics and anecdotes, *Eaarth* echoes the blend of deep concern and dogged optimism found in 350.org, the activist group founded by McKibben.

Michael E. Mann and Lee R. Kump, *Dire Predictions* (2008). In this colourful alternative to narrative-driven primers on climate change, researchers Mann and Kump present a smorgasbord of topics in short, digestible chunks, with each two-page spread focused on a single topic. Their message is bolstered with compelling photos and graphics.

Mark Maslin, *A Very Short Introduction to Global Warming* (Second Edition, 2009). Maslin zips through climate-change science and policy in this extended essay. Depth is limited, given the compact format, and graphics suffer in size and consistency. But Maslin touches on areas that others often omit, such as the world-views that inform people's positions on the topic.

George Monbiot, *Heat: How to Stop the Planet from Burning* (2006). Meticulously researched and convincingly argued, *Heat* lays out plans for reducing UK emissions to what Monbiot argues is a safe and equitable level (just 13% of what they are now) by 2030. Particularly valuable is the book's thorough analysis of the electricity grid and other parts of the energy infrastructure.

Gavin Schmidt and Joshua Wolfe, *Climate Change: Picturing the Science* (2008). Assembled in an engaging coffee-table-book style, this volume includes spectacular climate-related imagery from around the globe taken by Wolfe and other photographers, together with lucid essays by top researchers and communicators.

James Gustave Speth, *Red Sky at Morning* (2004). *Red Sky* offers a US-oriented discussion of climate change in the larger environmental context by a co-founder of the Natural Resources Defense Council and founder of the World Resources Institute. The book benefits greatly from Speth's first-hand knowledge of environmental work, and avoids the potential tangles of policy jargon.

Arctic and glaciers

Richard B. Alley, *The Two-Mile Time Machine* (2000). A personable book written by a geologist who's participated in ice-core expeditions in Greenland and Antarctica. Alley explains in lively, conversational prose how ice cores are retrieved and what they tell us. More than a decade on, Alley's explanations still serve as a great starting point on ice and climate.

Mark Bowen, *Thin Ice* (2005). Bowen's masterful, in-depth look at tropical glaciers and those who study them fills an important niche. The focus is on glaciologist Lonnie Thompson and his exploits in Peru's highest mountains, with more general background on climate change provided along the way. Himself a climber and physicist, Bowen digs deep into the science, but also brings tales of high-altitude exploits that will resonate with outdoor enthusiasts.

Andrew C. Revkin, *The North Pole Was Here* (2006). This book for young readers blends lavish photography with clear, engaging writing. It focuses on a three-day

visit to the high Arctic by Revkin, a veteran reporter on climate change at *The New York Times*. Full-page excerpts from the *Times* go back as far as a century.

Charles Wohlforth, *The Whale and the Supercomputer* (2004). One of the best books to date on what climate change is doing to the people, creatures and landscape of Alaska – and, by extension, high latitudes around the world. Wohlforth's narrative has the broad sweep of an epic, plus plenty of insightful details on researchers and everyday Alaskans pondering the future and grappling with the enormity of the changes already under way.

Historical

Paul Edwards, *A Vast Machine* (2010). A scholarly, timely and highly readable look at how computer models and climate data have come of age. Edwards argues that the two are now inseparable: without models, we'd never be able to process the torrent of global observations now at hand or read between the lines of clues to past climate. The perfect antidote to mindless model-bashing.

John Imbrie and Katharine Palmer Imbrie, *Ice Ages: Solving the Mystery* (1979). This classic book explains how our modern understanding of ice ages emerged. Though it's not the place to turn for the latest cutting-edge work in glaciology, its accounts of early figures such as Milankovitch are packed with detail.

Jeremy Leggett, *The Carbon War* (1999). As a former oil geologist and then a science advisor to Greenpeace, Leggett got a close-up look at the negotiations leading up to Kyoto. He shares the victories and agonies of climate-change activism and sheds light on many of the key players and processes, making no secret of whom he considers to be the good guys and bad guys. A book of great historical value.

Naomi Oreskes and Erik M. Conway, *Merchants of Doubt* (2010). A meticulously assembled dossier on a highly influential handful of scientists, well connected with top US officials and conservative think tanks, who over the course of decades helped sow uncertainty about climate change as well as acid rain, the ozone hole, and smoking. *Merchants of Doubt* makes a powerful, well-referenced and persuasive case.

Fred Pearce, *The Climate Files* (2010). A premier example of journalism's traditional role as "the first draft of history", *The Climate Files* emerged from the extended series on Climategate filed by Pearce for *The Guardian*. In his reflections on the trove of emails released in 2009, Pearce critiques the behaviour of researchers reluctant to release data, as well as observers who took the emails as evidence of a grand conspiracy to manipulate the science.

William K. Stevens, *The Change in the Weather* (1999). One of the best global-warming overviews of its time, this is worth seeking out for its thorough coverage of climate research in the 1990s. A long-time science writer for *The New York Times*, Stevens provides illuminating profiles of scientists (including sceptics) and how they arrived at their positions on climate change.

Spencer Weart, *The Discovery of Global Warming* (Revised and Expanded, 2008). A superb chronicle of how fragments of theories and observations gradually coalesced to form a new scientific discipline. The sparkling prose is concise and precise, giving a sense of the full story quickly and pleasurably. Just as impressive is the adjunct website, www.aip.org/history/climate, which includes all the material from the book and far more besides.

Paleoclimate

John Cox, *Climate Crash* (2005). A wonderfully written report on the science of abrupt climate change and its roots in clues such as glaciers and fossils. Cox conveys the excitement of the research chase, the difficulty of the work and the mixed signals it provides about our future. *Climate Crash* is one of the few places to find easily digestible treatments of such complex phenomena as Dansgaard–Oeschger events.

Brian Fagan, *The Little Ice Age* (2000) and *The Long Summer* (2003). Literate, highly entertaining accounts of the first two great climate epochs of the last millennium (the third being the warm-up since the mid-1800s). They're not so much books about science as vivid historical portraits of climate's impact on humanity.

William Ruddiman, *Plows, Plagues and Petroleum* (New Edition, 2010) and *Earth's Climate* (Second Edition, 2007). An accomplished paleoclimatologist, Ruddiman shook up his field by speculating that agriculture may have postponed the next ice age. *Plows, Plagues and Petroleum* outlines the hypothesis in lay-oriented language. It's a satisfying read and a great case study of a scientist creating and defending a theory. *Earth's Climate*, designed as a textbook to introduce students to climate history, is clear, colourful and accessible, free from equations and jargon but packed with helpful diagrams.

Policy

Andrew L. Dessler and Edward A. Parson, *The Science and Politics of Global Climate Change* (Second Edition, 2010). This soberly written guide covers the basics of climate change, the main points of contention and the approaches of policy experts to the topic. Striking a middle ground between the comprehensiveness of an IPCC report and the immediacy of *An Inconvenient Truth*, it's especially good for students and scholars needing a one-stop reference.

Gwynne Dyer, *Climate Wars* (2008). A veteran writer on international affairs, Dyer examines the geopolitical tensions stoked by the spectre of climate change. While sharing his own views along the way, Dyer draws on a wide range of science, policy, and military experts, often including generous excerpts from their own works.

James Hansen, *Storms of My Grandchildren* (2009). In his first book for a general audience, the eminent climate researcher gives first-person accounts from his vast trove of experience, from TV interviews to White House meetings and mountaintop protests, while outlining his prescription for dramatic cuts in greenhouse emissions. Hansen's earnestness and conviction enliven this compelling and profoundly serious book.

James Hogg and Richard Littlemore, *Climate Cover-Up* (2009). This book hits many of the same themes as Hogg's DeSmogBlog, particularly the roles of public relations and propaganda in manipulating the public dialogue on climate change and obscuring the level of agreement among key researchers.

Eli Kintisch, *Hacking the Planet* (2010). An entertaining, personable look at the embryonic world of geoengineering. Kintisch nimbly weaves portraits of the bold personalities, audacious plans and huge scientific and social question marks around the idea of engineering Earth's climate.

Aubrey Meyer, *Contraction and Convergence* (2001). The full story about one of the leading candidates for a post-Kyoto system of controlling greenhouse emissions. Meyer developed C&C more than a decade ago and makes the case for it with passion and conviction.

Roger Pielke Jr., *The Climate Fix* (2010). A political scientist who's long focused on climate issues, Pielke presents a number of suggestions for crafting global-warming policy that's mindful of societal realities (including people's craving to improve their economic lot) while staying rooted in science (but avoiding fear-driven appeals, which he argues can backfire). A very readable introduction to the issues around decarbonizing a market-based global economy.

Eric Pooley, *The Climate War* (2010). The most complete report to date on the epic evolution – or, as some would view it, devolution – of US climate change policy over the last twenty-plus years. Pooley combines thoroughness and perspective with a top journalist's knack for colourful storytelling and humanizing.

Oliver Tickell, *Kyoto2* (2008). A comprehensive explanation of the Kyoto2 plan, one of the leading proposals for a post-2012 world. Kyoto2 calls for catalysing the power of markets by placing a global cap on the total amount of carbon emitted, with gases regulated as close to their sources as possible. Clearly presented and worth pondering.

Energy and climate

Chris Goodall, *Ten Technologies to Save the Planet* (2008). Although he's no Pollyanna, Goodall musters a surprising amount of well-grounded optimism in covering this suite of promising tools, from super-efficient homes to electric cars and biochar. A superbly written and remarkably clear-eyed survey.

Richard Heinberg, *Powerdown* (2004) and *Peak Everything* (2010). An expert on sustainability, Heinberg has long been concerned about the potential for oil depletion. In *Powerdown*, Heinberg focused on the threat posed as oil and gas resources draw scarce, while *Peak Everything* expands the range of concerns to encompass water and food shortages, all goosed by a growing global population. Overall, Heinberg manages to be both pessimistic and pragmatic, envisioning big trouble ahead while lobbying for action that would at least stanch the bleeding.

James Howard Kuntsler, *The Long Emergency* (2005). A fascinating if depressing read, this book places climate change in the context of other global issues such as oil depletion, terrorism and disease. With a mix of snarkiness and genuine concern, Kuntsler explains why he believes that renewables, nuclear, tar sands and oil shale will fail to protect us from the massive impacts of diminished oil production.

Jeremy Leggett, *Half Gone* (2005), also published as *The Empty Tank*. An incisive book that explores how a drop in oil production (which Leggett sees as being imminent) could affect climate change. It's full of sharp observations on the workings of the fossil-fuel industry, benefiting from Leggett's own experience as an oil geologist. The book is let down only by the cloying opening and closing parables about "The Blue Pearl".

David J.C. MacKay, *Sustainable Energy – Without the Hot Air* (2009). This information-rich volume covers the gamut of options through which Britain (and the world) might craft a carbon-neutral future. Among its many strong suits, the book conveys the sense of scale needed to compare various forms of alternative energy. Despite the sophisticated content, MacKay's voice is remarkably conversational, and the book's clever structure allows one to read in depth or skim the highlights of each chapter.

Joseph J. Romm, *The Hype about Hydrogen* (2005) and *Hell and High Water* (2006). In *Hype about Hydrogen*, Romm – who worked for years at the US Department of Energy and now runs the Climate Progress blog – deftly punctures the bubble of optimism that surrounded "the hydrogen economy" at the dawn of the new millennium. Romm broadens his view in *Hell and High Water* to include the politics swirling around climate change in general and energy policies in particular, including the action-thwarting roles of sceptics and vested interests.

Personal action

Mike Berners-Lee, *How Bad are Bananas? The Carbon Footprint of Everything* (2010). An excellent guide to how much carbon gets emitted in doing what we do and buying what we buy. The eye-opening statistics are leavened with humour and juxtaposed against each other to give readers a useful and pragmatic sense of proportion.

Duncan Clark, *The Rough Guide to Green Living* (2009). Covering personal climate action as well as other forms of ethical consumerism, this UK-focused book provides useful, digestible advice on everything from home energy generation through to greener cars.

Chris Goodall, *How to Live a Low-carbon Life* (2007). If you're truly passionate about reducing your carbon footprint, this is the ultimate guide. In 326 pages, Goodall walks you through each lifestyle aspect in exhaustive detail – including a full page on how best to heat a teacup's worth of water – with an exacting eye for the subtleties that shorter guides miss or omit. No doubt too rigorous for some readers, but unsurpassed in its depth and commitment and quite readable to boot.

Fiction

John Barnes, *Mother of Storms* (1995). In this sci-fi thriller, set in 2028, nuclear warheads plough into the Pacific, destabilizing colossal amounts of methane clathrates. The greenhouse effect then runs amok, spawning cataclysmic hurricanes around the world, and the future rests on an uncommonly smart astronaut and his plan to screen out sunlight.

Michael Crichton, *State of Fear* (2004). The bestselling thriller that cheered climate-change sceptics across the world is an undeniably gripping tale in the Crichton mould, but hardly the place to turn for a well-rounded exploration of global warming science. See p.284 for more on Crichton.

Kim Stanley Robinson, *Forty Signs of Rain* (2004), *Fifty Degrees Below* (2005) and *Sixty Days and Counting* (2007). In this sci-fi trilogy, the collapse of warm North Atlantic currents results in an ice blitz. *Forty Signs of Rain* chronicles a cast of characters coming to grips with worsening climate, while *Fifty Degrees Below* focuses on one man living in snowbound Washington, DC, and a grand scheme to jump-start the Gulf Stream. The action – both interpersonal and meteorological – reaches a peak in *Sixty Days and Counting*.

Bruce Sterling, *Heavy Weather* (1995). This landmark blend of climate change and cyberpunk follows a group of rebellious, tech-savvy storm chasers facing off with epic tornadoes across the US Great Plains – just one manifestation of global warming c. 2031.

Websites

Given the enormous scope of the Web and the equally vast realm of climate-change science and debate, the sampling below is only the barest sliver of what's available online. We've focused on websites with a rich array of climate change content. For sites covering carbon offsetting and home energy, see "What can you do?" (p.365).

Scientific assessments

If you want undiluted, in-depth scientific facts about climate change, you can't do better than the assessment publications of relevant scientific institutions. The full text of each report since 1998 from the **IPCC** (see p.308) can be viewed online; for recent reports, full PDFs are also available, typically broken into chapters. The UK Climate Impacts Programme (**UKCIP**) produces scenarios every few years and sends a monthly e-bulletin, while the ongoing US Global Change Research Program (**USGCRP**) carries out a variety of assessments geared to the States. National-scale scenarios are also provided by **CSIRO** (Australia) and **Environment Canada**. **RAPID-WATCH** is a UK-funded, observationally oriented programme focusing on the present and future state of North Atlantic currents, whose evolution could affect large areas.

IPCC www.ipcc.ch
UKCIP www.ukcip.org.uk
US Global Change Research Program www.globalchange.gov
CCSP www.climatescience.gov
CSIRO www.climatechangeinaustralia.gov.au
Environment Canada www.ccsn.ca/index-e.html
RAPID-WATCH www.noc.soton.ac.uk/rapid/rw

Research centres

Some of the world's focal points for global warming research also provide extensive materials designed for the interested public. These labs include the UK Met Office **Hadley Centre** and the US National Center for Atmospheric Research (**NCAR**), which is operated by the University Corporation for Atmospheric Research (**UCAR**). UCAR's member universities and its international affiliates include many of the world's top academic centres studying global climate. Several branches of **NASA** and the National Oceanic and Atmospheric Administration (**NOAA**) focus on climate change and offer lay-friendly summaries of their work. The non-profit **Pew Center on Global Climate Change** works with leading experts to produce comprehensive, readable analyses tailored for policymakers and the public.

Hadley Centre www.metoffice.com/research/hadleycentre
UCAR/NCAR www.ucar.edu
UCAR members www.ucar.edu/governance/members/institutions.shtml
NASA Goddard Space Flight Center www.giss.nasa.gov
NOAA Climate Services www.climate.gov
NOAA Climate Dynamics and Prediction Group www.gfdl.noaa.gov/research/climate
NOAA Paleoclimatology Program www.ncdc.noaa.gov/paleo
Pew Center on Global Climate Change www.pewclimate.org

Climate change news

The following sites are among the best for keeping up with news on global warming and the environment in general. Among UK newspapers, **The Guardian** has delved into Climategate and many other aspects of climate change research and policy in unparallelled detail, with **The Independent** also relentless in covering climate change in recent years. **The New York Times** leads the US pack in the depth and frequency of its climate-change reporting. **The Huffington Post** and **The Daily Climate** are excellent online sources of both original coverage and aggregated news, and **Climate Central** offers top-notch articles and multimedia features. **Grist** is another standout, with content that's light-hearted but razor-sharp.

The Guardian www.guardian.co.uk/environment/climate-change
The Independent www.independent.co.uk/environment/climate-change
The New York Times www.nytimes.com
The Huffington Post www.huffingtonpost.com/news/climate-change
The Daily Climate www.dailyclimate.org
Climate Central www.climatecentral.org
Grist www.grist.org

Blogs

Though they're no substitute for peer-reviewed literature, blogs are invaluable for sorting through the ever-increasing stacks of global warming research and finding out what's on people's minds across the spectrum of thought. **Real Climate**, developed by an international team of top climate researchers, tackles complex science with clarity and insight. If you're looking for solid, accessible replies to common arguments against global warming (such as, "It hasn't warmed since 1998"), **Skeptical Science** is both comprehensive and pithy. News features accompany its handy set of more than 130 arguments and responses. Among the most popular of all blogs related to climate change is **Watts Up With That**, whose posts are a lodestone for sceptics but often critiqued by climate researchers. Another lively, frequently updated site with a loyal following is the unabashedly liberal **Climate Progress** (edited by Joe Romm, an energy expert with a doctorate in physics). Steve McIntyre critiques data sets and data access at **Climate Audit**, while climate whistleblower **Rick Piltz** keeps an eye on US government doings and **DeSmogBlog** dissects the PR strategies of climate sceptics. **Roger Pielke Jr.** maintains a thought-provoking blog focused on climate policy. Several other useful sites maintained by individual scientists and journalists, each with large helpings of climate-oriented content, are listed below.

Real Climate www.realclimate.org
Skeptical Science www.skepticalscience.com
Watts Up With That www.wattsupwiththat.com
Climate Progress www.climateprogress.org
Climate Audit www.climateaudit.org
Climate Science Watch (Rick Piltz) www.climatesciencewatch.org
DeSmog Blog www.desmogblog.com
Roger Pielke Jr.'s Blog rogerpielkejr.blogspot.com
Mark Lynas www.marklynas.org
George Monbiot www.monbiot.com
Chris Mooney and Sheril Kirshenbaum www.scienceblogs.com/intersection
John Fleck www.inkstain.net/fleck

Index

INDEX

INDEX